# Artificial Intelligence Applications in Agriculture and Food Quality Improvement

Mohammad Ayoub Khan
*University of Bisha, Saudi Arabia*

Rijwan Khan
*ABES Institute of Technology, India*

Pushkar Praveen
*Govind Ballabh Pant Institute of Engineering and Technology, India*

IGI Global
PUBLISHER of TIMELY KNOWLEDGE

A volume in the Advances in Computational Intelligence and Robotics (ACIR) Book Series

Published in the United States of America by
IGI Global
Engineering Science Reference (an imprint of IGI Global)
701 E. Chocolate Avenue
Hershey PA, USA 17033
Tel: 717-533-8845
Fax: 717-533-8661
E-mail: cust@igi-global.com
Web site: http://www.igi-global.com



Library of Congress Cataloging-in-Publication Data

Names: Khan, Mohammad Ayoub, 1980- editor. | Khan, Rijwan, 1981- editor. | Praveen, Pushkar, 1985- editor.
Title: Artificial Intelligence applications in agriculture and food quality improvement / Mohammad Ayoub Khan, Rijwan Khan, Pushkar Praveen, editors.
Description: Hershey, PA : Engineering Science Reference, [2022] | Includes bibliographical references and index. | Summary: "This book emphasizes artificial intelligence, machine learning and data analytics as a means to increase production in agriculture and food quality through the use of smart solutions improvements"-- Provided by publisher.
Identifiers: LCCN 2022007816 (print) | LCCN 2022007817 (ebook) | ISBN 9781668451410 (h/c) | ISBN 9781668451427 (s/c) | ISBN 9781668451434 (eISBN)
Subjects: LCSH: Agriculture--Data processing. | Artificial intelligence.
Classification: LCC S494.5.D3 A775 2022 (print) | LCC S494.5.D3 (ebook) | DDC 338.10285--dc23/eng/20220314
LC record available at https://lccn.loc.gov/2022007816
LC ebook record available at https://lccn.loc.gov/2022007817

This book is published in the IGI Global book series Advances in Computational Intelligence and Robotics (ACIR) (ISSN: 2327-0411; eISSN: 2327-042X)

British Cataloguing in Publication Data
A Cataloguing in Publication record for this book is available from the British Library.

# Advances in Computational Intelligence and Robotics (ACIR) Book Series

Ivan Giannoccaro
University of Salento, Italy

ISSN:2327-0411
EISSN:2327-042X

## Mission

While intelligence is traditionally a term applied to humans and human cognition, technology has progressed in such a way to allow for the development of intelligent systems able to simulate many human traits. With this new era of simulated and artificial intelligence, much research is needed in order to continue to advance the field and also to evaluate the ethical and societal concerns of the existence of artificial life and machine learning.

The **Advances in Computational Intelligence and Robotics (ACIR) Book Series** encourages scholarly discourse on all topics pertaining to evolutionary computing, artificial life, computational intelligence, machine learning, and robotics. ACIR presents the latest research being conducted on diverse topics in intelligence technologies with the goal of advancing knowledge and applications in this rapidly evolving field.

## Coverage

- Cyborgs
- Heuristics
- Neural Networks
- Brain Simulation
- Synthetic Emotions
- Robotics
- Artificial Intelligence
- Agent technologies
- Computational Intelligence
- Computer Vision

IGI Global is currently accepting manuscripts for publication within this series. To submit a proposal for a volume in this series, please contact our Acquisition Editors at Acquisitions@igi-global.com or visit: http://www.igi-global.com/publish/.

The Advances in Computational Intelligence and Robotics (ACIR) Book Series (ISSN 2327-0411) is published by IGI Global, 701 E. Chocolate Avenue, Hershey, PA 17033-1240, USA, www.igi-global.com. This series is composed of titles available for purchase individually; each title is edited to be contextually exclusive from any other title within the series. For pricing and ordering information please visit http://www.igi-global.com/book-series/advances-computational-intelligence-robotics/73674. Postmaster: Send all address changes to above address. 

## Titles in this Series

*For a list of additional titles in this series, please visit: http://www.igi-global.com/book-series/advances-computational-intelligence-robotics/73674*

***Applied AI and Multimedia Technologies for Smart Manufacturing and CPS Applications***
Emmanuel Oyekanlu (Drexel University, USA)
Engineering Science Reference • © 2022 • 300pp • H/C (ISBN: 9781799878520) • US $270.00

***Artificial Intelligence of Things for Weather Forecasting and Climatic Behavioral Analysis***
Rajeev Kumar Gupta (Pandit Deendayal Energy University, India) Arti Jain (Jaypee Institute of Information Technology, India) John Wang (Montclair State University, USA) Ved Prakash Singh (India Meteorological Department, Ministry of Earth Sciences, Government of India, India) and Santosh Bharti (Pandit Deendayal Energy University, India)
Engineering Science Reference • © 2022 • 277pp • H/C (ISBN: 9781668439814) • US $270.00

***Unmanned Aerial Vehicles and Multidisciplinary Applications Using AI Techniques***
Bella Mary I. Thusnavis (Karunya Institute of Technology and Sciences, India) K. Martin Sagayam (Karunya Institute of Technology and Sciences, India) and Ahmed A. Elngar (Faculty of Computers & Artificial Intelligence Beni-Suef University, Beni Suef City, Egypt)
Engineering Science Reference • © 2022 • 300pp • H/C (ISBN: 9781799887638) • US $270.00

***Principles and Applications of Socio-Cognitive and Affective Computing***
S. Geetha (VIT University, Chennai Campus, India) Karthika Renuka (PSG College of Technology, India) Asnath Victy Phamila (VIT University, Chennai, India) and Karthikeyan N. (Syed Ammal Engineering College, India)
Engineering Science Reference • © 2022 • 330pp • H/C (ISBN: 9781668438435) • US $270.00

***Applications of Computational Science in Artificial Intelligence***
Anand Nayyar (Duy Tan University, Da Nang, Vietnam) Sandeep Kumar (CHRIST University (Deemed), Bangalore, India) and Akshat Agrawal (Amity University, Guragon, India)
Engineering Science Reference • © 2022 • 284pp • H/C (ISBN: 9781799890126) • US $270.00

***Challenges and Applications for Hand Gesture Recognition***
Lalit Kane (School of Computer Science, University of Petroleum and Energy Studies, India) Bhupesh Kumar Dewangan (School of Engineering, Department of Computer Science and Engineering, O.P. Jindal University, Raigarh, India) and Tanupriya Choudhury (School of Computer Science, University of Petroleum and Energy Studies, India)
Engineering Science Reference • © 2022 • 249pp • H/C (ISBN: 9781799894346) • US $215.00

IGI Global
PUBLISHER of TIMELY KNOWLEDGE

701 East Chocolate Avenue, Hershey, PA 17033, USA
Tel: 717-533-8845 x100 • Fax: 717-533-8661
E-Mail: cust@igi-global.com • www.igi-global.com

# Table of Contents

**Preface** .......... xiv

**Chapter 1**
Artificial Intelligence-Based Sustainable Agricultural Practices .......... 1
*Usha Chauhan, Galgotias University, India*
*Divya Sharma, Galgotias University, India*
*Sharzeel Saleem, Galgotias University, India*
*Mahim Kumar, Galgotias University, India*
*Shaurya Pratap Singh, Galgotias University, India*

**Chapter 2**
Applications of Deep Learning in Agriculture .......... 17
*Padmesh Tripathi, IIMT College of Engineering, India*
*Nitendra Kumar, IIMT College of Engineering, India*
*Mritunjay Rai, Shri Ramswaroop Memorial University, India*
*Ayoub Khan, University of Bisha, Saudi Arabia*

**Chapter 3**
Freshness Grading of Agricultural Products Using Artificial Intelligence .......... 29
*Zeynep Elbir, Marmara University, Turkey*
*Berat Asrin Caferoglu, Marmara University, Turkey*
*Onur Cihan, Marmara University, Turkey*

**Chapter 4**
Precision Farming Using Image Processing and Machine Learning .......... 55
*Ayesha Naaz, Muffakham Jah College of Engineering and Technology, India*
*Sucharitha B., Muffakhjam Jah College of Engineering and Technology, India*

**Chapter 5**
The Use of Pesticide Management Using Artificial Intelligence .......... 74
*Sapna Katiyar, ABES Institute of Technology, Ghaziabad, India*

**Chapter 6**
Role of Deep Learning in Weed Detection ........ 95
*Kavita Srivastava, Institute of Information Technology and Management, GGSIP University, India*

**Chapter 7**
Evaluation of Machine Learning Techniques for Crop Yield Prediction ........ 112
*Divya Goel, J. C. Bose University of Science and Technology, YMCA, India*
*Payal Gulati, J. C. Bose University of Science and Technology, YMCA, India*
*Suman Kumar Jha, IIMT College of Engineering, India*
*Dr Nitendra Kumar, IIMT College of Engineering, India*
*Ayoub Khan, University of Bisha, Saudi Arabia*
*Priti Kumari, B. S. M. College of Engineering, India*

**Chapter 8**
An Extensive Analysis of Flying Ad-Hoc Network Applications and Routing Protocols in Agriculture ........ 129
*Ahona Ghosh, Maulana Abul Kalam Azad University of Technology, West Bengal, India*
*Caprio Mistry, Brainware University, India*
*Mousumi Biswas, Brainware University, India*

**Chapter 9**
Deep Learning-Based Plant Disease Detection Using Android Apps ........ 148
*Mohd Shoaib, Department of Computer Engineering, Zakir Husain College of Engineering and Technology, Aligarh Muslim University, India*
*Firoz Ahmad, Department of Computer Engineering, Zakir Husain College of Engineering and Technology, Aligarh Muslim University, India*
*Mohd Ammad Rehman, Department of Computer Engineering, Zakir Husain College of Engineering and Technology, Aligarh Muslim University, India*

**Chapter 10**
A Sustainable Approach of Artificial Neural Network for Prediction of Irrigation, Pesticides, Fertilizers, and Crop Yield ........ 169
*Pandurangan Prakash, Sathyabama Institute of Science and Technology, India*
*G. Elanthendral, Sathyabama Institute of science and Technology, India*
*V. Vedanarayanan, Sathyabama Institute of Science and Technology, India*
*J. Aravindh Kumar, Saveetha School of Engineering SIMATS (Deemed), India*
*Antony V. Samrot, Mahsa University College, Malaysia*
*R. Jaiganesh, SRM Institute of Science and Technology, India*

**Chapter 11**
Impact of Artificial Intelligence and Machine Learning in the Food Industry: A Survey ........ 190
*Archana Sharma, ABES Institute of Technology, India*
*Kajol Mittal, ABES Institute of Technology, India*
*Sunil Kumar, Institute of Management Studies (University Courses Campus), India*
*Utkarsh Sharma, Indraprastha Institute of Information Technology, Delhi, India*

*Prashant Upadhyay, ABES Institute of Technology, India*

**Chapter 12**
Applications of Machine Learning in Food Safety ........ 216
*Rakesh Mohan Pujahari, ABES Institute of Technology, India*
*Rijwan Khan, ABES Institute of Technology, India*

**Chapter 13**
Artificial Intelligence for Improving Food Quality ........ 241
*Avinash Kumar Sharma, Department of CSE, ABES Institute of Technology, AKTU, Ghaziabad, India*
*Shivani Singh, Department of CSE, ABES Institute of Technology, AKTU, Ghaziabad, India*
*Shivam Kumar, Department of CSE, ABES Institute of Technology, AKTU, Ghaziabad, India*

**Chapter 14**
The Role of Machine Learning and Computer Vision in the Agri-Food Industry ........ 257
*Aswathy Ravikumar, Mar Baselios College of Engineering and Technology, India*

**Chapter 15**
Artificial Intelligence-Based Food Calories Estimation Methods in Diet Assessment Research ....... 276
*Naimoonisa Begum, Muffakamjah College of Engineering, Hyderabad, India*
*Ankur Goyal, Chandigarh University, India*
*Sachin Sharma, Koneru Lakshmaiah Education Foundation, India*

**Compilation of References** ........ 291

**About the Contributors** ........ 330

**Index** ........ 334

# Detailed Table of Contents

**Preface**........................................................................................................................................ xiv

**Chapter 1**
Artificial Intelligence-Based Sustainable Agricultural Practices........................................................... 1
*Usha Chauhan, Galgotias University, India*
*Divya Sharma, Galgotias University, India*
*Sharzeel Saleem, Galgotias University, India*
*Mahim Kumar, Galgotias University, India*
*Shaurya Pratap Singh, Galgotias University, India*

A world hunger crisis has risen since 2019 when COVID-19 hit the world. This pandemic has shifted this generation back in time, and now it is very important to be involved in new techniques that are effective in terms of better yield with less toxins. With the rate at which the population is growing, it is expected that by the year 2050, the world population would cross 9 billion. This exponential rise would require the food production to rise by 70 to 80%. This is a matter of concern for agriculture and food industries. As the world is in the fourth industrial revolution, it is the need of the hour to embed artificial intelligence and machine learning algorithms with agriculture. This research aims to accumulate different methodologies that are present and come up with a critical analysis. These methodologies have the capability to increase the yield, predict the diseases, and even increase the safety and help enhance traceability.

**Chapter 2**
Applications of Deep Learning in Agriculture ........................................................................................ 17
*Padmesh Tripathi, IIMT College of Engineering, India*
*Nitendra Kumar, IIMT College of Engineering, India*
*Mritunjay Rai, Shri Ramswaroop Memorial University, India*
*Ayoub Khan, University of Bisha, Saudi Arabia*

Today's era is the era of technologies. Technologies have widely been employed in each and every field. The field of agriculture is not untouched with the technologies, and in several segments of agriculture; it has been employed at large. Deep learning techniques and its variants like convolutional neural networks (CNN), recurrent neural networks (RNN), generative adversarial network (AGN), and their various subcategories like AlexNet, ImageNet, visual geometry group (VGG), etc. have widely been employed in many sectors of agriculture in order to increase the quality and quantity of production. In this chapter, some applications of deep learning have been explored.

**Chapter 3**
Freshness Grading of Agricultural Products Using Artificial Intelligence.......................................... 29
*Zeynep Elbir, Marmara University, Turkey*
*Berat Asrin Caferoglu, Marmara University, Turkey*
*Onur Cihan, Marmara University, Turkey*

This chapter presents a deep learning model that can be used to determine the levels of freshness of three agricultural products: strawberries, lemons, and tomatoes. For this purpose, YOLO, a state-of-the-art object detection algorithm is utilized. The data for training, validation, and testing are collected from online sources, and by applying image augmentation techniques, a sufficient number of images are obtained. Test results show that the model is performing quite well, and the speed of the model is fast. These results are promising and can be utilized to reduce a significant amount of agricultural waste and increase customer satisfaction once it is utilized by online groceries.

**Chapter 4**
Precision Farming Using Image Processing and Machine Learning................................................... 55
*Ayesha Naaz, Muffakham Jah College of Engineering and Technology, India*
*Sucharitha B., Muffakhjam Jah College of Engineering and Technology, India*

Agriculture in general is plagued by numerous problems that can be solved using modernization techniques and help the farmers making them aware of problems related to crop yield, nutrient status, crop disease, etc. Precision agriculture is a promising approach to address these issues. The objective of the chapter is design and deployment of farmer-friendly precision farming using multispectral imaging techniques for providing advanced, smart, and connected agricultural management solutions by integrating high-precision hardware, software, and data. Remote sensing is a useful tool for monitoring spatio-temporal variations of crop morphological and physiological status and supporting practices in precision agriculture. Of the multiple technologies used for remote sensing, multispectral and hyperspectral remote sensing are widely in demand. The data is processed employing deep learning to allow multispectral image classification based on spectral-spatial features. Applying data analytics on the mapped data can provide suggestions/notifications/alerts to the farmers.

**Chapter 5**
The Use of Pesticide Management Using Artificial Intelligence......................................................... 74
*Sapna Katiyar, ABES Institute of Technology, Ghaziabad, India*

The population of the world may reach almost 10 billion by 2050, and currently, approximately 37.7% of land is used for the production of crops. Agriculture is a major source of revenue for any country. Globally, automation in agriculture is in demand. Innovation and integration of technologies contributes for challenges faced by farmers with enlarged revenue and employment opportunities. Artificial intelligence has brought a revolution in agriculture. Crop wellbeing is important as it is a crucial factor that relates all parameters directly; therefore, crop health examination is mandatory. Premature detection of pests also reduces the quantity and frequent use of pesticides, but human intervention in process makes it time consuming and expensive. Time and techniques to use the pesticides in large farmland using AI along with computer vision and IoT converts traditional processes into smart agriculture. This chapter presents the assessment and implementation of an intelligent system for pesticide management.

**Chapter 6**
Role of Deep Learning in Weed Detection ............................................................................ 95
*Kavita Srivastava, Institute of Information Technology and Management, GGSIP University, India*

Deep learning, robotics, AI, and automation have lots of applications that are beneficial to society at large. In fact, nearly every sector, such as transportation, industries, manufacturing, healthcare, education, retail, and home automation, are adopting AI, machine learning, IoT, and robotics to their advantage. Of course, agriculture is no exception. The chapter starts with an introduction to the applications of deep learning in agriculture. Next, a comprehensive survey of the research work done in recent years is provided. It is followed by the description of various techniques of deep learning (DL). The next section briefly describes the traditional ways of weed detection and removal. Next, the architecture of deep learning for weed detection and removal is presented along with the associated code. Further, the chapter goes on to discuss the pros and cons of this approach. Finally, the chapter concludes by citing the important points discussed in this study.

**Chapter 7**
Evaluation of Machine Learning Techniques for Crop Yield Prediction........................................... 112
*Divya Goel, J. C. Bose University of Science and Technology, YMCA, India*
*Payal Gulati, J. C. Bose University of Science and Technology, YMCA, India*
*Suman Kumar Jha, IIMT College of Engineering, India*
*Dr Nitendra Kumar, IIMT College of Engineering, India*
*Ayoub Khan, University of Bisha, Saudi Arabia*
*Priti Kumari, B. S. M. College of Engineering, India*

The agricultural segment is a major supporter of the Indian economy as it represents 18% of India's GDP, and it gives work to half of the nation's work power. The farming segment is required to satisfy the expanding need for food because of the increasing populace. Therefore, to cater to the ever-increasing needs of people of the nation, yield prediction is done prior. The farmers are also benefited from yield prediction as it will assist the farmers to predict the yield of crops prior to cultivating. There are multiple parameters that affect the yield of crops like rainfall, temperature, fertilizers, pH level, and other atmospheric circumstances. Thus, considering these factors, the yield of a crop is thus hard to predict and becomes a challenging task. In this chapter, the dataset of different states producing different crops in different seasons is considered; further, after preprocessing the data, the authors applied machine learning algorithms, and their results are compared.

**Chapter 8**
An Extensive Analysis of Flying Ad-Hoc Network Applications and Routing Protocols in Agriculture ........................................................................................................................... 129
*Ahona Ghosh, Maulana Abul Kalam Azad University of Technology, West Bengal, India*
*Caprio Mistry, Brainware University, India*
*Mousumi Biswas, Brainware University, India*

Agriculture and irrigation are sources of man's potential. For the sake of cost-effectiveness and the betterment of agricultural professionals, UAVs (unmanned aerial vehicles) can be deployed for surveillance, utilization of pesticides and insecticides, and detection of bioprocessing errors. With the proper collaboration and coordination of the clusters of UAVs forming a network, linked with a ground infrastructure or satellite will exceed the competencies of a single UAV system. However, one of the vital design issues FANETs deal with is in selecting the accurate routing protocol which is a prerequisite for the creation of FANET. In this chapter, the authors discuss the routing protocols of FANET in different platforms and different strategic manners. The open research challenges have been discussed and possible solutions have been attempted to be drawn from the conclusion. The main contribution lies in suggesting the most suitable routing protocol for each particular agriculture application based on the mobility model and requirement.

**Chapter 9**
Deep Learning-Based Plant Disease Detection Using Android Apps ............................................ 148
*Mohd Shoaib, Department of Computer Engineering, Zakir Husain College of Engineering and Technology, Aligarh Muslim University, India*
*Firoz Ahmad, Department of Computer Engineering, Zakir Husain College of Engineering and Technology, Aligarh Muslim University, India*
*Mohd Ammad Rehman, Department of Computer Engineering, Zakir Husain College of Engineering and Technology, Aligarh Muslim University, India*

Disease detection in plants is crucial for preventing losses in yield and agricultural productivity. Historically, disease identification has been supported by agriculture extension organizations, which were difficult to access from villages. Farmers have to go to their field and manually monitor plant disease. The aim of this work is to develop an Android application that provides an easy-to-use platform for farmers to identify diseases in their crops. The mobile application will help to take responsive action according to the disease detected in their plants and can be easily used by anyone who is interested in analyzing the disease of the plants. This work reports on the classification of 26 diseases in 14 crop species using 54,306 images from PlantVillage dataset using a convolutional neural network approach. The models used are Inception-v3 and MobileNet. The correct prediction of the correct crop-diseases pair in 38 classes decides the criteria for performance measurement. The most accurate model achieves an overall accuracy of 96.32%.

**Chapter 10**
A Sustainable Approach of Artificial Neural Network for Prediction of Irrigation, Pesticides, Fertilizers, and Crop Yield ............................................................................................................ 169
*Pandurangan Prakash, Sathyabama Institute of Science and Technology, India*
*G. Elanthendral, Sathyabama Institute of science and Technology, India*
*V. Vedanarayanan, Sathyabama Institute of Science and Technology, India*
*J. Aravindh Kumar, Saveetha School of Engineering SIMATS (Deemed), India*
*Antony V. Samrot, Mahsa University College, Malaysia*
*R. Jaiganesh, SRM Institute of Science and Technology, India*

Food demand increases as world population expands; the world populace can reach 9.9 billion by 2050. India will lead as the world's largest population by 2024. Agriculture therefore, should be fruitful and economical for subsistence. Organic methods of agriculture are still effective for healthier crops. However, production shrinks as it depends wholly on manual labour. Since traditional farming, agriculture saw various revolutions and developments. Currently, in the era of Agriculture 5.0, precision agriculture principles using artificial intelligence, machine learning and IoT are being used. India still relies heavily on manual work. Educational level, inadequate training and indigent farmers, make India incompetent. Technology may lead to sustainable agriculture, which means integration of plant and animal production that leaves unshakable benefits on the environment, farmers and society, essential for the climate change and disaster-prone world. Machine learning techniques that can possibly cater various agricultural challenges faced by famers in India are reviewed.

**Chapter 11**
Impact of Artificial Intelligence and Machine Learning in the Food Industry: A Survey.................. 190
*Archana Sharma, ABES Institute of Technology, India*
*Kajol Mittal, ABES Institute of Technology, India*
*Sunil Kumar, Institute of Management Studies (University Courses Campus), India*
*Utkarsh Sharma, Indraprastha Institute of Information Technology, Delhi, India*
*Prashant Upadhyay, ABES Institute of Technology, India*

In recent years, the food sector or industry has escalated to prominence as the most important industry to receive widespread attention. It encompasses various industrial activities related to food production, distribution, processing, preparation, preservation, transportation, and packaging. Machine learning (ML) is a subpart of artificial intelligence (AI), and it is widely used in the food sector for industrial automation and predictive modeling with the world's growing demand and population. AI assists in improving package shelf life, menu selection, food cleanliness, and safety. Because of AI and machine learning, smart agriculture, drones, and robotics in the area of the food sector are becoming the need of the modern era. This chapter discusses how AI and machine learning have the potential to be used in the food business to save money while simultaneously increasing resource efficiency. It highlights the food industry's achievements and challenges with specific attention to the role of machine learning and artificial intelligence.

**Chapter 12**
Applications of Machine Learning in Food Safety............................................................................ 216
*Rakesh Mohan Pujahari, ABES Institute of Technology, India*
*Rijwan Khan, ABES Institute of Technology, India*

Food safety has a major correlation with health related to the public. Machine learning can be a great help for large volume and emerging data sets to enhance the safety of the food supply and minimise the impact of food safety incidents. Pathogen genomes which are food borne and unique data streams, transactional, including text, and trade data, have ample emerging applications initiated by a machine learning approach, like prediction of antibiotic resistance, source related pathogens, and detection of food borne outbreak and also assessment of risk. In this chapter, a gentle introduction of machine learning in the pretext of food safety and a detailed overview of various developments and applications has been enumerated.

**Chapter 13**
Artificial Intelligence for Improving Food Quality............................................................................ 241
*Avinash Kumar Sharma, Department of CSE, ABES Institute of Technology, AKTU, Ghaziabad, India*
*Shivani Singh, Department of CSE, ABES Institute of Technology, AKTU, Ghaziabad, India*
*Shivam Kumar, Department of CSE, ABES Institute of Technology, AKTU, Ghaziabad, India*

There is no doubt about the fact that artificial intelligence (AI) has stepped into improving quality of food throughout the world. Artificial intelligence makes it possible to be done for machine to profit as a material or substance of fact examining facts from different data sources and make conclusions, and can perform most humanly tasks with more than the limits of act without any fault. Due to the interference of humans in the production and packaging of food product, there is lack in maintaining the demand supply chain and also in safety of food to prevail over these issues of food industry. Automation is the best possible solution. It wraps up an enormous amount of streaming data, often noted as "big data," which brings fresh and new opportunities to watch agricultural and food processes. Besides sensors, big data from social media is becoming important for the food industry. During this review, the authors present a summary of artificial intelligence (AI) and its role in shaping further of agri-food systems.

**Chapter 14**
The Role of Machine Learning and Computer Vision in the Agri-Food Industry ............................ 257
*Aswathy Ravikumar, Mar Baselios College of Engineering and Technology, India*

The food industry has been unable to control the demand-supply cycle and has also fallen short on food safety due to human engagement. Food production and distribution activities may be controlled more efficiently while also improving operational competence using an artificial intelligence-based solution. The food industry's future is entirely reliant on drones and also witty and robotic farming, thanks to AI and machine learning. Smart farming (soil monitoring, pest monitoring, and fertilizer management), food processing (production, processing, marketing, and customer feedback), and food safety, among other topics, need a detailed evaluation of machine learning applications in the agrifood business. It is vital to monitor production lines to ensure that the manufacturing process and products fulfill the required quality standards. A plethora of data may now be created throughout the production process due to growing digitization.

**Chapter 15**
Artificial Intelligence-Based Food Calories Estimation Methods in Diet Assessment Research....... 276
*Naimoonisa Begum, Muffakamjah College of Engineering, Hyderabad, India*
*Ankur Goyal, Chandigarh University, India*
*Sachin Sharma, Koneru Lakshmaiah Education Foundation, India*

The standard of healthy intake of food is the necessity for keeping a balanced diet to prevent the obesity problem and many other health problems in humans. Obesity is increasing at an alarming speed and keeping people's health at risk. Mankind needs to have careful control on their daily intake of calories by choosing healthier foods, which will be the most fundamental method in preventing obesity and ill health. Even though the packaging of food comes with calorie and nutrition labels, it might not be very favorable for the reference of people. Thus, the scientists to help people started using AI-based techniques and methodologies to know the ways of determining their daily calorie intake of their food. This chapter proposes a review of various AI-based food calorie estimation methodologies in diet assessment which are suggested to help the normal people and patients so that normal people and doctors could succeed to fight against diet-based health conditions.

**Compilation of References** .......................................................................................................... 291

**About the Contributors** ................................................................................................................ 330

**Index** .............................................................................................................................................. 334

# Preface

Nowadays, technology is an important part of the food production and distribution in food industry. The standard of healthy food is the necessity for keeping a balanced diet to prevent the obesity problem and many other health problems in humans. The rate at which obesity is increasing at an alarming speed, is keeping people's health at risk. Technologies, such as equipment, drones, robots, and 3D printing, improve food quality and reduce the cost of producing food, while artificial intelligence (AI) and machine learning (ML) technologies address a wide range of automated acceleration and processing processes, saving revenue, eliminating human error, minimizing waste, happier consumers, more efficient and automated operations, and different orders in the various food businesses such as restaurants, bars, cafes, and food manufacturers.

The agricultural sector must be prudent to increase food security. Food is an essential element of human life, and it can be described as the final product of agricultural activity, which is produced after farmers grind their various products. Agriculture is essential to any country in the global economy. Because the food business is so important to both the global economy and the global economy, smart solutions based on Artificial Intelligence (AI) are desperately needed to ensure product quality and food security. The agricultural sector is constantly under pressure to improve crop production due to population growth. As a result, increasing production without the use of smart solutions is not possible in the same country. This necessitates the use of Artificial Intelligence applications. We focus on AI, ML, and data analysis in this book.

In recent years, the food industry has risen to prominence as the most important industry for widespread attention. It covers a wide range of industrial activities related to food production, distribution, processing, processing, storage, transportation, and packaging. ML is a branch of AI and is widely used in the field of industrial automation and modeling for growing demand and population. AI helps to improve package shelf life, menu selection, food hygiene, and safety. Thanks to AI and mechanical learning, clever agriculture, drones, and robots in the food industry are becoming a necessity in modern times. Mankind needs to carefully control the daily calorie intake by choosing healthy foods, which will be the most important way in preventing obesity and poor health. Although food packaging comes with calorie labels and nutritious foods, it may not be very good at targeting people. Therefore, scientists are assisting individuals in adopting AI-based strategies and methods so they can calculate how many calories they consume every day. The overall calorie amount of any food item is not justified by the calorie ratio based on the item as a whole; rather, its calorie balance might vary significantly depending on its size, quantity, and contents. Many recent studies have used a variety of machine learning techniques, tools, and algorithms for accurate nutrition. A person's nutritional status can be determined by considering various factors such as diet, physical health, nutritional knowledge, body digestive rate, and physical activity.

For designing accurate personal nutrition information and developing the goal of leading a healthier and healthier life, there are many advanced computational techniques such as AI, ML, and deep learning, promising strategies to provide an integrated framework.

This book addresses how AI and machine learning have the potential to be applied in the food industry to save money and increase resource efficiency concurrently. It examines the successes and difficulties of the food sector, with a focus on the use of machine learning and artificial intelligence.

## ORGANIZATION OF THE BOOK

This book contains 15 chapters. A brief description of each of the chapters is as follows:

Chapter 1, titled "Artificial Intelligence-Based Sustainable Agricultural Practices", presents the research aims to accumulate the different methodologies that are present and come up with a critical analysis. These methodologies have the ability to increase the yield, predict diseases and even increase safety and help enhance traceability.

Chapter 2, titled "Applications of Deep Learning in Agriculture", presents the deep learning techniques which emerged three decades earlier and its variants like Convolutional neural Network (CNN), Recurrent Neural Network (RNN), generative adversarial network (AGN) and their various subcategories like AlexNet, ImageNet, visual geometry group (VGG), etc. have widely been employed in many sectors of agriculture to increase the quality and quantity of production. In this article, some applications of deep learning have been explored.

Chapter 3, titled "Freshness Grading of Agricultural Products Using Artificial Intelligence", presents a deep learning model that can be used to determine the levels and the freshness of three agricultural products: strawberries, lemons, and tomatoes. For this purpose, YOLO, a state-of-the-art object detection algorithm, is utilized.

Chapter 4, titled "Precision Farming using Image Processing and Machine Learning", presents the design and deployment of farmer-friendly precision farming using multispectral imaging techniques for providing advanced, smart and connected agricultural management solutions by integrating high-precision hardware, software, and data. Remote sensing is a useful tool for monitoring spatio-temporal variations of crop morphological and physiological status and supporting practices in precision agriculture. Of the multiple technologies used for remote sensing, multispectral and hyperspectral remote sensing are widely in demand. The data is processed employing deep learning to allow multispectral image classification based on spectral-spatial features. Applying data analytics on the mapped data, this can provide suggestions / notifications / alerts to the farmers.

Chapter 5, titled "The Use of Pesticide Management Using Artificial Intelligence", presents the assessment and implementation of an intelligent systems for Pesticide Management. Crop wellbeing is important as it is a crucial factor which relates all parameters directly; therefore, crop health examination is mandatory. Premature detection of pests also reduces the quantity and frequent use of pesticides, but human intervention in this process makes it time consuming and expensive. Time and techniques to use pesticides in large farmland using AI along with Computer Vision and IoT converts the traditional process into Smart Agriculture.

Chapter 6, titled "Role of Deep Learning in Weed Detection", presents the applications of deep learning in agriculture and a comprehensive survey of the research work done in recent years is provided. It is followed by the description of various techniques of Deep Learning (DL). The next section briefly describes the traditional ways of weed detection and removal. Next, the architecture of deep learning for weed detection and removal is presented along with the associated code. Furthermore, the chapter goes on to discuss the pros and cons of this approach.

Chapter 7, titled "Evaluation of Machine Learning Techniques for Crop Yield Prediction", presents the dataset of different states producing different crops in different seasons is considered. After pre-processing the data, the machine learning algorithms are applied and their results are compared. The farmers are also benefited from yield prediction as it will assist the farmers to predict the yield of crops prior to cultivating. There are multiple parameters that affect the yield of crops like rainfall, temperature, fertilizers, pH level, and other atmospheric circumstances. Thus, considering these factors, the yield of a crop is thus hard to predict and becomes a challenging task.

Chapter 8, titled "An Extensive Analysis of Flying Ad-Hoc Network Applications and Routing Protocols in Agriculture", presents the routing protocols of FANET in different platforms and different strategic manners. Open research challenges have been discussed and possible solutions have been attempted to be drawn from the conclusion. The main contribution lies in suggesting the most suitable routing protocol for each particular agriculture application based on the mobility model and requirement.

Chapter 9, titled "Deep Learning-Based Plant Disease Detection Using Android App", presents the classification of 26 diseases in 14 crop species using 54,306 images from PlantVillage dataset using a Convolutional Neural Network approach. The models used are Inception-v3 and MobileNet. The correct prediction of the correct crop diseases pair in 38 classes decides the criteria for performance measurement. The aim of this work is to develop an Android application which provides an easy -to-use platform for farmers to identify diseases in their crops. The mobile application will help to take responsive action according the disease detected in their plants and can be easily used by anyone who is interested to analyze the disease of the plants.

Chapter 10, titled "A Sustainable Approach of Artificial Neural Network for Prediction of Irrigation, Pesticides, Fertilizers, and Crop Yield", presents the machine learning aspect of precision agriculture. It is the integration of plant and animal production that leaves deep-rooted beneficial effects on the environment, farmers, and society, which is essential for the present climate change and disaster-prone world.

Chapter 11, titled "Impact of Artificial Intelligence and Machine Learning in Food Industry: A Survey", presents how AI and ML have the potential to be used in the food business to save money while simultaneously increasing resource efficiency. It highlights the food industry's achievements and challenges, with specific attention to the role of machine learning and artificial intelligence. It also shows artificial intelligence's application in the food sector for industrial automation and predictive modeling of the world's growing demand and population.

Chapter 12, titled "Applications of Machine Learning in Food Safety", presents the gentle introduction of machine learning in the pretext of food safety and a detailed overview of various developments and applications. Machine learning can be of great help for large and emerging data sets to enhance the safety of food supply and minimize the impact of food safety incidents.

Chapter 13, titled "Artificial Intelligence for Improving Food Quality", presents the Artificial intelligence (AI), and their troublesome role in shaping further of agri-food systems. It wraps up in an enormous amount of streaming data, often noted as "big data", which brings fresh and new opportunities to watch agricultural and food processes. Besides sensors, big data from social media is too becoming important for the food industry.

Chapter 14, titled "The Role of Machine Learning and Computer Vision in the Agri-Food Industry", presents the smart farming (soil monitoring, pest monitoring, and fertilizer management), food processing (production, processing, marketing, and customer feedback), and food safety, among other topics, need a detailed evaluation of machine learning applications in the agri-food business. It is vital to monitor production lines to ensure that the manufacturing process and products fulfill the required quality standards. A plethora of data may now be created throughout the production process due to growing digitization.

Chapter 15, titled "Artificial Intelligence-Based Food Calories Estimation Methods in Diet Assessment Research", presents the review of various AI based food calorie estimation methodologies in diet assessment which are suggested to help normal people and patients so that normal people and doctors could succeed to fight against diet-based health conditions. Even though the packaging of food comes with calorie and nutrition labels, it might not be still very favorable for the reference of people. Thus, the scientists to help people started using AI Based techniques and methodologies to know the ways of determining their daily calorie intake of their food.

*Mohammad Ayoub Khan*
*University of Bisha, Saudi Arabia*

*Rijwan Khan*
*ABES Institute of Technology, India*

*Pushkar Praveen*
*Govind Ballabh Pant Institute of Engineering and Technology, India*

# Chapter 1
# Artificial Intelligence-Based Sustainable Agricultural Practices

**Usha Chauhan**
*Galgotias University, India*

**Divya Sharma**
*Galgotias University, India*

**Sharzeel Saleem**
*Galgotias University, India*

**Mahim Kumar**
*Galgotias University, India*

**Shaurya Pratap Singh**
*Galgotias University, India*

## ABSTRACT

*A world hunger crisis has risen since 2019 when COVID-19 hit the world. This pandemic has shifted this generation back in time, and now it is very important to be involved in new techniques that are effective in terms of better yield with less toxins. With the rate at which the population is growing, it is expected that by the year 2050, the world population would cross 9 billion. This exponential rise would require the food production to rise by 70 to 80%. This is a matter of concern for agriculture and food industries. As the world is in the fourth industrial revolution, it is the need of the hour to embed artificial intelligence and machine learning algorithms with agriculture. This research aims to accumulate different methodologies that are present and come up with a critical analysis. These methodologies have the capability to increase the yield, predict the diseases, and even increase the safety and help enhance traceability.*

DOI: 10.4018/978-1-6684-5141-0.ch001

## INTRODUCTION

To feed 9-10 billion people by 2050, global food production is estimated to need to rise by 60–110% (Krishna, 2016; Rockstrom et al., 2017). As a result, agriculture's long-term viability is critical to ensuring food security and to eradicate hunger for the world's ever-increasing population. Furthermore, a well-written system involving traceability has become a necessity for controlling quality in the food chain as a result of various food safety scandals and accidents in the food business, such as bovine spongiform encephalopathy and dioxin in poultry (Ben-Ayed et al., 2013). Furthermore, weather and climate change circumstances, as well as long-term water management due to shortage, will be major issues in the coming years. For these reasons, a deliberate shift away from the existing paradigm of increased agricultural output and toward agricultural sustainability is urgently required. Helping farmers and stakeholders make better decisions by adopting sustainable agriculture practises, particularly through the use of digital technologies such as the Internet of Things (IoT), Artificial Intelligence (AI), and cloud computing, is a critical choice for anticipating efficient solutions. Additionally, AI components (machine and deep learning algorithms) are frequently integrated with location intelligence technology. The purpose of this research is to highlight the most important uses of artificial intelligence and machine learning in the agri-food industry.

AI technology has just been available for use in the agri-food industry. In reality, AI approaches contribute significantly to understanding a model's identification, service generation, and decision-making processes as well as support for various agri-food applications and supply chain stages. In agriculture, the main purpose of AI is to give accuracy and anticipating decisions in order to increase productivity while preserving resources.(Patel et al., 2021)

The Food demand is expected to rise from 60% to 97% percent by 2050 as the world's population grows (Suchithra & Pai, 2020). Thus, AI has been used to meet this food demand in areas such as supply chain management, food sorting, production development, food quality enhancement, and adequate industrial hygiene.(Ahumada & Villalobos, 2009; Ben Ayed et al., 2017; Kamilaris & Prenafeta-Boldú, 2018) Food safety has been identified as one of the most pressing challenges in the food business, prompting the development of smart packaging technologies to meet the requirements of the food supply chain. The state of foods is monitored through intelligent packaging, which provides details on the food's quality while in storage and transportation (Gaurav et al., 2019). Another study looked into intelligent packaging as a method for reducing food waste, and found that there have been roughly 45 recent improvements in the field of optical systems for freshness monitoring.

## Methodology

### AI in Agriculture and Food Industry

AI-based technologies contribute in the improvement of efficiency in all fields and the management of difficulties faced by numerous industries, including crop yield, irrigation, soil content sensing, crop monitoring, weeding, and crop establishment in the agricultural sector.(Ahumada & Villalobos, 2009) The agriculture sector is experiencing a crisis as the world population grows, but AI has the potential to provide a much-needed answer. Farmers have been able to generate more output with less input and increase the quality of their output, as well as ensure a faster time to market for their harvested crops, which is possible because of AI-based technology solutions.

Disease detection: The image sensing technique guarantee that the plant leaf images are divided into surface areas such as background, area with disease, and area without disease. After that, the contaminated or sick region is further analysed. This also aids in the detection of pests and the detection of vitamin insufficiency. Figure 1 depicts the sequence in detail.(Dharmaraj & Vijayanand, 2018)

*Figure 1. Flow diagram of disease identification*

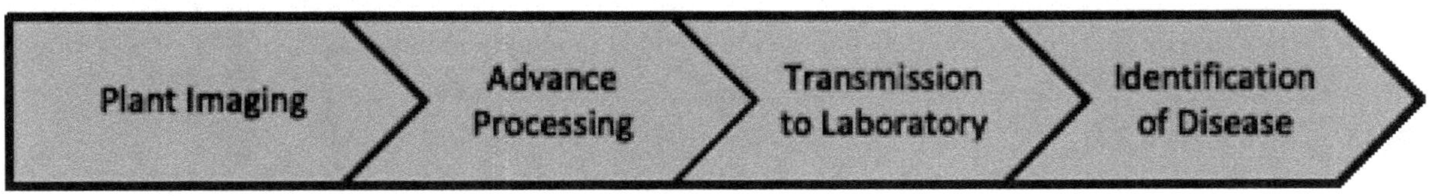


Skills and workforce: artificial intelligence allows farmers to collect large amounts of data from government and open source websites, evaluate it, and gives farmers with solutions to many problematic concerns, as well as a smarter irrigation system that results in higher output for farmers. Farming will be discovered to be a blend of technology and biological skills in the coming years as a result of artificial intelligence, which will provide a better outcome in terms of quality, but it will also reduce their losses and workloads. AI in agriculture can be used to ease the procedures, eliminate risks, and enable farmers with a more simple and coherent farming experience.(Talaviya et al., 2020)

Maximizing the output: upcoming technologies have benefited in the accurate crop selection and have improved the decision of hybrid seed options that are suitable to the wants of farmers. It has been adopted by determining how seeds react to different weather conditions and soil types. Plant diseases are much less likely as a result of this data collection. We can now meet market trends, yearly outcomes, and customer needs, allowing farmers to maximise agricultural returns more efficiently.(Talaviya et al., 2020)

Use of weather forecasting: Farmers can use Artificial Intelligence to analyse weather conditions by using weather forecasting algorithms, that helps to plan the type of crop that can be grown and determine the time as when seeds should be sown.(Jain, 2020)

Analysing health of crops by drones: Drone-based Arial imaging solutions for crop health monitoring have been introduced by SkySqurrel Technologies. This method, involves the drone collects data from agricultural fields, which is subsequently sent to a computer via USB drive and examined by experts. (Jain, 2020)

This company analyses the taken photographs with algorithms and provides a full report on the farm's current health. It assists farmers in identifying pests and bacteria, allowing them to utilise pest management and other ways in a timely manner.

Agricultural Robotics: Robots that can perform several duties in agricultural fields are being developed by AI businesses. When compared to people, these robots are trained to remove weeds and harvest crops at a faster rate with bigger volume. These robots are programmed to inspect crop quality and detect weeds while picking and packaging crops at the same time. These robots can also deal with the difficulties that agricultural labour faces.(Jain, 2020)

It has been observed that the artificial intelligence is playing a critical role in food industry and with days passing by application of Artificial intelligence is increasing in the food industry has due to many reasons like food safety, food storing, predication and classification parameters, quality control. Expert system, like as ANN, adaptive neuro-fuzzy inference system (ANFIS) and machine learning are the popular technique that are being used or utilized in food industries. Before AI was incorporated There was a lot of efforts on training people about agriculture that will increase the crop yield and quality. After AI has been embedded the agriculture industry is growing at more faster rate because it is easier to train a system rather than training a human.

*Figure 2. This block diagram explain the different types of uses of AI in food industry*

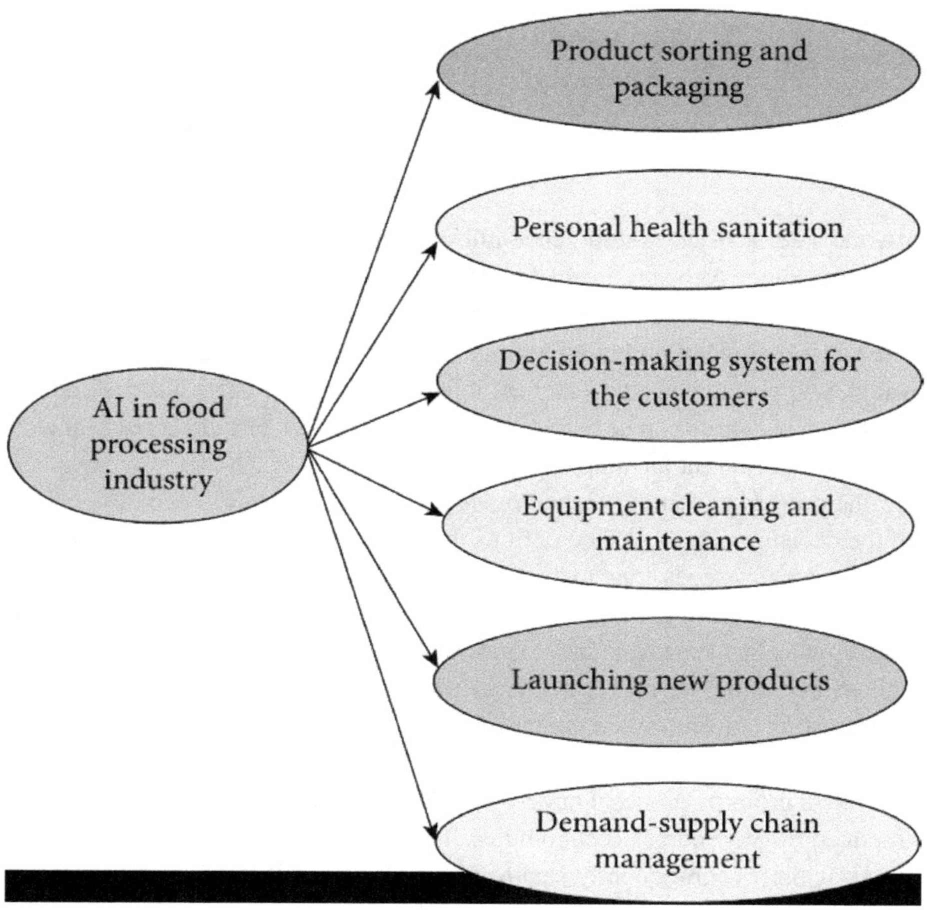


## Crop and Soil Monitoring through AI

Frist of all we have to go through the ground. For the quantity and quality of the crop Micro and macronutrients are very important factors. Once the crops are present in soil, so we have to monitor the soil as well as crop for efficient growth and quality. It's very important to understand the environment and the growth of crop on regular basis to do whatever needed for crop growth (D N & Choudhary, 2021).

The traditional method of measuring the soil quality and crop quality is by the human but this is not that much accurate nor that much timely. In place of traditional method now we can use drones to capture the images data and AI for the intelligent monitoring of crop and soil conditions (Talaviya et al., 2020).

To analyse and interpret this data to we can use Visual sensing AI:

- Health of crop can be tracked
- Prediction are data based so it will accurate
- To detect the diseases and it is more accurate than manual identification

*Figure 3. Drone monitoring in agriculture*

Observing crop maturity: AI can help farmers also to give the data about the maturity of the crop just like whatever the crop is that is ready for go to market or not. If we are taking the example of that type of farmers who are doing farming of vegetables or fruits. they can easily check with the help of AI that their crop is ready for sell or not and the same manner they can able to detect any type of diseases in the crop. In the bellow figure we can see that the AI is detecting the tomato that it is ripped or not. it is doing that with the help of color of the tomato. it is detecting the color through the live image captured and by that it is giving the information that the tomato is ripped or not. In the same manner it can do for any crop (Talaviya et al., 2020).

*Figure 4. Representation of a ripe/not-ripe fruit*

In the previous sections we have observed that the with the help of AI we can easily predict the environment and soil condition as well as we can determine the maturity of the crop but what about the, but when we are moving to agriculture condition that are less predictable compare to the previous ones?

With the help of image recognition technology which is based on deep learning, we can now able to automate detection of plant diseases and pests. It is working on the image classification detection and segmentation method with the help of this we build the module and its work is to keep track of plant health.(Geetha et al., 2020)

*Figure 5. Disease classification based on image processing*

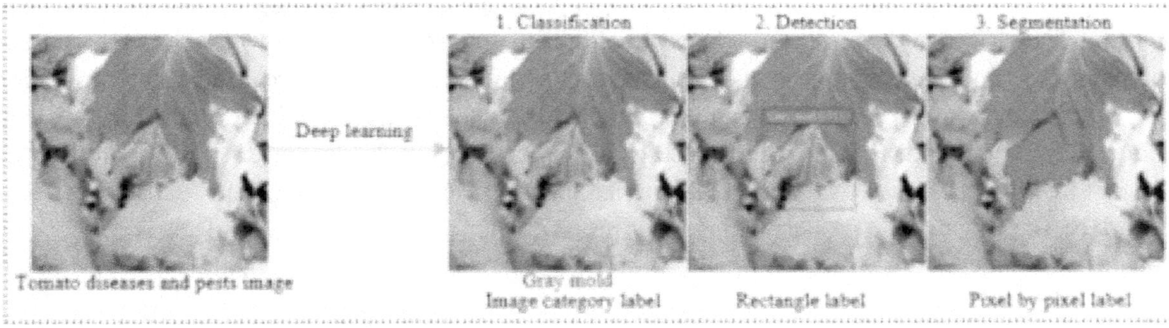


Automatic Weeding: If we are thinking that the Intelligent sprayers are the only AI getting into weeding then it is totally wrong. There are too many AI using robot which are going to use in removing the unwanted plants.

Now from here we are going to detect weed with the same manner we are detecting insects. but for the larger help of farmer the we have to use the AI for both purpose detecting weed as well as removing it.(Zhang et al., 2017)

*Figure 6. Weed detection using image analysis and AI*

Difficulty faced by farmers while using AI: Farmers have a tendency to think of AI as something that only applies in the digital world. They may be unable to see how it can assist them in working on the actual land. This isn't because they're fearful of the unknown or conservative. Their resistance stems from a lack of awareness of how AI tools can be applied in the real world. Because Agri-Tech vendors fail to adequately explain why their solutions are valuable and how they should be deployed, new technologies sometimes appear complex and excessively expensive. This is what occurs in agriculture when artificial intelligence is used. Although AI can be beneficial, technology companies must still do a lot of work to assist farmers in properly implementing it.(Zhang et al., 2018)

Agriculture includes a variety of processes and phases, the majority of which are performed manually. AI can help with the most complex and routine jobs by supplementing existing technology. When integrated with other technology, it can gather and evaluate massive data on a digital platform, determine the best course of action, and even initiate that action (Ali et al., 2018a).

*Figure 7. Function of AI in information management and implementation*

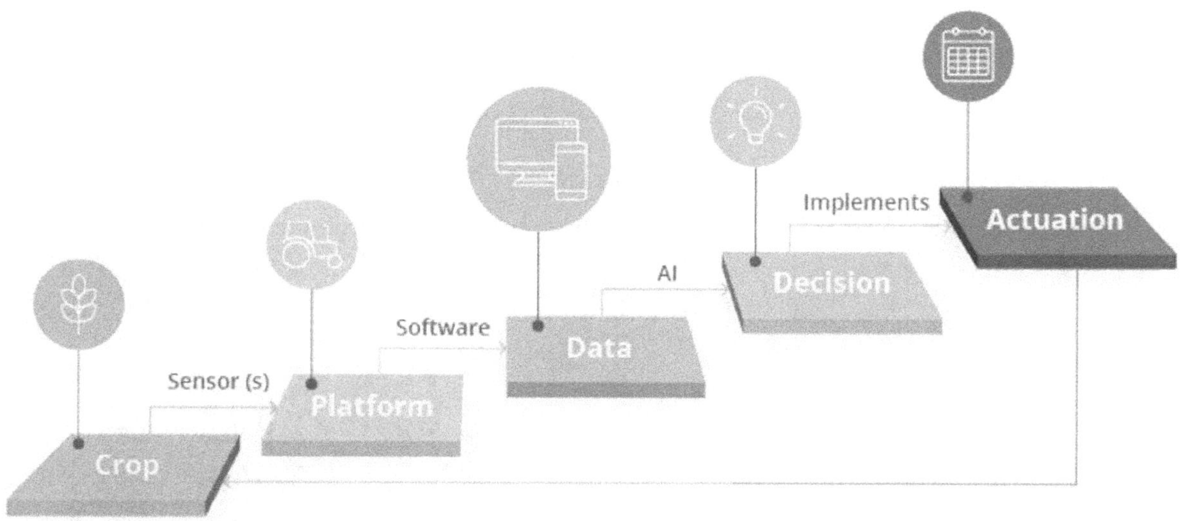


AI can be used for further analysis of the market in the following ways

*Figure 8. Processes involved that would benefit AI*

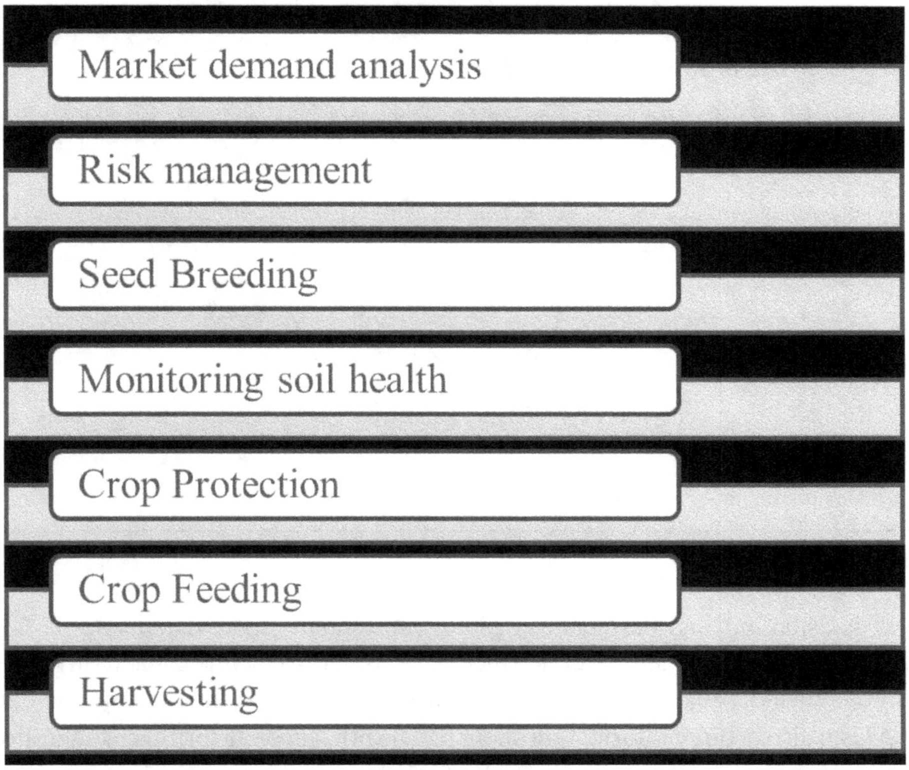


Analyzing market demand: I can simplify crop selection by analyzing market demand and assisting farmers in identifying the most profitable product.

Risk management: Forecasting and predictive analytics can help farmers reduce errors in business processes and lower the chance of crop failure.

Seed breeding: I can assist develop crops that are less prone to disease and better adaptable to environmental conditions by gathering data on plant growth.

Monitoring soil health: AI systems that monitor soil health can conduct chemical soil analysis and provide precise estimations of missing nutrients.

Protecting crops: AI can detect and even anticipate diseases in plants, identify and eradicate weeds, and offer appropriate pest control.

Feeding crops: AI can be used to determine the best irrigation and fertilizer treatment times, as well as anticipate the best agronomic product combination.

Harvesting: Harvesting can be automated with AI, and the ideal time for harvesting can even be predicted.

## How AI Helps in Different Challenges

*Figure 9. Overview of use of AI in agricultural practices*

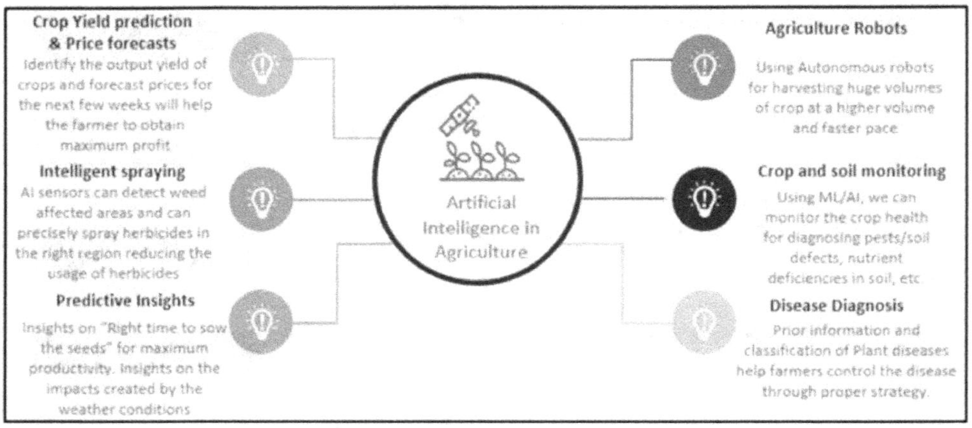


AI helps better decision making: Farmers can gather substantially more data using AI than they could without it, and they can do so much faster. Farmers may use AI to handle critical difficulties including analyzing market demand, projecting pricing, and calculating the best time to sow and harvest.

However, AI can also acquire information about soil health, generate fertilizer suggestions, check the weather, and track the availability of product. All of this helps farmers make better judgments at every stage of the crop-growing process (Al-Jarrah et al., 2015).

Cost savings from AI: Precision agriculture, for example, can assist farmers in growing more crops with fewer resources. Precision agriculture enabled by artificial intelligence (AI) could be the next big thing in agriculture. Precision farming combines the greatest soil management strategies, variable rate technologies, and data management strategies to assist farmers increase yields while lowering costs.

Farmers can use AI to get deep insights into their fields, enabling them to spot areas that require irrigation, fertilizer, or pesticide application. Furthermore, modern farming techniques such as vertical agriculture may assist improve food output while reducing resource consumption. Herbicide use is reduced, harvest quality is improved, earnings are increased, and significant cost savings are realized as a result.

Artificial intelligence can help with labor shortages: Agricultural work is difficult, and labor shortages are nothing new in this business. With the use of automation, farmers can address this problem. Farmers can save money by using self-driving tractors, smart irrigation and fertilization systems, vertical farming software, and AI-based harvesting robots. AI-driven farm tools are faster, tougher, and more accurate than any human farm worker.

Farmers must realize that artificial intelligence is really a more advanced version of basic technologies for processing, acquiring, and analyzing field data. For AI to function, it requires a proper technological infrastructure. As a result, even those farms with some technology in place may find it difficult to progress (Akhtar et al., 2016).

This is a problem for software companies as well. Farmers should be approached gradually, with simpler technology such as an agricultural trading system being offered first. It will be reasonable to step up and offer something additional, including AI features, once farmers have been adapted to a less difficult solution.

Agriculture in developing countries differs from that in Western Europe and the United States. Artificial intelligence agriculture could be beneficial in some areas, but it may be difficult to sell in areas where agricultural technology is not widely used. It is likely that farmers will require assistance in implementing it (Behmann et al., 2015).

As a result, tech firms interested in doing business in rising agricultural economies may need to be strategic. They'll have to give training and continuing support for farmers and agribusiness owners who are willing to try out new ideas in addition to delivering their products.

Precision agriculture create several legal difficulties that often go unanswered because there are no defined policies and regulations surrounding the use of AI not just in agriculture but in general. Farmers may face major challenges as a result of privacy and security threats like as cyberattacks and data leaks. Unfortunately, many farmers are at risk from these dangers.

Big data as a tool for making well-informed decisions: The real purpose of data collection and production is to put it to good use. Data analytics in agriculture can lead to large production gains and significant cost reductions. Farmers can acquire credible advice based on well-sorted reliable details on crop demands by merging AI and big data. As a result, guessing will be eliminated, allowing for more accurate farming.

Robotics and automation to reduce manual work: One of the most common challenges in farming is a manpower shortage, which can be solved with artificial intelligence, automated tractors, and the Internet of Things. Because these technologies are more accurate and so eliminate errors, they have the potential to be cost-effective. AI, automated tractors, and the Internet of Things, when combined, are the key to precision agriculture.

Robotics is a less well-known but quickly increasing technology. Agricultural robots are already performing laborious tasks like harvesting fruits and vegetables. Robots have a number of benefits over farmworkers. They can work for longer periods of time, are more exact, and are less likely to make mistakes.

IoT devices for data collection and analysis: Farmers can monitor, measure, and save data from fields in real time using IoT sensors and other technologies (such as drones. Farmers may acquire more accurate information faster by combining AI agricultural technologies with IoT sensors and software.

Artificial intelligence expectations vs. reality for sustainable farming: The advantages of artificial intelligence in agriculture are unquestionable. Farm employees can use their time for more strategic activities that demand human intellect by using smart farming instruments and vertical farming systems to accomplish small, repeatable, and time-consuming activities. It's crucial to note, though, that, unlike a tractor, AI cannot be purchased and started. AI isn't something that can be touched. It's a collection of technologies that have been programmed to work together (Singh et al., 2020).

Artificial intelligence is simply simulation of reasoning; it is data-driven learning and problem-solving. AI is only the next step in the evolution of smart farming, and it requires the use of other technologies to function properly. To put it another way, farmers will need a technological infrastructure before they can maximize the benefits of AI.

## Agricultural Lifecycle

The following process is involved in lifecycle of agriculture

*Figure 10. Process diagram of cycle of agriculture*

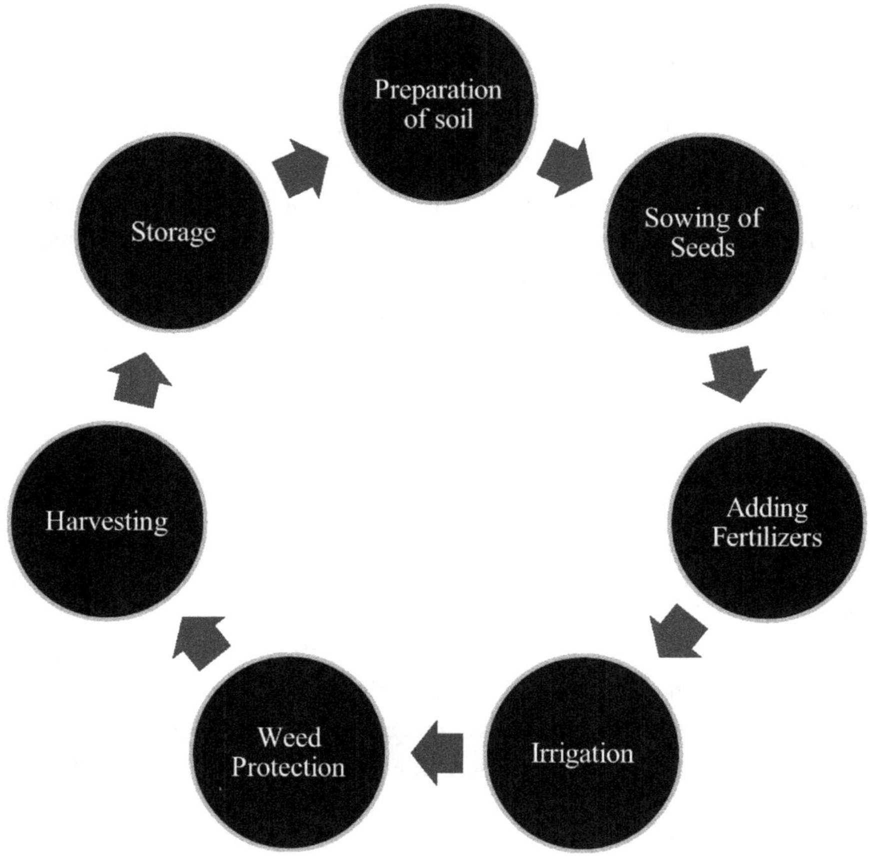


Preparation of soil: It is the first stage of agriculture where farmers have to prepare the soil accordingly for sowing seeds. In this process first of all we break large soil in to clumps and then remove debris like as roots, sticks and rocks. the fertilizers are added according to the type of crop as well as according to the type of soil also.

Sowing of seeds: after preparing the soil we are going for sowing the seeds process in this stage we have to take care of the distance between the two seeds and in how much depth we are planting it. farmers have to also keep eye on the climatic conditions like as rainfall, temperature humidity these are going to play very critical role in sowing the seeds.

Adding Fertilizers: To maintain the soil fertility it is an important method to add fertilizers to it. if farmers are not mixing the fertilizers, then the quantity of the crop will get reduced and it will cause financial damage to the farmers. Farmers started using fertilizers because the fertilizers contain potassium, nitrogen, phosphorus etc. If we are going to know what is fertilizers? then the answer comes is fertilizers are nothing but nutrients applied to the filed to give necessary elements naturally in soil.it is very important stage because by the this stage only the quality and quantity of crop is depend.

Irrigation: this stage is responsible for maintaining the humidity and moist of the soil. If the amount of water is more then it also harmful and the if amount of water is more than it is also harmful. if it can't be done properly the it can cause heavy damage to crops.

Weed protection: we can easily see that there are too many unwanted plants near the crop or on the boundary of the crop they are known as the weeds. farmer needs to protect the crop from the weed because it decreases the yields and increase the production cost and decrease the crop quality.

Harvesting: in this process we gather the crop from the field we can simply say that it is a post-production process which include cleaning sorting packing and cooling. In this process there is a lot of labor needed to do it.

Storage: This in this stage products are kept in that type of place. Where it can't get damaged by different types of climate changes. this phase also includes the packing of the product. Packing should be good so it can't get damaged. This phase also includes the transportation.

## COMPARITIVE ANAYLSIS

*Table 1. Comparison of different research works*

| **Authors** | *Algorithm/System* | *Findings* |
|---|---|---|
| F. K. Shaikh, M. A. Memon, N. A. Mahoto, S. Zeadally and J. Nebhen | RESNET, GRU, and CNN(Shaikh et al., 2022) | The authors have used an embedded system which is helpful in soil management, they have monitored the soil moisture and based on that different conclusion have been stated |
| F. Bu Xi.Wang | Deep Reinforcement learning(Bu & Wang, 2019) | Deep learning models by policy gradient methods have been proposed which uses Markov decision process. |
| H.Bhardwaj P., Tomar, A.Sakalle, U. Sharma | Computer vision based farming and use of AI robots (Bhardwaj et al., 2021) | The authors have used a weather monitoring system embedded with computer vision-based use of disease prediction; this enables early detection of diseases in crops. |
| A.R.Zanella, E. d. Silva, L. C. Albini | Data security using IOT and AI (Rettore de Araujo Zanella et al., 2020) | The authors have developed a system that uses internet protocol to prevent hijacking of UAV and smart robots used in agriculture. |
| V.S. Magomadov | DL based model with six convolutional kernels(Magomadov, 2019) | The author has developed a 6-layer convolutional model with a test accuracy of 95% and train accuracy of 96% |
| L. Jun, S.Zhiqi, Z. Anting, C. Yueting | Blockchain and IoT based smart ecosystem (Lin et al., 2018) | The authors have proposed a smart ecosystem that incorporates blockchain that gives and extensive overview as ERP legacy system embedded with IoT. |
| M. Naresh and P. Munaswamy | WSN system using ARM processor (Naresh & Munaswamy, 2019) | The author has used ARM processor LPC2148 with humidity sensors to gather data and it is displayed on LCD screen and further this data is uploaded to cloud for predictive analysis |
| S. Saleem and R. Rana | Locusts repellent using basic electronic components(Saleem & Rana, 2021) | The authors have developed an ultrasonic locust repellent circuit that is low cost and is helpful in driving away these insects from the agriculture field. |
| P.K.Srivastava, J.shiney, P.shan | Plant leaf disease identification using image processing (Srivastava et al., 2021) | The authors have classified the disease based on abiotic and biotic factors and stated the symptoms as rusts, mold, smuts, distortion. |
| K.Muruganandam & U. Chauhan | Used soil moisture sensor for developing IOT-based irrigation sytem (Muruganandam & Chauhan, 2021) | The authors have used IoT for solving the problem also a systematic rule-based model has developed. |

## CONCLUSION

All the paper that has been reviewed are last 4 years shows methodologies that are beneficial in terms of agriculture field protection, high yield, disease prediction and even pest repellents. Mostly the authors focused on the livestock management and data protection when it comes to artificial intelligence, very few authors have developed a robust system for smart agricultural practices.

Future scope-The future scope of this research can be doing a more extensive and efficient research in this field and develop and more effective model. Further a comparative study can also be made.

## REFERENCES

Ahumada, O., & Villalobos, J. R. (2009). Application of planning models in the agri-food supply chain: A review. *European Journal of Operational Research, 196*(1), 1–20. doi:10.1016/j.ejor.2008.02.014

Akhtar, P., Tse, Y. K., Khan, Z., & Rao-Nicholson, R. (2016). Data-driven and adaptive leadership contributing to sustainability: Global agri-food supply chains connected with emerging markets. *International Journal of Production Economics, 181*, 392–401. doi:10.1016/j.ijpe.2015.11.013

Al-Jarrah, O. Y., Yoo, P. D., Muhaidat, S., Karagiannidis, G. K., & Taha, K. (2015). Efficient machine learning for big data: A review. *Big Data Research, 2*(3), 87–93. doi:10.1016/j.bdr.2015.04.001

Ali, M., Deo, R. C., Downs, N. J., & Maraseni, T. (2018a). Multi-stage hybridized online sequential extreme learning machine integrated with Markov Chain Monte Carlo copula-Bat algorithm for rainfall forecasting. *Atmospheric Research, 213*, 450–464. doi:10.1016/j.atmosres.2018.07.005

Behmann, J., Mahlein, A. K., Rumpf, T., Römer, C., & Plümer, L. (2015). A review of advanced machine learning methods for the detection of biotic stress in precision crop protection. *Precision Agriculture, 16*(3), 239–260. doi:10.100711119-014-9372-7

Ben Ayed, R., Ennouri, K., BenAmar, F., Moreau, F., Triki, M. A., & Rebai, A. (2017). Bayesian and phylogenic approaches for studying relationships among table olive cultivars. *Biochemical Genetics, 55*(4), 300–313. doi:10.100710528-017-9802-0 PMID:28466379

Ben-Ayed, R., Kamoun-Grati, N., & Rebai, A. (2013). An overview of the authentication of olive tree and oil. *Comprehensive Reviews in Food Science and Food Safety, 12*(2), 218–227. doi:10.1111/1541-4337.12003

Bhardwaj, H., Tomar, P., Sakalle, A., & Sharma, U. (2021). *Artificial Intelligence and Its Applications in Agriculture With the Future of Smart Agriculture Techniques*.. doi:10.4018/978-1-7998-1722-2.ch002

Bu, F., & Wang, X. (2019). A smart agriculture IoT system based on deep reinforcement learning. *Future Generation Computer Systems, 99*, 500–507. doi:10.1016/j.future.2019.04.041

D N, V., & Choudhary, D. S. (2021). An AI solution for Soil Fertility and Crop Friendliness Detection and Monitoring. *International Journal of Engineering and Advanced Technology, 10*(3), 172–175. doi:10.35940/ijeat.C2270.0210321

Dharmaraj, V., & Vijayanand, C. (2018). Artificial Intelligence (AI) in Agriculture. *International Journal of Current Microbiology and Applied Sciences, 7*(12), 2122–2128. doi:10.20546/ijcmas.2018.712.241

Gaurav, V., Choudhary, S., Singh, A., & Agarwal, S. (2019). *Autonomous crop irrigation system using artificial intelligence. International Journal of Engineering and Advanced Technology* , 8).

Geetha, G., Samundeswari, S., Saranya, G., Meenakshi, K., & Nithya, M. (2020). Plant Leaf Disease Classification and Detection System Using Machine Learning. *Journal of Physics: Conference Series, 1712*(1), 012012. doi:10.1088/1742-6596/1712/1/012012

Jain. (2020). *Artificial Intelligence in Agriculture: Using Modern Day AI to Solve Traditional Farming Problems*. Data Science Bloagthon.

Kamilaris, A., & Prenafeta-Boldú, F. X. (2018). Deep learning in agriculture: A survey. *Computers and Electronics in Agriculture, 147*, 70–90. doi:10.1016/j.compag.2018.02.016

Krishna, K. R. (2016). *Push Button Agriculture: Robotics, Drones, Satellite-Guided Soil and Crop Management*. Apple Academic Press.

Lin, J., Shen, Z., Zhang, A., & Chai, Y. (2018). Blockchain and IoT based food traceability for smart agriculture. *Proceedings of the 3rd International Conference on Crowd Science and Engineering (ICCSE 2018)*, 3. 10.1145/3265689.3265692

Magomadov, V. S. (2019). Deep learning and its role in smart agriculture. *Journal of Physics: Conference Series, 1399*(4), 044109. doi:10.1088/1742-6596/1399/4/044109

Muruganandam, K., & Chauhan, U. (2021). Soil Moisture Sensor Nodes in IoT-Based Drip Irrigation System for Water Conservation. In Smart Sensors for Industrial Internet of Things. Internet of Things. Springer. doi:10.1007/978-3-030-52624-5_13

Naresh, M., & Munaswamy, P. (2019). Smart Agriculture System using IoT Technology. *International Journal of Recent Technology and Engineering, 7*(5).

Patel, G. S., Rai, A., Das, N. N., & Singh, R. P. (Eds.). (2021). *Smart Agriculture: Emerging Pedagogies of Deep Learning, Machine Learning and Internet of Things* (1st ed.). CRC Press. doi:10.1201/b22627

Rettore de Araujo Zanella, A., da Silva, E., & Carlos Pessoa Albini, L. (2020). Security challenges to smart agriculture: Current state, key issues, and future directions. *Array, 8*, 100048. doi:10.1016/j.array.2020.100048

Rockstrom, Williams, & Daily. (2017). Sustainable intensification of agriculture for human prosperity and global sustainability. *Ambio, 46*(1), 4–17.

Saleem, S., & Rana, R. (2021). Ultrasonic Locust Repellent Circuit. *2021 International Conference on Computational Intelligence and Knowledge Economy (ICCIKE)*, 547-549. 10.1109/ICCIKE51210.2021.9410712

Shaikh, Memon, Mahoto, Zeadally, & Nebhen. (2022). Artificial Intelligence Best Practices in Smart Agriculture. *IEEE Micro, 42*(1), 17-24. . doi:10.1109/MM.2021.3121279

Singh, R., Srivastava, S., & Mishra, R. (2020). AI and IoT Based Monitoring System for Increasing the Yield in Crop Production. *2020 International Conference on Electrical and Electronics Engineering (ICE3),* 301-305. 10.1109/ICE348803.2020.9122894

Srivastava, Shiney, & Shan. (2021). Plant Disease Prediction using Image Processing and Soft Computing Algorithms: A Review. *2021 International Conference on Computational Intelligence and Knowledge Economy (ICCIKE),* 143-148. . doi:10.1109/ICCIKE51210.2021.9410728

Suchithra, M. S., & Pai, M. L. (2020). Improving the prediction accuracy of soil nutrient classification by optimizing extreme learning machine parameters. *Information Processing in Agriculture*, *7*(1), 72–82. doi:10.1016/j.inpa.2019.05.003

Talaviya, T., Shah, D., Patel, N., Yagnik, H., & Shah, M. (2020). Implementation of artificial intelligence in agriculture for optimisation of irrigation and application of pesticides and herbicides. *Artificial Intelligence in Agriculture.*, *4*, 58–73. Advance online publication. doi:10.1016/j.aiia.2020.04.002

Zhang, C., Li, M., & Guo, P. (2017). An interval multistage joint-probabilistic chance-constrained, programming model with left-hand-side randomness for crop area planning under uncertainty. *Journal of Cleaner Production*, *167*, 1276–1289. doi:10.1016/j.jclepro.2017.05.191

Zhang, C., Liu, J., Shang, J., & Cai, H. (2018). Capability of crop water content for revealing variability of winter wheat grain yield and soil moisture under limited irrigation. *The Science of the Total Environment*, *631*, 677–687. doi:10.1016/j.scitotenv.2018.03.004 PMID:29539596

# Chapter 2
# Applications of Deep Learning in Agriculture

**Padmesh Tripathi**
https://orcid.org/0000-0001-9455-1652
*IIMT College of Engineering, India*

**Nitendra Kumar**
https://orcid.org/0000-0001-7834-7926
*IIMT College of Engineering, India*

**Mritunjay Rai**
*Shri Ramswaroop Memorial University, India*

**Ayoub Khan**
https://orcid.org/0000-0002-0398-3722
*University of Bisha, Saudi Arabia*

## ABSTRACT

*Today's era is the era of technologies. Technologies have widely been employed in each and every field. The field of agriculture is not untouched with the technologies, and in several segments of agriculture; it has been employed at large. Deep learning techniques and its variants like convolutional neural networks (CNN), recurrent neural networks (RNN), generative adversarial network (AGN), and their various subcategories like AlexNet, ImageNet, visual geometry group (VGG), etc. have widely been employed in many sectors of agriculture in order to increase the quality and quantity of production. In this chapter, some applications of deep learning have been explored.*

DOI: 10.4018/978-1-6684-5141-0.ch002

# INTRODUCTION

## Smart Farming

With the applications of technologies in agriculture, the term smart farming has emerged. Though, the majority of farmers is not aware with this term as they are not familiar with the technologies being used. But, the future of agriculture lies in the technology based farming. Smart farming talks about the applications of technologies like location systems, internet of things, robots, sensors, artificial intelligence, etc. in farming. Using these technologies, a farmer can increase both the quantity and quality of crops (Tyagi, 2016).

Smart farming/agriculture is a technology that depends on its employment on the application of artificial intelligence (AI) and internet of things (IoT) in cyber-physical farm management (Bacco et al., 2019). Smart farming talks of many issues associated with the crop production like observing the changes of climate factors, soil moisture (Kumar et al., 2021), soil characteristics, etc. IoT technology is able to link various remote sensors such as ground sensors, robots, and drones, as this technology allows devices to be linked together using the internet to be operated automatically (AlMetwally et al., 2020). IoT based Smart farming technology are beneficial in improvement of product quality, irrigation and plant protection, disease prediction, fertilization process control, etc. (Adamides et al., 2020). With the applications of IoT, all the agricultural equipment and devices are connected together to take precise decisions infertilizer supply and irrigation (Kumar and Periasamy, 2021). Darwin et al. (2021) have surveyed the applications of deep learning in smart farming.

Some technologies applied in farming are listed below:

- Location systems – GPS, satellites, etc.
- Sensors – for light soil, moisture, water, temperature management, etc.
- Robots - for spaying the pesticides
- Drone cameras – for monitoring the status of crops
- Precise plant nutrition an precision irrigation
- Software platforms
- Optimization and analytics platforms

With the rising population worldwide, the rise in food production is essential. Not only food production is essential but also it is essential to maintain the availability and nutritional quality worldwide. Smart farming is essential to deal with the challenges encountered in agricultural production in terms of productivity, food security, environmental impact, and sustainability (Gebbers and Adamchuk, 2010).

A report of the Food and Agriculture Organization (FAO 2017) revealed that approximately 20–40% of crops are lost per annum due to pests and diseases and as a result of lack of good monitoring of the state of the crop. Hence, the use of smart farming technologies allows monitoring of weather factors, fertility status, and also determining the exact amount of fertilizers necessary for crop growth.

## Deep Learning (DL)

Last two decades have witnessed a new technology called deep learning (DL) which made things simpler in dealing with a set of large data. It is a branch of yet another powerful technology called machine learning. Artificial neural networks are the backbone of deep learning. As neural network is simply impersonator of human brain, DL is also an impersonator of human brain. The beauty of deep learning is that one doesn't need to program everything.

A neural network (NN) is a circuit or network of biological neurons, or in a contemporary sense, an artificial neural network (ANN), made up of artificial neurons or nodes (Hopfield, 1982). According to that, a NN consists of a biological NN, composed of biological neurons, or an ANN, for unraveling hitches related to AI (artificial intelligence). Weights between different nodes are vital in NN. The networks of

*Figure 1. Machine learning versus deep learning*

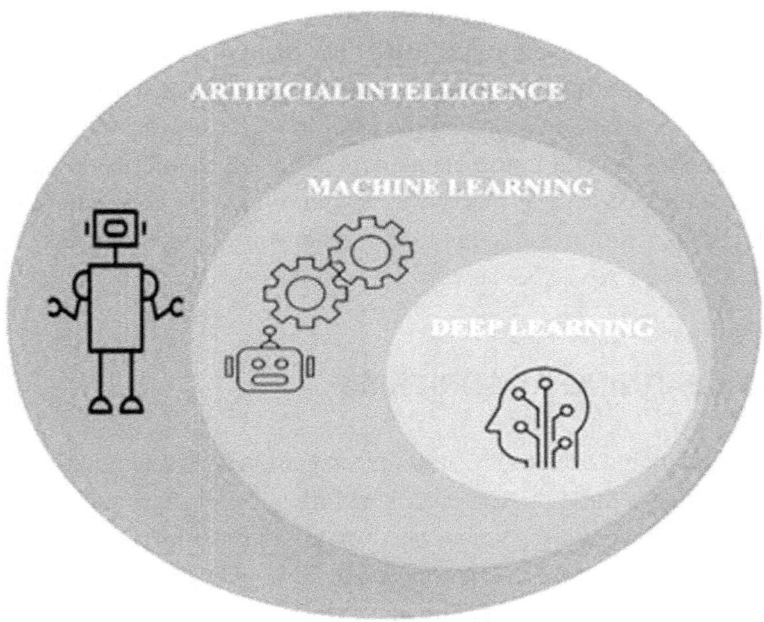


*Figure 2. Structure of neural network*

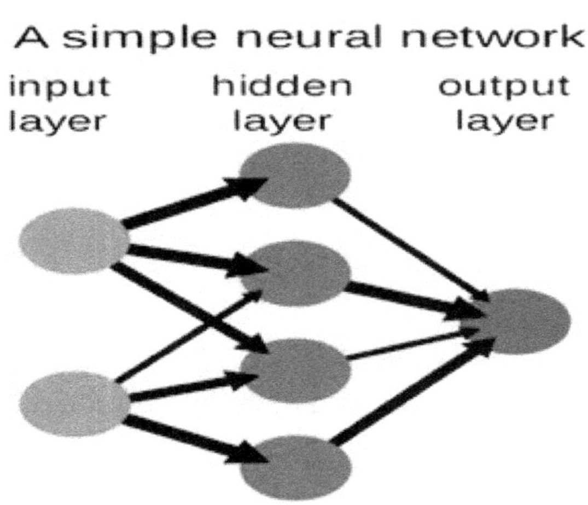

the biological neurons are exhibited in ANN as weights between diverse nodes. Both the positive and negative weights have echoes. Excitatory connections are reflected by a positive weight; on the other hand, inhibitory connections are reflected by a negative weight. Each input is reformed by a weight and added up. This action is known as linear combination. Ultimately, an activation function controls the amplitude of the output. These artificial networks can be applied for adaptive control, predictive modeling and the applications in which they can be trained via a dataset.

DL deals with the "deeper" NNs that offer a hierarchical depiction of the data via numerous convolutions. Due to this, deep learning enjoys higher learning proficiencies and hence, higher precision and performance (Kamilaris and Prenafeta - Boldú, 2018). One of the best important features of DL is feature learning (Le Cun et al., 2015). In DL, complex models are used and these complex models permit immense parallelization. That is the reason why DL can resolve the multifaceted problems in more efficient and fast manner (Panand Yang, 2010). The main disadvantage of deep learning is that it needs large data sets. If there are large datasets, the complex models used in deep learning have abilities to escalate the classification accuracy or lessen the error in regression problems.

Deep learning comprises several components depending on the network architectures employed. Components consists of encode/decode schemes, convolutions, fully connected layers, pooling layers, gates, activation functions, memory cells, etc. and the network architecture consists of Unsupervised Pre-trained Networks(UPN), RNN, CNN, etc.

## COMMON DEEP LEARNING ALGORITHMS

Several DL algorithms are being used today. But, the very often used and popular algorithms are CNN, RNN and GAN. In addition to these three main algorithms, several other sub categories of deep learning algorithms like: Deep convolution generative adversarial networks (DCGAN) (Suárez, et al 2017), Visual Geometry Group network (VGGNet) proposed by Simonyan and Zisserman (2014), Long Short-Term Memory (LSTM) networks proposed by Fan, et al. (2014), etc. can be derived from these three.

### Convolutional Neural Networks (CNN)

CNN is a DL algorithm made up of various convolutional layers, fully connected layers and pooling layers, that has widely been used in the field of speech recognition, natural language processing, face recognition, etc. (Razavian, et al., 2014). The structure made up of the pooling layers plus the convolutional layers are used for feature extraction while the fully connected layers are used as a classifier. The computation process for convolution and pooling layers are shown in figure 3 and figure 4 while flow chart for CNN is shown in figure 5.

### Recurrent Neural Networks (RNN)

RNN is an algorithm widely used for sequential data. The specificity of this algorithm is that it is first algorithm which remembers its input, because of an internal memory, which brands it effortless appropriate for machine learning. In deep learning, RNN has been widely employed and produced amazing results in last two decades. RNN can recall significant things regarding the received input, which permits them in guessing precisely about coming next. This is the reason why RNN is the favorite algorithm in deep

*Figure 3. Convolution layer computation process*

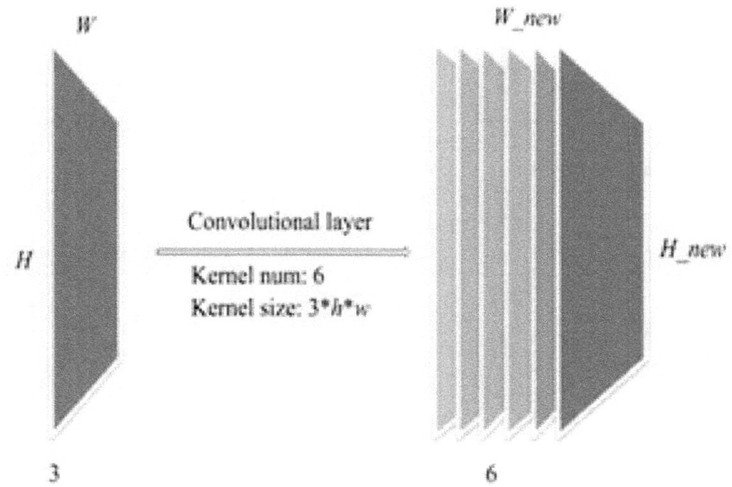


*Figure 4. Pooling layer computation process*

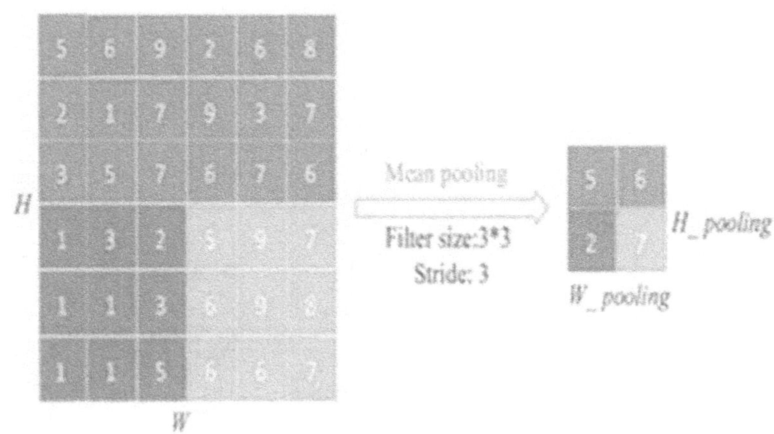


learning for sequential data like audio, video, time series, speech, weather, text, language, financial data, etc. RNN may be regarded as back propagation NN where the output is applied in place of the input of subsequent network (LeCun, 2015).
Figure 6 represents the flow chart for RNN programming.

## Generative Adversarial Networks (GAN)

GAN learns to generate new data having the same statistics for a given training data set. The central notion behind GAN is the "indirect" training using the discriminator. Discriminators are a type of neural networks having the ability to communicate the amount of the reality of an input, which itself is also being restructured dynamically (Vanilla, 2020). Two networks: generative and discriminative are employed here. The generative network generates contenders while the discriminative network performs evaluations

*Figure 5. Flow chart for CNN*

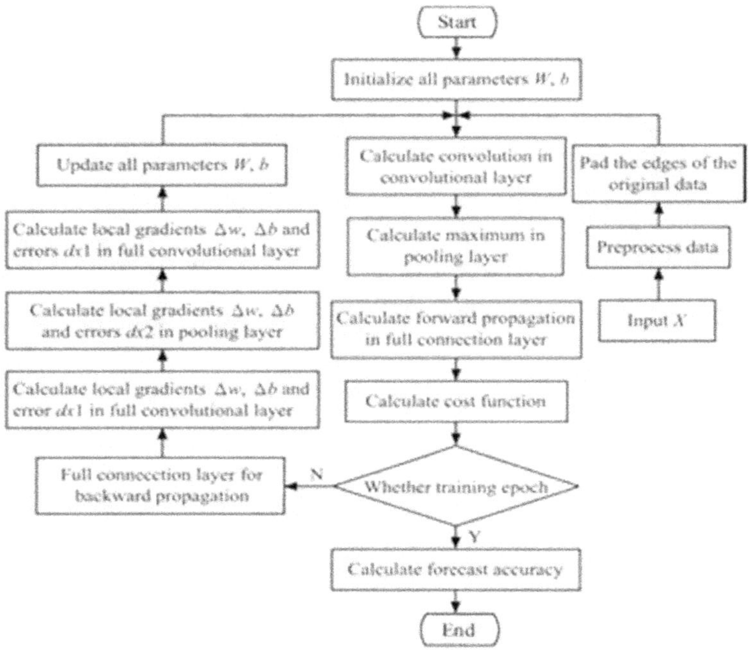


*Figure 6. Flow chart of RNN programming*

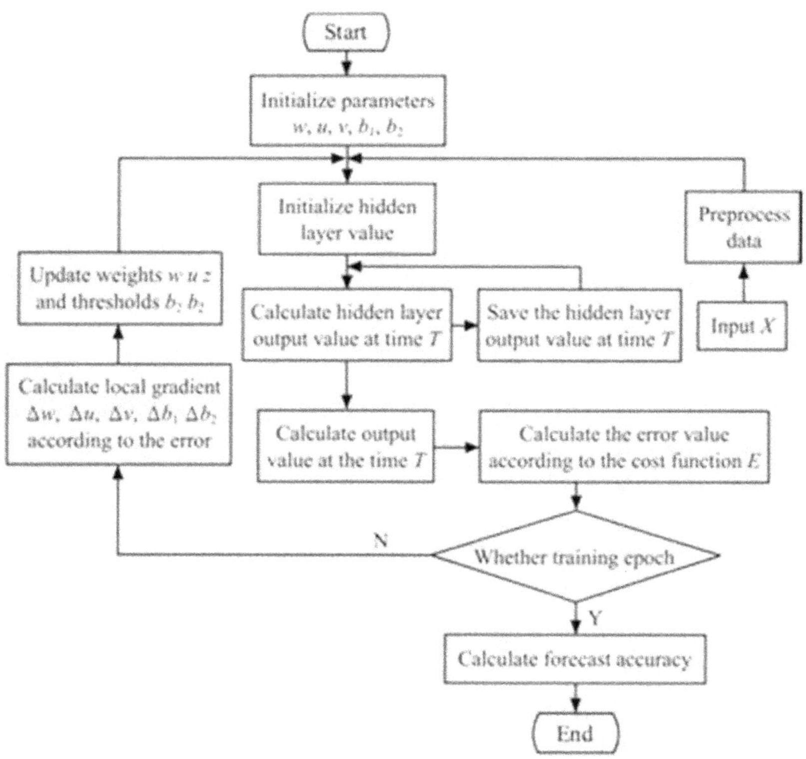

*Figure 7. Generative adversarial networks structure*

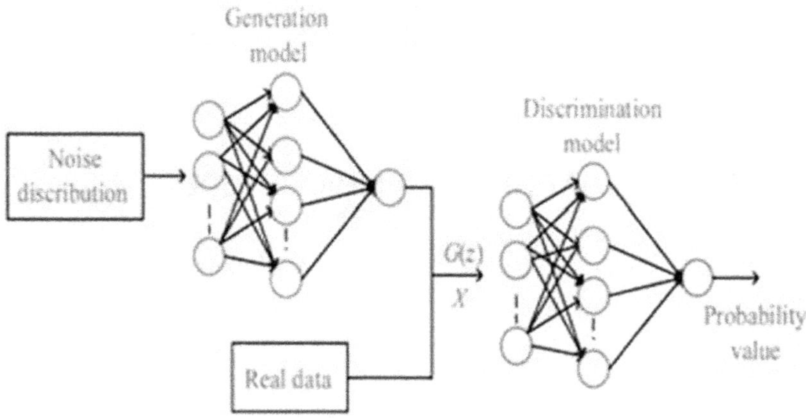


*Figure 8. Flow chart for generative adversarial network*

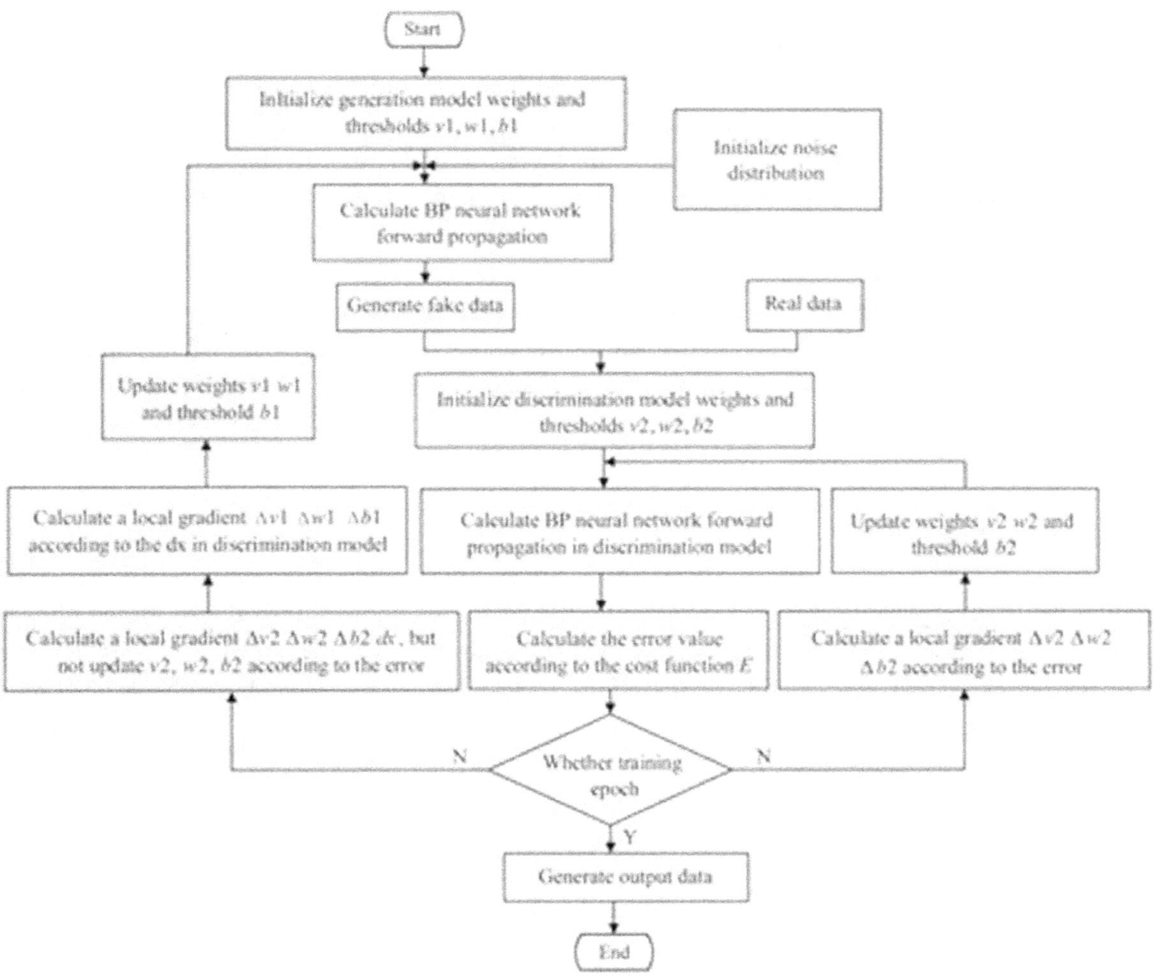

of them. For the discriminator, an identified dataset works as the preliminary training data. Training it include expressing it with samples from the training dataset, until it reaches to an adequate accuracy.

Figure 7 represents generative adversarial network structure for a DNN which generates the fake data $G(z)$ and x is a real data set.

## APPLICATIONS OF DEEP LEARNING IN AGRICULTURE

DL algorithms CNN, RNN and GAN have widely been used in many fields of agriculture. Below are listed some applications:

### Plant Disease Detection

A huge amount of time is consumed in detecting plant diseases when performed manually. With the emergence of artificial intelligence, this problem has been easily handled able using image processing. Pattern recognition and leaf image classification have played vital role in detecting the plant diseases. Based on these, several models have been developed for detecting the plant diseases. An innovative deep learning framework established by the Berkley Vision and Learning Centre (BVLC) has been employed to shape a model for detecting the plant diseases.

Sladojevic, et al. (2016) proposed a technique on the basis of leaf image classification employing the deep convolution network. Caffe was employed to execute the deep CNN training. This model resulted in recognition of 13 plant diseases. In addition, this model attained precision between 91% and 98%.

Chen, et al. (2020) suggested an approach to identify plant disease in which VGG (visual geometric group) Net pre-trained on ImageNet and inception module were chosen. The accuracy level of more than 91.82% was achieved.

Hassan, et al. (2021) proposed a model using Shallow Convolutional Neural Networks. They used shallow visual geometric group (VGG) incorporating Random Forest (RF) and shallow VGG incorporating Xgboost to identify the diseases on corn, potato and tomato data sets. They observed that Shallow VGG incorporating Xgboost resulted the uppermost accuracy of 94.47%, 98.74%, and 93.91% respectively in corn, potato and tomato datasets.

### Weather Forecasting

Weather forecasting plays vital role in agriculture. If farmers can know the weather of future, they can plan accordingly. Several deep learning techniques have employed in weather forecasting in last few years.

Biswas, et al. (2014), employed CBR (case-based reasoning) and an RNN based NARXNet model in weather forecasting and compared the accuracy performance. They concluded that NARXNet out-performed with an accuracy of 93.95%.

Salman, el al. (2015) compared the deep learning algorithms CN, RNN and CRBM (Conditional Restricted Boltzmann Machine) and found them useful in weather prediction.

Zaytar et al. (2016) engaged Long-Short Term Memory (LSTM) in weather forecasting and compared with the other models and concluded that LSTM performed better than other models.

## Crop Yield Prediction

Predicting the Crop yield is a thought-provoking task in agriculture. It plays a vital role in decision making at field, regional and global levels. There are several parameters like crop, soil, environment, meteorology based on which crop yield is affected. Therefore, it becomes essential to understand properly the relation between the yield and these parameters.

You, et al. (2017) employed DL algorithms on remote sensing data for prediction of the crop yield. They employed CNN and LSTM network using a dimensionality reduction method incorporating Gaussian Process component. The results obtained showed that the approach employed outperformed.

Khaki and Wang (2019) devised a technique on basis of DNN (deep NN) and found that the approach achieved RMSE = 12% of average yield.

Khaki, et al. (2020) presented a model comprising of CNNs and RNNs based on management practices and environmental data to predict the crop yield and compared with RF, DFNN and LASSO. Their model achieved a RMSE of 9% of average yield and outperformed all other models used.

## Plant Classification

Plant classification is another challenging task in agriculture.

Arafat, et al. (2021) proposed an automated plant classification model using DL techniques, image preprocessing, and ResNet18 to identify plants. The model was evaluated on two datasets: GRANDYMU and SWEDISH leaf dataset. On the first dataset, achieved accuracy was 99% and on second it was 99.9%

Wagle, et al. (2022) proposed a model based on compact CNN and AlexNet incorporating transfer learning and employed them on 9 species of plants of the Plant Village dataset for classification. The classification accuracy was found to be 99.45%.

## Weed Identification

Weeds are the unwanted plants that spread rapidly and unenviably, and influence on crop yields as well as crop quality. Therefore, it becomes essential for farmers to use resources to reduce weeds. Several techniques have been developed for identification of weeds.

Tang, et al. (2017) projected a method in which K-means feature learning was combined with CNN and identification accuracy of 92.89% was achieved.

Jiang, et al. (2020) projected an approach on basis of CNN feature based GCN (graph convolutional network). The approach was tested over four datasets and achieved accuracy was 97.80%, 99.37%, 98.93% and 96.51% respectively.

Osorio, et al. (2020) proposed a method on basis of Mask - region based CNN and F-1 score was achieved as 94%.

## Plant Classification

AlexNet, a pre-trained CNN technique is extensively employed for classification of plants. Yalcin (2017) proposed a method in which for extraction of image features, a pre-trained CNN was used and was compared with machine learning algorithms. It was concluded that the proposed method outperformed.

Gyires-Tóth, et al. (2019) employed CNN for feature learning and for classification, they employed fully connected layers incorporating log softmax output. Pre-trained models on ImageNet were employed and transfer learning was used. Offered method outperformed every other method.

## Plant Phenotyping

Plant phenotyping is the valuation of complex plant traits like tolerance, resistance, growth, development, physiology, ecology, architecture, yield, and individual quantitative constraints which are the basis for complex trait valuation.

Namin, et al. (2017) projected a novel DL technique for recognition of plant phenotype in which combination of CNN and LSTM was performed. CNN was employed for the feature extraction and then LSTM was employed on obtained output. The accuracy was achieved as 93%..

Ubbens & Stavness (2017) developed Deep Plant Phonemics, an open-source DL technique. This technique affords pre-trained NN for various plant phenotyping works, and a simple stage that can be applied by researchers and scientists of plants to train models for their own phenotyping applications.

## Leaf Area Index (LAI) Estimation

Plant canopies are characterized by LAI which is a quantity without dimension. Several agriculture models require the leaf area index. LAI and its changing aspects are extensively used by scientists for estimation of carbon cycles, vegetation status and environment, etc.

Chai, et al. (2012) developed a model for estimation of time series LAI. This model was based on NARX (Nonlinear Autoregressive model process with eXogenous input) and was called as NARXNN. It demonstrated as a talented technique for time series LAI.

Liu, et al. (2021) proposed a model for estimation of LAI employing multimodal data fusion & DNNs. They applied it on maize data set to estimate maize LAI. The model produced best results with relative RMSE of 12.78%.

## Soil Moisture Estimation

Soil moisture is a vibrant hydrological attribute for climate change, meteorology and precision agriculture. But, soil moisture in farmland is affected by many factors. Hence, it is a difficult task to estimate the soil moisture. Neural networks are very useful to overcome this situation.

Lu, et al. (2017) employed NARX model to estimate moisture. They compare this model with JAXA, LPRM and GLDAS and checked the validity. The model continued as stable in direct validation in both the seasons frozen and unfrozen.

## Land Classification

Land classification generally includes classification of large areas of land. It is vital for purposes like disaster risk assessment, food security, land use and landcover (LULC), and agriculture. DL techniques have widely been used in land classification.

Lu, et al. (2017) employed deep CNN method to classify the land. The precision of 90% was achieved.

## Land Cover Classification

One of the exciting tasks in agriculture is the Land cover classification. In classification of land cover, LSTM, a variant of RNN is widely employed.

Rußwurm et al. (2017) employed the LSTM in classification of land cover and concluded that it outperformed all mono-temporal models based on SVM, CNN and standard RNN.

Gharbia, et al. (2021) employed two CNN models namely VGG-16 and AlexNet on Landsat dataset for land cover classification. When tested on this dataset, the classification accuracy was 74.8% using AlexNet and 90.2% using VGG-16. When tested on data with seven times of original data set, accuracy achieved using AlexNet was 90.0% and using VGG-16was 94.6%.

## DISCUSSION

Most of the contemporary developments made by scientists and researchers in field of agriculture are researchers are thoroughly associated with production and other fragments of agriculture for the purposes of refining yields of crops, removing or at least dropping plant diseases, advancing automated an mechanized up-to-date agro-industry and agriculture. Deep learning provides better performance almost in all the related works here. Data classification and image recognition are two main aspects that are dealt with deep learning. These can be performed in four different steps. First step is the data collection and data preprocessing, second step is neural networks training, third step is model testing and fourth and final step is analysis of results. Data collection and pre-processing is vital for any experiment or analysis. Better the data better the result. Here, data collection is done with the modern technologies like radar, satellites, drones, internet of things, etc. with combination of deep learning and offers datasets of images or other forms with high quality. It has been observe that the data obtained in such manner has greatly increased the application of deep learning in agriculture. For neural network training, several algorithms like AlexNet, K-means feature learning, FCNN, etc., play vital role in the applications where high precision is required. These have been used extensively in agriculture and proved their utility. DL-based models have also been trained based on field sensory data. For Model testing, it is always beneficial to use new data. Result analysis, interpretation and performance evaluation are another vital aspects. Performances of models have been measured using performance accuracy, RMSE, and $R^2$value.

Several deep learning models have been proposed by researchers in field of agriculture. Still, a lot is remained to be explored in future. For example - exploitation of time dimension for higher performance prediction or classification is possible using other approaches for LSTM and other RNN models. Some approaches can be applied for commercialization purposes. The approaches incorporating DetectNet CNN and Faster Region-based CNN would be tremendously advantageous for automatic robots that remove weeds or collect crops. Additionally, automatic robots could be used in the estimation of the expected yields of crops. Similarly, using new approaches, other issues can be resolved in future.

## CONCLUSION

In this article, it has been witnessed that deep learning has enormous applications in diverse fields of agriculture like: land classification, land cover classification, plant classification, plant disease detection, weed identification, weather forecasting, soil moisture estimation, crop yield prediction, estimation of leaf area index, etc. Several things are still worth to be explored in agriculture using DL techniques.

## REFERENCES

Fan, Y., Qian, Y., Xie, F., & Soong, F. K. (2014). *TTS synthesis with bidirectional LSTM based recurrent neural networks. Proc. Interspeech.*

Gebbers, A. V. I. (2010). Precision agriculture and food security. *Science, 327*(5967), 828–831.

Hopfield, J. J. (1982). Neural networks and physical systems with emergent collective computational abilities. *Proceedings of the National Academy of Sciences of the United States of America, 79*(8), 2554–2558. doi:10.1073/pnas.79.8.2554 PMID:6953413

Kamilaris, A., & Prenafeta-Boldú, F. X. (2018). Deep learning in agriculture: A survey. *Comp. Elec. Agri., 147*, 70–90. doi:10.1016/j.compag.2018.02.016

Le Cun, Y., Bengio, Y., & Hinton, G. (2015). Deep learning. *Nature, 521*(7553), 436–444. doi:10.1038/nature14539 PMID:26017442

LeCun, Y., Bengio, Y., & Hinton, G. (2015). Deep learning. *Nature, 521*(7553), 436–444. doi:10.1038/nature14539 PMID:26017442

Pan, S. J., & Yang, Q. (2010). A survey on transfer learning. *IEEE Transactions on Knowledge and Data Engineering, 22*(10), 1345–1359. doi:10.1109/TKDE.2009.191

Razavian, A. S., Azizpour, H., Sullivan, J., & Carlsson, S. (2014). *CNN features off-the-shelf: an astounding baseline for recognition. In IEEE Conference on Computer Vision and Pattern Recognition Workshops*. CVPRW.

Simonyan, K., & Zisserman, A. (2014). Two-stream convolutional networks for action recognition in videos. *Advances in Neural Information Processing Systems, 1-4*, 568–576.

Suárez, P. L., Sappa, A. D., & Vintimilla, B. X. (2017). Infrared image colorization based on a tripletdcgan architecture. *IEEE Conference on Computer Vision and Pattern Recognition Workshops (CVPRW).*

Tyagi, A. C. (2016). Towards a second green revolution. *Irrigation and Drainage, 65*(4), 388–389. doi:10.1002/ird.2076

# Chapter 3
# Freshness Grading of Agricultural Products Using Artificial Intelligence

**Zeynep Elbir**
https://orcid.org/0000-0002-6330-2773
*Marmara University, Turkey*

**Berat Asrin Caferoglu**
https://orcid.org/0000-0002-9176-6428
*Marmara University, Turkey*

**Onur Cihan**
https://orcid.org/0000-0002-5729-2417
*Marmara University, Turkey*

## ABSTRACT

*This chapter presents a deep learning model that can be used to determine the levels of freshness of three agricultural products: strawberries, lemons, and tomatoes. For this purpose, YOLO, a state-of-the-art object detection algorithm is utilized. The data for training, validation, and testing are collected from online sources, and by applying image augmentation techniques, a sufficient number of images are obtained. Test results show that the model is performing quite well, and the speed of the model is fast. These results are promising and can be utilized to reduce a significant amount of agricultural waste and increase customer satisfaction once it is utilized by online groceries.*

## INTRODUCTION

In the rapidly developing world of the twenty-first century, food delivery and freshness control have become a major issue because of the fast increase in the population. Thanks to modern technologies, people started to spend less time grocery shopping and take advantage of the comfort brought by online shop-

DOI: 10.4018/978-1-6684-5141-0.ch003

ping. Among the online orders, fruit and vegetable orders constitute a significant percentage. However, in online fruit and vegetable shopping, the end consumers cannot apply their choices and preferences when selecting goods, and customer satisfaction is highly dependent on products that are chosen by the supplier. Therefore, to provide customers better shopping experience, online groceries should determine the freshness and quality of the product in a systematic, consistent, and objective way and price them accordingly. To achieve this, it is necessary to evaluate the freshness level with measurement methods, and physical properties, in other words, the analysis should be quantitative. Using quantitative analysis, end consumers will be able to do fruit and vegetable grocery shopping online with the freshness level they prefer, leading to more comfortable and healthy shopping experiences. In this way, customer satisfaction is expected to increase. Furthermore, by detecting the fruits and vegetables that are soon to be wasted, the grocery can rearrange their price and prevent them to be wasted in the store. This not only is advantageous for the grocery but also helps to reduce wastage.

Among the studies about measuring the freshness of fruits and vegetables in the literature, some studies damage and waste agricultural goods during the measurement process. On the other hand, studies that do not damage or waste the goods are insufficient when applied alone. Therefore, analyses and test methods need to be applied and brought together harmoniously and systematically to obtain a useful tool. This research aims to bring these analysis methods together with the use of deep learning technologies and contribute to the existing literature with an innovative approach. In this chapter, the authors consider the analysis of the color, shape, texture, and size properties of agricultural goods are considered in the deep learning model and provide a method to measure the freshness of agricultural products in an objective way.

The freshness levels of agricultural products can be measured using chemical and physical test methods. For instance, for freshness detection of asparagus, the concentration of asparagine amino acid in the tip is a significant indicator while the fluorescence of chlorophyll is important for the detection of the freshness of broccoli. Nevertheless, chemical freshness analysis methods are not practical and not suitable in all conditions. Therefore, chemical tests are not commonly used among groceries and consumers. Furthermore, the freshness perception of consumers includes various quantitative and qualitative factors, such as size, shape, color, firmness, smell, and texture (Péneau, 2005). These factors cannot be generalized to cover all agricultural products at one time. To the best of the authors' knowledge, there does not exist a comprehensive method that uses all the listed features to estimate the freshness levels of goods in the literature. On the other hand, analysis by physical inspection may damage the items and make them inconsumable. As a result, physical methods are not widely used in the freshness analysis of agricultural products.

## Visual Analysis

The methods based on visual analysis are advantageous since they do not damage the product in the process of freshness grading. There are four main types of visual analysis methods in the literature: analysis of the color, brightness, shape/size, and the defective area.

### Color Analysis

The color analysis method is the fastest and easiest way to comment on the freshness of fruits and vegetables. For instance, color changes can be used as a maturity marker for a banana fruit. To be clearer, the dominant color is green for an unripe banana; a mix of green and yellow for a slightly ripe banana, yellow for a ripe banana, dominant brown and black or white (due to mold) for a rotten banana. To apply the color analysis method, the color scale of each item needs to be properly determined by experts in the agricultural field. Figure 1 illustrates a color scale of banana fruit starting with unripe and ending with rotten.

*Figure 1. A color scale for banana*

There are two ways to apply the color analysis method in the literature. The first way is to analyze the color using the human eye alone. Nonetheless, the result of the test varies from person to person, that is, the method is subjective and inconsistent. Therefore, this method is not practical. Another way to analyze the color is to measure the light which is reflected from the product's surface with a device called a colorimeter and determine the color of the item by measuring the absorbance of the material. This method gives more reliable and objective results compared to the first one. On the other hand, the main drawback of this method is its price. That is why, this method is used by agricultural field experts, but not by daily consumers or groceries, and therefore it is not a cost-effective solution to the freshness detection problem.

## Brightness Analysis

The brightness of some fruits and vegetables is related to their juiciness and freshness. For instance, mandarin and orange are bright when they are fresh and their brightness significantly decreases over time (Barrett, Beaulieu, & Shewfelt, 2010). Like the previously mentioned color analysis method, a gloss meter can be used to determine the brightness of the product. However, it is not a practical and applicable solution for most applications due to its price.

## Shape/Size Analysis

Another visual analysis method is the evaluation of the shape/size. Standard and characteristic knowledge about the physical appearance is important to comment on the quality and freshness of the fruits and vegetables. In this context, items with abnormal shape and size tend to be damaged by external factors in general. Furthermore, most of the items with abnormal shapes have bad tastes and are not preferred by consumers. There are various size and shape scales developed by the experts in the literature. Still, these scales are not widely known by consumers (Mendoza, Dejmek, & Aguilera, 2010).

### Defected Area Analysis

Defected area analysis is another estimation method that can be used to grade agricultural products. Tolerance rates are determined for each agricultural product with defects (Verma, 2015; Nunes, Emond, Rauth, Dea, & Chau, 2009; Mitcham, Cantwell, & Kader, 1996). Note here that these tolerance levels are not known by the groceries or consumers and there is no practical tool to determine the freshness level of fruits and vegetables with such a method of analysis.

## Analysis by Touch and Punching/Force Application

Ripeness and freshness levels of agricultural products are directly related to their softness or firmness (Xu, He, Hauser, & Gerling, 2019). For instance, the maturity level of a peach can be measured by considering its softness, since it becomes softer as they become rotten. On the other hand, this analysis method is not applicable to all agricultural products. Besides, applying force during the analysis process may lead to deformity or even food wastage. Since deformed products cannot be sold at the same price as a perfect product, this method is not very practical in daily life.

### Analysis by Touch

Consumers do not tend to prefer too soft or too firm products in general and the firmness of the agricultural products can be checked by touching. On the other hand, the fundamental problem of this analysis method is lack of hygiene and being subjective. Besides, some consumers may damage the product by touching which leads to more wastage at the end of the day.

### Analysis by Punching/Force Application

In the context of the punching method, the reaction of the object to an applied force is measured and the firmness of the object's surface can be determined via measurement. However, the object is subject to regional damages in this method as well. A device called a penetrometer is used in the application. The other method is the firmness test through deformation. In this test method, the degree of firmness is determined by measuring the force at the point where the shape of the object begins to deform due to the applied force. Mostly, the product becomes inconsumable because of the deformation (Mitcham, Cantwell, & Kader, 1996; Kemikoğlu & Özen, 2018).

From the above discussion, one can conclude that existing freshness analysis methods are either expensive or not applicable in practice. As a result, these methods are not suitable for continuous use and are not preferred by groceries or consumers. On the other hand, some of these freshness analysis methods are subjective. The main objective of this chapter is to provide a reliable and cost-effective method to measure the freshness of agricultural products in an objective way. As can be seen from the above discussion, the integration of existing methods in the literature to online fruit and vegetable sales is not easy since some are expensive and some damage the items. In classical online grocery shopping, the customers cannot apply their freshness preferences which means they are dependent on the selection of grocery employees. This may lead to customer dissatisfaction and limit the number of online sales. Therefore, letting the customers choose fruits and vegetables according to their preferences is vital for online groceries.

In this chapter, a useful mix of the existing methods in literature will be used to create a freshness analysis method with high accuracy. In this context, this research presents an innovative approach by scaling freshness levels with the help of deep learning methods to make the consumer more comfortable in online shopping. The method proposed in this research can be also integrated into the grocery section of markets, letting them sell the fruits and vegetables with accurate pricing according to their freshness levels. The research offers a cost-effective system based on artificial intelligence that allows end-consumers to apply their preferences more easily and with peace of mind during online fruit and vegetable shopping. Detection of accurate freshness levels of fruits/vegetables without harming them will also provide a decrease in food wastage by detecting products that are approaching their expiration date.

## Analysis by Artificial Intelligence

Valentino, Cenggoro, & Pardamean (2021) proposed a new convolutional neural network model for detecting the freshness levels of fruits that are growing in Indonesia. In their research, fruits are classified into two categories: fresh and rotten. Similarly, Ananthanarayana, Ptucha, & Kelly (2020) considered only two grading classes for three different fruits and adjusted environment lighting in order to increase the detection performance. Kuznetsova, Maleva, & Soloviev (2020) developed a deep learning model that is used to detect the harvest time of the apples which can be used in agricultural applications. In this model, they indicate that 9.2% of the apples are not recognized. It is stated that the developed system can be applied to not only apples but also other spherical products such as orange. They applied various filtering methods in order to minimize the effects of external factors such as shadows and branches. Anas, Jaya, & Nurjanah (2021) utilized the YOLO object detection algorithm to determine fish freshness. Researchers used physical specifications of the fish (eyes, gills, texture, and flesh) for the quality analysis. Three fish species are used during the research including Rastrelliger, Euthynnus affinis, and Chanos chanos. In this system, the tiny version of YOLO version 2 is used due to detection speed concerns. Ohali (2010) developed a grading system with 3 levels for dates using a classification algorithm based on backpropagation neural networks (BPNN). Azarmdel, Jahanbakhshi, Mohtasebi, & Muñoz (2020) used artificial neural networks and support vector machines for freshness grading of mulberries. They considered three grading levels (unripe, ripe and overripe) for white and red mulberries and proposed a future work plan for a sorting system according to their freshness levels. Mathew & Mahesh (2022) used multiple object detection algorithms (SSD, YOLOv3, YOLOv4, and YOLOv5) to detect bacterial spots on the leaf of bell paper plant. The authors stated that YOLOv5 has given the best result in terms of detection accuracy. Kuznetsova, Maleva and Soloviev (2021) used YOLOv3 and YOLOv5 to detect apples in orchards. According to the results of the research, YOLOv5 performed a better performance than YOLOv3 for detection of the apples. Yao, Qi, Zhang, Shao, Yang, & Li (2021) proposed a real-time method for detecting defects in kiwifruits based on YOLOv5. The researchers used an enhanced version of YOLOv5 and their model which performed 9% better than the original version in terms of mean average precision. Han, Jiang, & Zhu (2022) proposed a method to reduce the uneven impacts of human senses on selecting cherry fruits according to their qualities. The authors utilized YOLOv5 along with a flood filling algorithm to classify cherry fruits. The flood filling algorithm was applied to the image dataset to remove environmental effects on the images. Amemiya et al. (2021) proposed a support system for the prediction of the appropriate harvest time of Shine Muscat grapes using YOLOv5. The harvest time of the products is determined according to the colors and the ratios of pest grains, diseased grains, and normal grains of the grape bunch. The authors indicated that the proposed model has 90% accuracy.

## MAIN FOCUS OF THE CHAPTER

### Issues, Controversies, Problems

Under the influence of the Covid-19 pandemic, there is a significant increase in daily grocery shopping over the internet (Alaimo, Fiore, & Galati, 2020). In online shopping, the fresh grocery section has its place. Customers desire to buy fresh and high-quality items from these traders. However, some unfortunate situations that result in consumer dissatisfaction may occur since the consumers only know the name of the item they are buying over the internet, but they do not know the quality and freshness level. Therefore, consumers cannot apply their preferences and tastes in online shopping for fruits and vegetables which leads to consumer dissatisfaction. To prevent this, agricultural products must be classified according to a freshness scale evaluated using a quantitative classification that is acceptable by anybody. Besides that, purchasing agricultural products without knowing their freshness level and shelf life causes a significant amount of waste in the long term. About 30-50% of the products that have been bought from the market are thrown away by the consumer (Verma, 2015).

In the light of the above, the main objective of this research is to present a model that can be used to determine the freshness level of agricultural products by taking the color, brightness, and shape/size into account. The accuracy of the proposed model is shown to be higher than the existing studies in the literature and the time required to determine the freshness level of a product is very low.

## SOLUTIONS AND RECOMMENDATIONS

In this study, three popular and preferred fruits are chosen for freshness detection: strawberry, lemon, and tomato. Before building a deep learning model for this purpose, a dataset containing a sufficient number of images with different freshness levels must be prepared. To this end, the authors utilized a Python has a framework called *Icrawler* which consists of web crawlers and supports different types of media such as images and videos. It is a mini framework because it is a tiny and flexible downloading tool that allows multithreaded processes. In this research, the following keywords are used to search images on the internet for various conditions of agricultural products: "unripe, slightly ripe, ripe, overripe, and rotten". In this way, large datasets have been collected for each category of agricultural products under consideration. Data without labels have no meaning since labels are the information of classes/categories and the location of the object in the image. Therefore, all data must be properly labeled before training the deep learning algorithm. In this research, *MakeSenseAI* and *LabelImg* are used as labeling tools. MakeSenseAI is a website that allows users to label their data without any installation whereas LabelImg is a program written in Python which contains useful features such as checking the labels by selecting the image and its label file after the labeling process is done.

The freshness detection model contains a deep learning algorithm that analyzes the physical appearances of goods such as color, brightness, shapes, and surface defects. For this purpose, the authors utilize

YOLO (You Only Look Once), which is a fast and accurate deep learning algorithm in the context of requirements (Redmon, 2015). Originally, YOLO is an object detection algorithm that combines the concepts of classification and localization. Classification predicts the class of a specific object in an image, and localization determines the object's location in the image. YOLO is an algorithm that uses neural networks to provide real-time object detection. The first version of YOLO was released in 2015 and the most current version is 5. Each version of the algorithm includes several subversions which have tradeoffs between speed (i.e., frames per second) and accuracy. YOLO algorithm has superior performance over other object detection algorithms in terms of both speed and accuracy, and therefore it can be referred to as a state-of-the-art algorithm.

YOLO algorithm looks at the image for one time and divides up the image into a $S \times S$ grid (Escriva, & Laganiere, 2019). Each cell in this grid is responsible for predicting 5 bounding boxes (Li, Su, Geng, & Yin, 2018), the rectangles that enframe the objects. For all predictions, there are confidence scores that show how certain the prediction of the bounding box is. The algorithm chooses the best class among these predictions within a threshold and the final bounding boxes are weighted by the predicted probabilities. While other popular object detection algorithms process the image multiple times, YOLO uses only one neural network for the entire image and looks at the image only once. In this wise, YOLO is faster than most of the object detection algorithms (Redmon, 2015; Almog, 2020).

After constructing the dataset, labeling the data therein and selecting the deep learning algorithm, the next step is to train the model for object detection. In this research, two different versions of the YOLO algorithm are used: YOLOv4 and YOLOv5. In the training stage of YOLOv4, an open-source neural network framework written in C and CUDA called *Darknet* is utilized. The beneficial properties of Darknet are fast training, ease of installation and use, and support for both CPU and GPU computation (Redmon, 2015; Redmon, 2016). Darknet framework takes labeled images (images and their labels), configuration file, and a pre-trained weights file (to train new weights file faster) as inputs. The configuration file is a text file that contains information about layers, number of classes, learning rate, training image size, number of maximum iterations, number of images that are taken into GPU once at a time, and the number of subdivisions. In short, the configuration file determines the speed, accuracy, and how iterations are done. It is important to stay within the memory limits of the GPU for continuous training. If limitations are not properly adjusted in the config file, it may cause an interruption in the training. The weights file contains the weights of the features of the model and therefore it should be computed properly. The Darknet framework calculates the weights using labeled images and returns a new weights file as output. On the other hand, YOLOv5 was trained using *PyTorch* which is a machine learning framework based on the Torch library. It is easily applicable to the image processing and computer vision. There are several yaml files to customize the training process. The first of them is "data. yaml" file, which contains the class information and the image set (train, validation and test set) locations that will be used in training. YOLOv5 has five subversions, therefore another yaml file is used to determine the sub-type to be trained. Lastly, there is a yaml file for hyperparameters which determines the augmentations such as shear and flip.

YOLOv4 consists of two submodels: Tiny and Normal. On the other hand, there are five submodels in YOLOv5: Nano, Small, Medium, Large, and XLarge. As the size of the submodel gets larger, the accuracy increases, and the inference time increase which results in a decrease in FPS (frame per second). Therefore, the choice of the submodel is crucial and depends on the restrictions of the application. For instance, a powerful computer can handle the XLarge model. However, since most development cards such as Raspberry Pi have less computation power, the Nano submodel should be a better choice. In this research, the tiny subversion of YOLOv4 with the size of 416x416 pixels, and the small subversion of YOLOv5 with the size of 640 pixels for the long side (the size of the short side is handled automatically) are used to have high compatibility in all kinds of systems, a high FPS operation, and a faster training process. The size of the training model is important to be able to catch small objects in the images since as the size increases, the percentage of detection rate of tiny objects increases accordingly.

YOLO algorithms have three main parts in their architecture: the backbone, neck, and head (Nepal, & Eslamiat, 2022). The backbone of an object detection algorithm is used for feature extraction from images and the neck part improves the feature extraction by collecting the feature maps from different layer stages. The last part of the YOLO object detection algorithm is the head. The head part is used for generating bounding boxes and class prediction. The fundamental architecture of YOLOv4 and YOLOv5 algorithms are shown in Figures 2 and 3, respectively.

*Figure 2. YOLOv4 architecture*

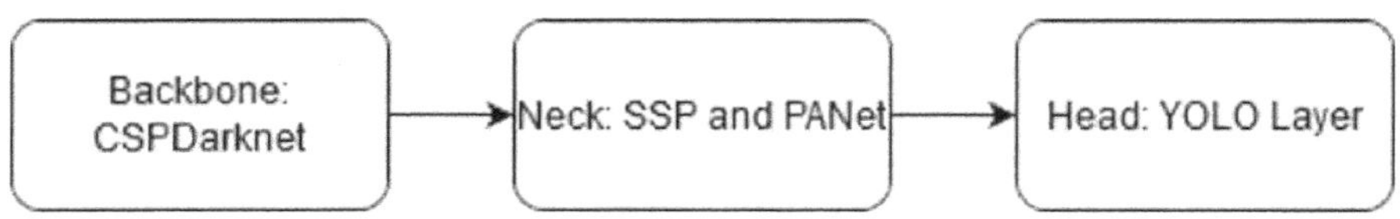


*Figure 3. YOLOv5 architecture*

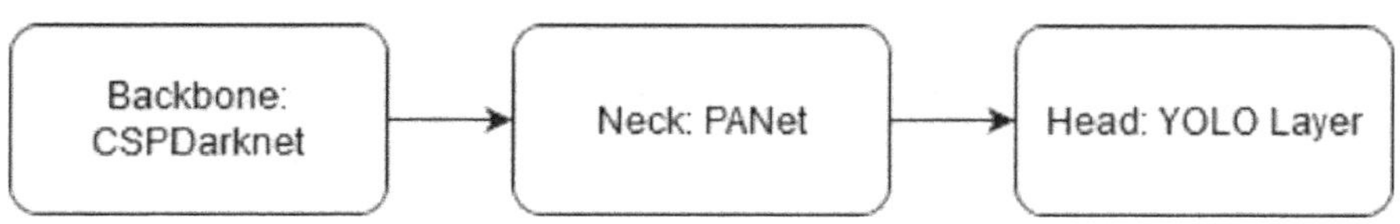


After training the deep learning model, OpenCV, which is a computer vision library is used in the image processing part of object detection. By utilizing this library, one can draw bounding boxes on the images. Figure 4 illustrates the bounding box placement on the images of agricultural products.

*Figure 4. a) A ripe tomato, b) a rotten strawberry, c) an unripe lemon*

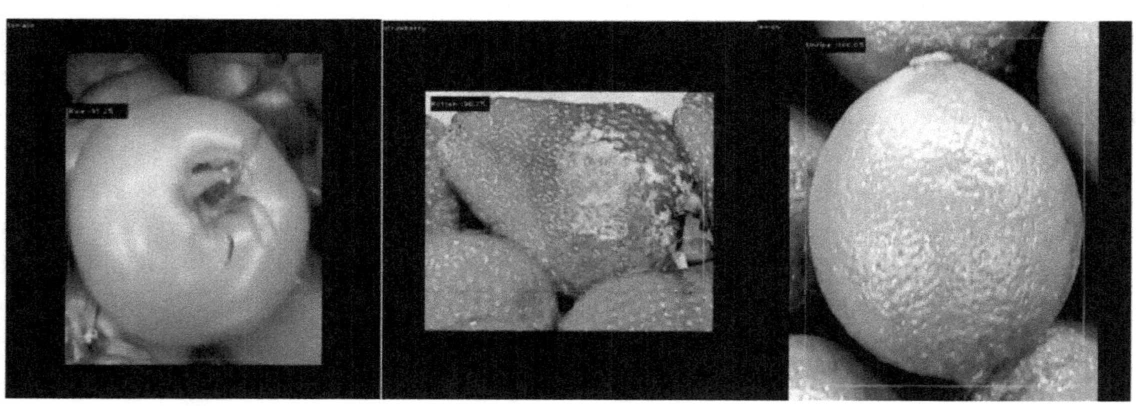

For each agricultural product under consideration, freshness detection models have been individually trained. In order to automatically determine the model to be used for freshness detection, an object detection model is constructed. Figure 5 shows how the objects are detected by the model.

*Figure 5. Object detection of a) tomato and b) lemon*

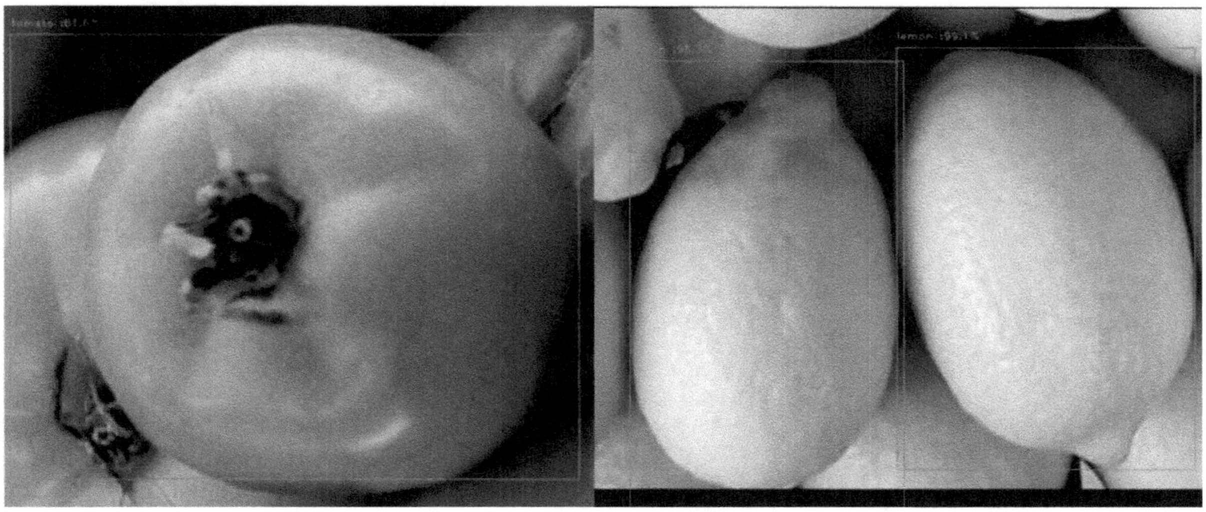

For YOLOv4, it is usually sufficient to train the model for 2000 iterations for each class, as reported in the documentation of the Darknet framework. Also, the accuracy of the trained model can be calculated using a method called Mean Average Precision (mAP). This method scales the accuracy level from 0% to 100%. Another method to calculate accuracy is the loss function. Figure 6 depicts the evolution of the mAP and loss values for the training phase of a model.

*Figure 6. The mAP and loss values as a function of iteration number*

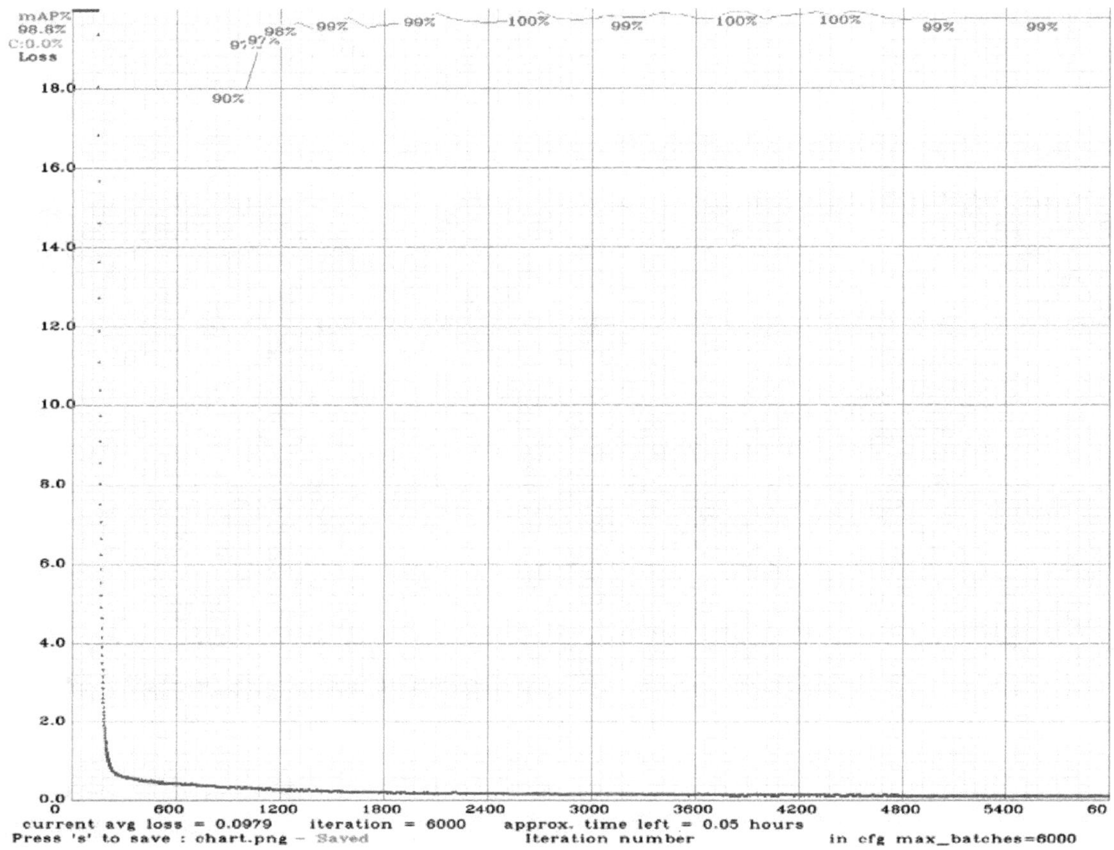


For YOLOv5, the training process works slightly differently. YOLOv5 uses an early stopping point as a training process stopping criteria. At this specific point, training should be stopped in order to have the best model without experiencing underfitting and overfitting. The early stopping point is determined according to a parameter called *patience*. The patience parameter changes according to the value variations of the mAP value, that is, if the difference between the precisions of consecutive iterations is less than a threshold, the patience increases. In other words, the patience value is the limit where the training will be stopped, and it corresponds to the early stopping point. Patience should be determined before the training process starts. If not, then the default value will be used to determine the early stopping point. In addition to the mAP chart, various charts are given as output to understand and examine the training process once YOLOv5 training is completed. The confusion matrix is one of the most useful charts that can be used to evaluate the performance of the model. The confusion matrices of the freshness detection models are given in Tables 4-7 (see Appendix). For a successful model, the diagonal elements of the confusion matrix are expected to be close to 1. A diagonal element that is equal to one corresponds to a perfect prediction for that specific class.

Figure 7 illustrates the localization and classification of a lemon image. In classification, the type of the product is detected without determining the exact location. On the other hand, finding the location of the product is called localization.

*Figure 7. Localization and classification of a lemon photo*

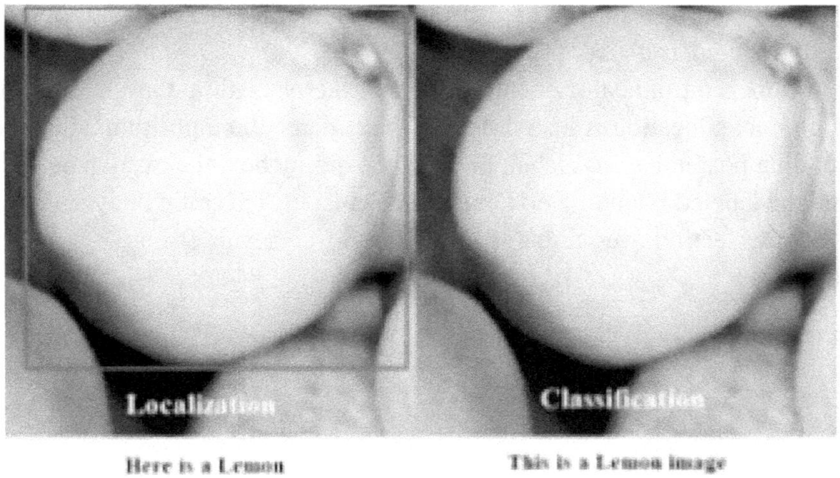


For object detection, the performance of the algorithm can be evaluated using the precision value. The precision value, denoted by *p*, is the ratio of true-positive (*TP*) detections to the sum of true and false-positive (*FP*) detections, i.e.,

$$p = \frac{TP}{TP + FP} \tag{1}$$

Figure 8 illustrates the ground truth and predicted bounding boxes for a photo consisting of four tomatoes.

*Figure 8. The ground truth and predicted bounding boxes*

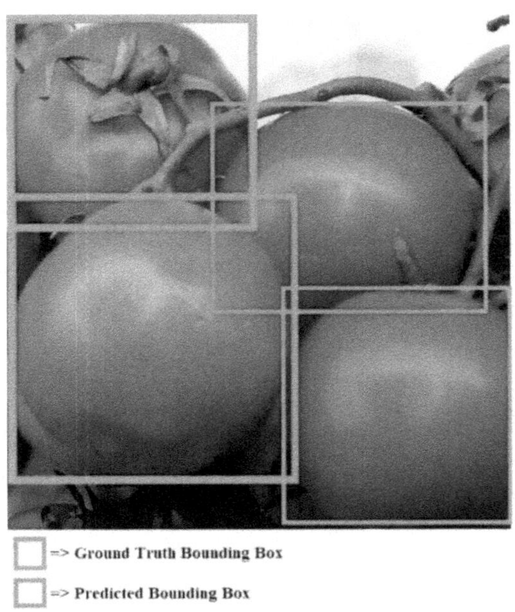

Green boxes show the original positions of the labels whereas two blue boxes are the predictions of the algorithm. Since the number of true predictions is 2, and the number of false predictions is 0, the precision value is computed as p=2/(2+0)=1.

Precision is an important indicator of the success of the algorithm. On the other hand, success in determining the product's location is also crucial. Object detection algorithms make their predictions in terms of a bounding box and a class label. For each bounding box, the overlap between the predicted bounding box and the labeled bounding box must be calculated. This ratio between the areas of overlap and union is called intersection over union or IoU in short (see Figure 9).

*Figure 9. IoU calculation*

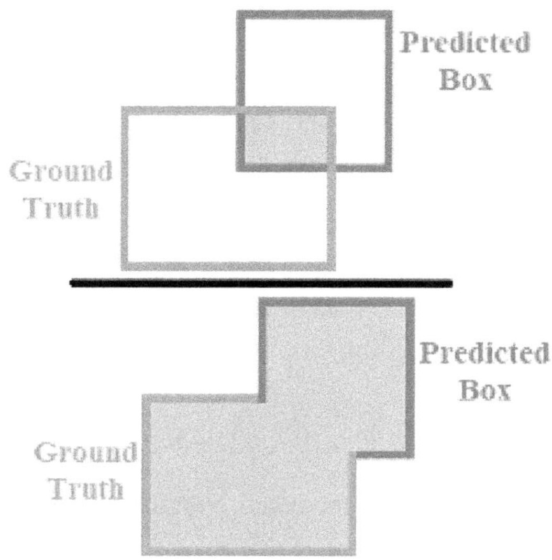


IoU value can be computed as

$$\mathrm{IoU} = \frac{A_o}{A_u} \tag{2}$$

where $A_o$ and $A_u$ are the areas of overlap and union, respectively. A detection is assumed to be successful if *IoU* value is greater than a certain threshold and unsuccessful vice versa. For instance, if *IoU* threshold is 0.5, and the calculated *IoU* value for a prediction is 0.9, then the estimation must be categorized as True Positive (*TP*). On the other hand, if the *IoU* value of prediction is 0.3, it should be categorized as False Positive (*FP*) as illustrated in Figure 10.

*Figure 10. False positive and true positive predictions*

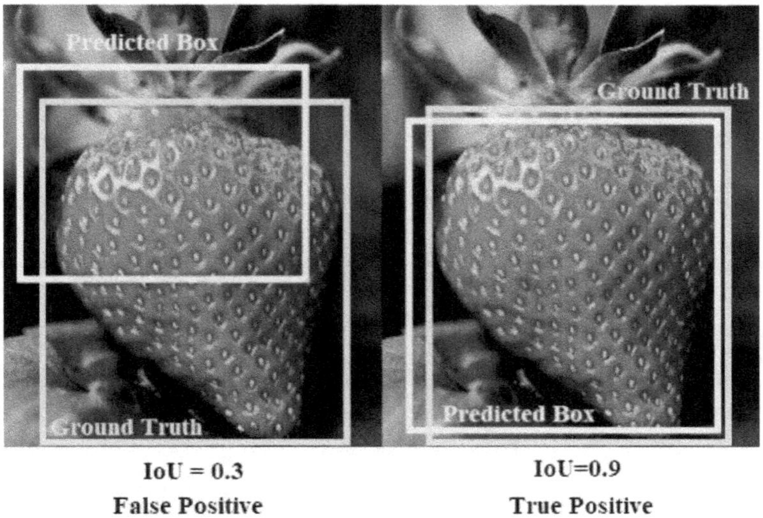


The recall value, denoted by *r*, measures how well the positives are founded and can be computed from the following formula

$$r = \frac{TP}{TP + FN} \tag{3}$$

where *FN* is the number of False Negative predictions. The mean Average Precision (m*AP*) is the area under the precision-recall curve calculated as

$$mAP = \int_0^1 p\left(r\right)^{\circ} \cdot dr \tag{4}$$

Predictions with higher *mAP* values are assumed to be more successful than prediction with lower values. On the other hand, smaller loss values correspond to better results. As reported in the documentation of Darknet, the final average loss takes values in the range between 0.05 (for a small model and dataset) and 3.0 (for a big model and dataset) (Bochkovskiy, 2020).

In the training stage of the model, two important problems may occur: under-fitting and over-fitting. Under-fitting is the regression problem with a function that is too simple to meet the needs. Consider a regression problem where a second-order polynomial is required to separate the classes. Once linear regression is applied, the resulting function will not be successful in separating classes (see Figure 11a). Under-fitting problems have a high bias. Over-fitting is again a regression problem where the function that separates the classes is too complex. Consider the same regression problem as in the previous example. If a tenth-order polynomial is chosen for regression, it will be very complex to meet the needs in real life (see Figure 11c). In over-fitting, the training error is low, and the testing error is high. Furthermore, an over-fitting problem brings a high variance in the solution. From Figure 11, one can conclude that training should be stopped in order to avoid overfitting.

*Figure 11. Under-fitting and over-fitting*

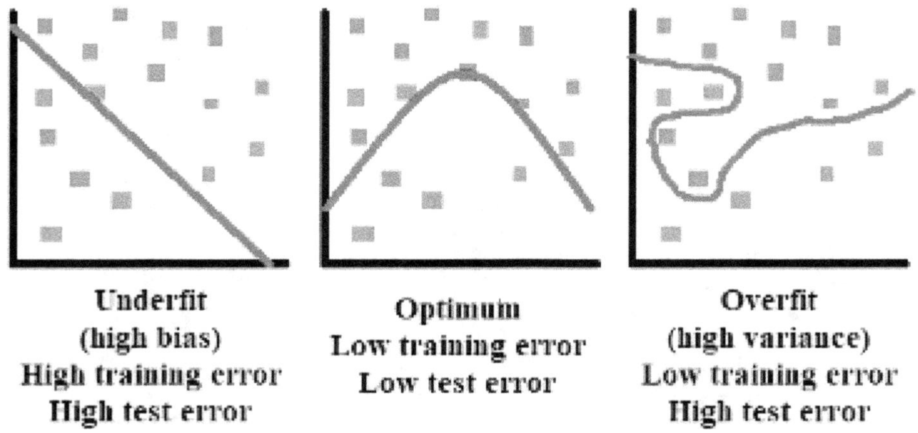


The optimal point to abort the training phase is called the sweet spot (IBM, 2021). Figure 12 depicts the errors in the validation and training sets as functions of the number of iterations. As can be seen from the figure, both errors decrease as the number of iterations increases until the sweet spot (or the optimum point). However, after the sweet plot, the error in the validation set increases due to over-fitting.

*Figure 12. The sweet spot*

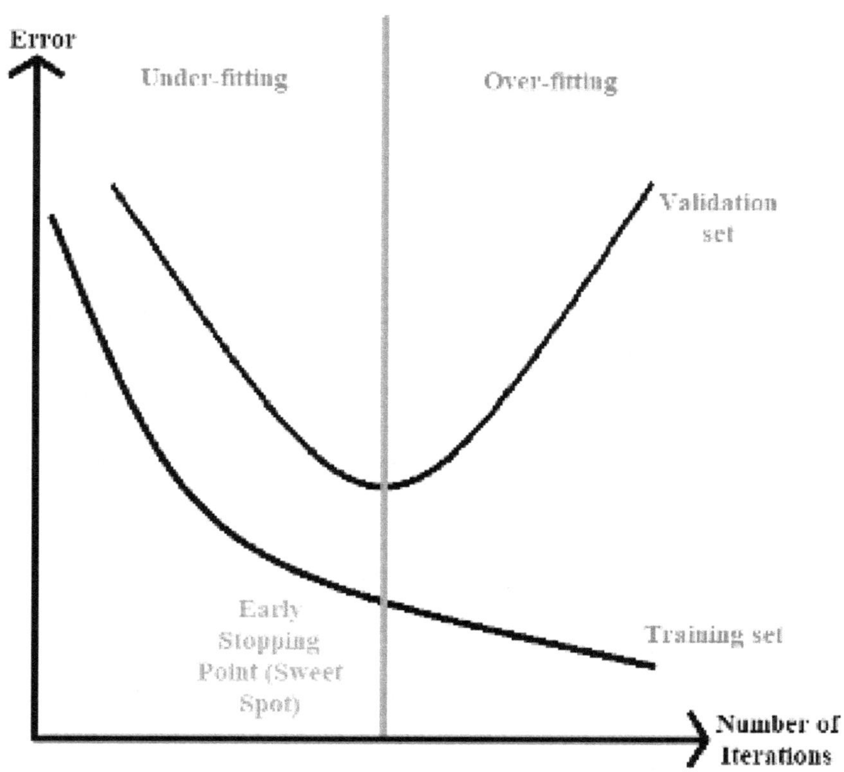

This point is also called an early stopping point to reduce the complexity of the model by eliminating less relevant inputs (IBM, 2021). According to Bochkovskiy (2020), the ideal training chart for YOLOv4 is shown in Figure 13.

*Figure 13. Ideal training graph (Bochkovskiy, 2020)*

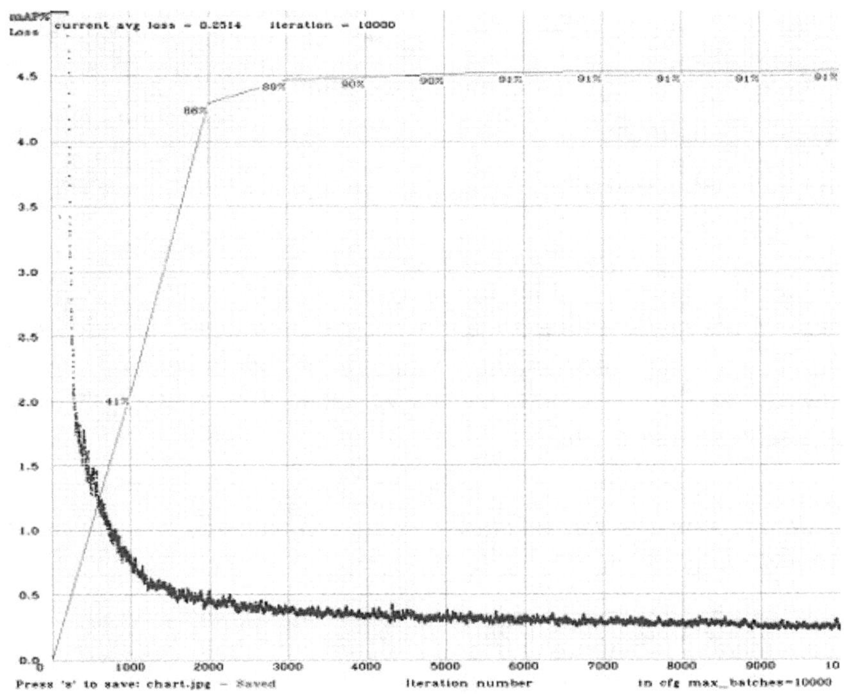


In object detection, the algorithm generates bounding box predictions with different confidence scores for the object. A non-filtered object detection algorithm might have two or more bounding boxes with different confidence scores for each object or might have wrong predictions as well. In order to avoid this, the non-maximum suppression filtering algorithm can be used. The non-maximum suppression algorithm calculates the overlap between the object detection proposals one by one and selects the best one (Sambasivarao, 2019). Figure 14 illustrates the bounding boxes proposed by the object detection algorithm and the result of the non-maximum suppression algorithm.

*Figure 14. Application of the non-maximum suppression algorithm*

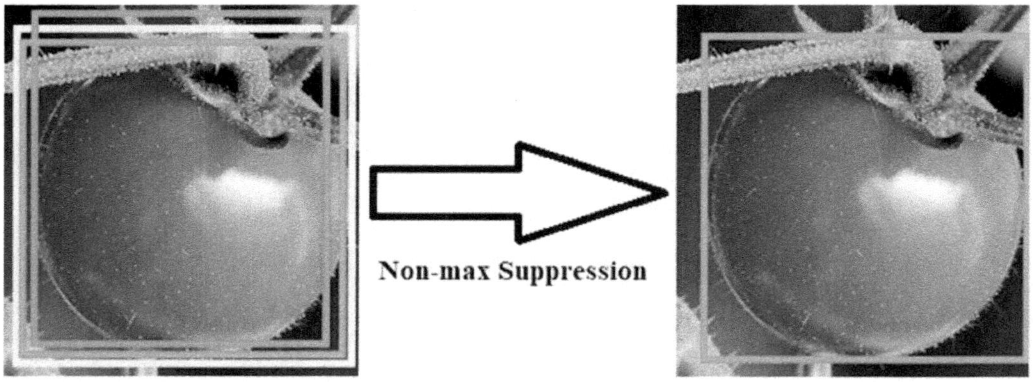


In order to increase the robustness of the freshness detection algorithm proposed in this study, data augmentation is used. Data augmentation is the generation of new data from existing ones by applying a noise effect. A data augmentation tool called *imgaug* is utilized to obtain distorted versions of the existing data in the dataset. Figure 15 shows the images generated by applying Gaussian noise, motion blur, and cut-out effects to a tomato image.

*Figure 15. The tomato picture a) before augmentation b) Gaussian noise augmentation c) motion blur augmentation d) cut-out augmentation*

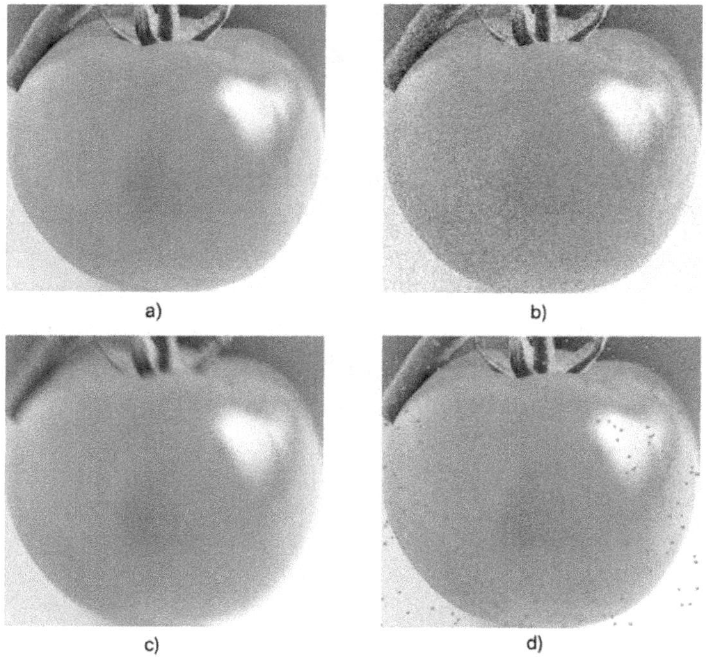


Figure 16 illustrates the steps of the freshness detection procedure considered in this research. First, an image is accepted as an input to the YOLO object detection algorithm. The algorithm detects the location and the type of the agricultural product, if any. In the next step, the image is forwarded to the freshness grading model related with the type of the agricultural product determined in the first step. After a successful implementation of these combined models, the freshness level of the product is aligned with a bounding box and a label in the output.

*Figure 16. Freshness detection procedure diagram*

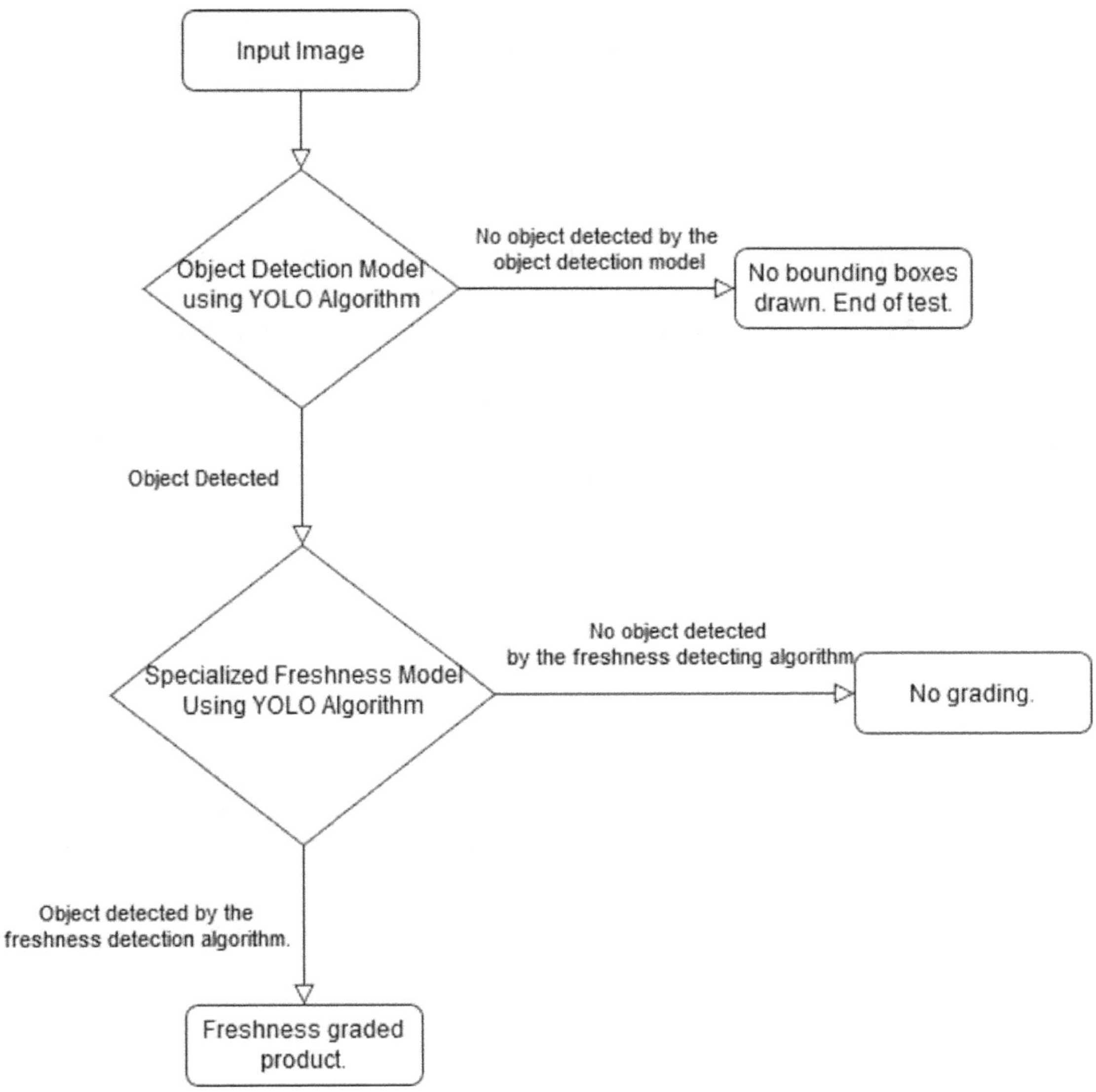


## RESULTS AND DISCUSSIONS

The object detection and freshness models are tested with the images in the test sets which do not appear in the training stages of the models. Therefore, the tests are reliable to evaluate the success of the models. For object detection, the performance of the model is tested for 58 strawberry, 50 tomato, and 52 lemon images. The detection performance is calculated using the following formulation:

$$\text{Detection percentage}(\%) = \frac{n_d}{n_t} \cdot 100 \tag{5}$$

where $n_d$ and $n_t$ are the numbers of detected products and the total number of products, respectively. For the successfully detected products, the error of classification is calculated using the following equation.

$$e_c = \frac{n_i}{n_d} \cdot 100 \tag{6}$$

where $e_c$ is the error in product classification and $n_i$ is the number of incorrect classifications. Once the products are detected via the object detection algorithm, freshness grading tests are performed separately for each product. The products are classified into 5 freshness levels: unripe, slightly ripe, ripe, over-ripe, and rotten. Since these classes are ordinal, the success of level detection model can be evaluated by defining indices for each class and computing the error as the difference between these indices. The indices of the classes are as follows: 0 (unripe), 1 (slightly ripe), 2 (ripe), 3 (overripe), and 4 (rotten). The estimation error and the absolute estimation error can be expressed as

$$e_c = \text{actual index} - \text{estimated index} \tag{7}$$

$$e_a = |e_e| \tag{8}$$

where $e_e$ is the estimation error and $e_a$ is the absolute estimation error. For instance, if the freshness detection algorithm classifies a slightly ripe strawberry as ripe, the estimation error will be computed as $e_e$=1–2= -1 which means the error in prediction is 1 class in the direction of increasing ripeness. Once the errors are computed for all test data, the total error is computed by the following equation

$$e_t = \frac{\sum e_a}{n_d} \cdot 100 \tag{9}$$

where $e_t$ is the total estimation error.

In order to compare the performance of this research with the existing ones in the literature, average precision, denoted by $p_a$, is defined as

$$p_a = \frac{1}{N} \sum_{i=0}^{N} p_i \tag{10}$$

where $p_i$ is the precision of the $i$-th test and N is the amount of test data. Detailed results of the object detection algorithm are given in Caferoglu, Elbir and Cihan (2022). The test results of the object detection models are summarized in Table 1.

*Table 1. Summary of the results of the object detection models*

| Model Name | Detection Percentage | Wrong Detection Percentage | Average Precision | Average Precision @0.5 (Training Result) |
|---|---|---|---|---|
| **YOLOv4 Object Detection** | 74.47% | 2.14% | 97.25% | 89% |
| **YOLOv5 Object Detection** | 92.02% | 0% | 100% | 93.1% |

As can be seen from Table 1, YOLOv4 algorithm is able to detect 74.47% of the products in the dataset and among these detections, only 2.14% of them are predicted incorrectly. On the other hand, the detection rate of YOLOv5 algorithm is 92.02% which is much higher than that of YOLOv4. Furthermore, the prediction of the fruit type has a success of 100%, which shows the efficacy of YOLOv5. The test results show that while YOLOv5 misses some of the objects in the test dataset, once an object is detected, the objects are classified perfectly.

For freshness detection of lemon, strawberry, and tomato, results are summarized in Table 2.

*Table 2. Summary of the results of the freshness level detection model*

| Model Name | Detection Percentage | Error | Average Precision (Testing Result) | Average Precision Using 0.5 IOU Threshold (Training Result) |
|---|---|---|---|---|
| **YOLOv4 Lemon Freshness** | 80.58% | 6.67% | 88.37% | 81.00% |
| **YOLOv5 Lemon Freshness** | 94.63% | 6.11% | 90.93% | 70.80% |
| **YOLOv4 Strawberry Freshness** | 87.88% | 2.30% | 98.57% | 87.00% |
| **YOLOv5 Strawberry Freshness** | 83.84% | 3.61% | 98.33% | 87.20% |
| **YOLOv4 Tomato Freshness** | 88.24% | 6.67% | 92.05% | 79.00% |
| **YOLOv5 Tomato Freshness** | 91.18% | 4.30% | 98.86% | 85.50% |

From Table 2, it can be concluded that for both versions, the detection rate is greater than 80% and the class shifting error is no more than 6.67%. Furthermore, by taking the average of the precision of each test data, the average precision value is calculated using (10). It is seen that the average precision values are higher than at least 90.93%. On the other hand, the average precision values when IOU threshold of 0.5 is used are greater than 70.80%.

Table 5 shows the confusion matrix for the lemon freshness YOLOv5 model. As can be seen from the table, most of the lemons that have overripe freshness are detected by the algorithm as rotten which is an exception that occurred only for lemon. The reasons behind this problem are 1) insufficient training data in the overripe class, 2) the overripe and rotten lemon data in the dataset have similar physical properties. In order to improve the success of the model, the amount of training data corresponding to the overripe and rotten classes should be increased.

The model developed in this research consists of two stages: detection of the product type, and detection of the freshness level of that product. For instance, given an image containing a tomato, the object detection model determines the product type. Then, the tomato freshness grading model is executed. The freshness level is the output of the model. While the object detection model has three different classes: lemon, strawberry, and tomato, the freshness models have five: unripe, slightly ripe, ripe, overripe, and rotten. In total, the developed model has fifteen combinations. On the other hand, the model developed by Ananthanarayana et al. (2020) contains only one stage and uses the SSD (Single Shot Detector) algorithm with MobileNetV2 as the backbone. Their model determines the freshness level of three different agricultural products (apples, bananas, and oranges) by considering two freshness levels (fresh and rotten) and the results are summarized in Table 3.

*Table 3. Summary of the results of Ananthanarayana et al. (2020)*

| Ground Truth Class | Average Precision Using 0.5 IOU Threshold |
|---|---|
| **Fresh Apples** | 94.71% |
| **Fresh Bananas** | 26.74% |
| **Fresh Oranges** | 57.71% |
| **Rotten Apples** | 90.34% |
| **Rotten Bananas** | 25.82% |
| **Rotten Oranges** | 76.89% |

Despite having only two classes for each fruit in their model, the average precision values of the models developed by Ananthanarayana et al. (2020) are less than those in this research. As can be seen from Tables 2 and 3, both YOLOv4 and YOLOv5 perform well for all fruits models under consideration. In other words, in terms of detection and true prediction YOLO algorithms produce better predictions compared to SSD. Moreover, the models developed in this chapter have five different freshness scales while showing a better detection rate, but the SSD models in Ananthanarayana et al. (2020) have only two. Please notice that the SSD object detection algorithm does not show the same performance on different fruits makes it less reliable in agricultural applications.

The method offered in this chapter is applying freshness detection in two steps. Once the agricultural product type is determined in the first step, the corresponding freshness grading model is utilized in the second step. In order to calculate all the freshness levels of the objects contained in the image, they must be located perfectly, i.e., the object detection rate must be as high as possible. As can be seen from Table 1, YOLOv5 detects a greater number of objects in the images compared to YOLOv4. Therefore, in the object detection manner, version 5 is better. Once the product type is determined correctly, the individual freshness models continue to investigate the image frame. For freshness detection, YOLOv5 performs a higher true prediction and detection rate for lemon and tomato, but on the contrary, YOLOv4 has s smaller error and a slightly higher detection rate for strawberries. To sum up, overall results show that YOLOv5 is a better choice in terms of detection rate and true freshness prediction for most products. Note that it is also possible to combine these models to increase the reliability of the freshness detection system.

## CONCLUSION AND FUTURE RESEARCH DIRECTIONS

In this chapter, a deep learning model that can be utilized to grade the freshness of strawberries, lemons, and tomatoes, is proposed. To this end, the authors used YOLO - a state of art object detection algorithm. Test results show that the performance of the model is very high compared to existing studies in the literature and the speed of the model is fast. The authors believe that these promising results can be utilized to reduce a significant amount of agricultural waste and increase customer satisfaction once it is utilized by online groceries. While the deep learning model is trained by post-harvest data, it is possible to adapt the model to pre-harvest applications by training the model with pre-harvest images.

Density is an important specification for the freshness analysis of fruits and vegetables (Khoshnam, Namjoo, Golbakshi, & Dowlati, 2015). In the lifecycle of a fruit or vegetable, the density varies since the amount of water in it decreases over time. It is possible to compute the density of an agricultural product by estimating its volume using image processing techniques and obtaining its weight using computer vision algorithms once it is placed on a digital scale. With the density information, the optimum density values of the products can be determined, and the model can be trained by using both photos and density values. Therefore, the success of the freshness detection model can be improved using density information.

The data used for training consist of pre-harvest images of the agricultural products and therefore it performs well for pre-harvest applications. The model can be used for pre-harvest applications once it is trained by a data set containing post-harvest images. By this means, it can be integrated into autonomous drones that are picking the agricultural products from trees, plants, or vines. The drones with freshness detection capability can pick the right product and the right time, yielding an increase in income and a reduction in agricultural waste. Finally, the model can be improved to determine plant diseases along with detecting the freshness level.

## ACKNOWLEDGMENT

This work was supported by The Scientific and Technological Research Council of Turkey (TUBITAK) 2209/A Programme.

## REFERENCES

Al Ohali, Y. (2011). Computer vision based date fruit grading system: Design and implementation. *Journal of King Saud University - Computer and Information Sciences, 23*(1), 29–36. doi:10.1016/j.jksuci.2010.03.003

Alaimo, L. S., Fiore, M., & Galati, A. (2020). How the covid-19 pandemic is changing online food shopping human behaviour in Italy. *Sustainability*, *12*(22), 9594. doi:10.3390u12229594

Alexey. (n.d.). *darknet: YOLOv4 / Scaled-YOLOv4 / YOLO - Neural Networks for Object Detection (Windows and Linux version of Darknet)*. Academic Press.

Almog, U. (2020). *YOLO V3 explained*. Towards Data Science. https://towardsdatascience.com/yolo-v3-explained-ff5b850390f?source=rss----7f60cf5620c9---4&ref=morioh.com&utm_source=morioh.com

Amemiya, T., Akiyama, K., Leow, C. S., Buayai, P., Makino, K., Mao, X., & Nishizaki, H. (2021). Development of a support system for judging the appropriate timing for grape harvesting. *2021 International Conference on Cyberworlds*. 10.1109/CW52790.2021.00040

Ananthanarayana, T., Ptucha, R., & Kelly, S. C. (2020). Deep Learning based Fruit Freshness Classification and Detection with CMOS Image sensors and Edge processors. *IS&T International Symposium on Electronic Imaging*, *2020*(12), 172–172. doi:10.2352/ISSN.2470-1173.2020.12.FAIS-172

Anas, D. F., Jaya, I., & Nurjanah. (2021). Design and implementation of fish freshness detection algorithm using deep learning. *IOP Conference Series. Earth and Environmental Science*, *944*(1), 012007. doi:10.1088/1755-1315/944/1/012007

Azarmdel, H., Jahanbakhshi, A., Mohtasebi, S. S., & Muñoz, A. R. (2020). Evaluation of image processing technique as an expert system in mulberry fruit grading based on ripeness level using artificial neural networks (ANNs) and support vector machine (SVM). In Postharvest Biology and Technology (Vol. 166, p. 111201). Elsevier BV. doi:10.1016/j.postharvbio.2020.111201

Barrett, D. M., Beaulieu, J. C., & Shewfelt, R. (2010). Color, flavor, texture, and nutritional quality of fresh-cut fruits and vegetables: Desirable levels, instrumental and sensory measurement, and the effects of processing. *Critical Reviews in Food Science and Nutrition*, *50*(5), 369–389. doi:10.1080/10408391003626322 PMID:20373184

Bochkovskiy, A., Wang, C.-Y., & Liao, H.-Y. M. (2020). *YOLOv4: Optimal speed and accuracy of object detection*. https://arxiv.org/abs/2004.10934

Caferoglu, B. A., Elbir, Z., & Cihan, O. (2022). *GitHub repository*. https://github.com/BeratAsrin/Freshness_Detection_of_Agricultural_Products_Using_Artificial_Intelligence

Design of a Sensor to Detect Fruit Freshness. (2018). *International Journal of Scientific and Technological Research, 4*(1), 1–6.

Fu, Y. (2020). *Fruit Freshness Grading Using Deep Learning*. Auckland University of Technology.

Han, W., Jiang, F., & Zhu, Z. (2022). Detection of Cherry Quality Using YOLOV5 Model Based on Flood Filling Algorithm. *Foods*, *11*(8), 1127. doi:10.3390/foods11081127 PMID:35454714

IBM Cloud Education. (n.d.). *What is overfitting?* Retrieved February 20, 2022, from https://www.ibm.com/cloud/learn/overfitting

Kader, A. A. (2008). Flavor quality of fruits and vegetables. *Journal of the Science of Food and Agriculture*, *88*(11), 1863–1868. doi:10.1002/jsfa.3293

Kemiklioğlu, E., & Özen, Ö. (2018). Design of a Sensor to Detect Fruit Freshness. *International Journal of Scientific and Technological Research*, *4*(1), 1–6.

Khoshnam, F., Namjoo, M., Golbakshi, H., & Dowlati, M. (2015). Physical and Mechanical Changes in Ripening Melon Fruits. *Yuzuncu Yil University Journal of Agricultural Sciences*, *26*(2), 135–144.

Kuznetsova, A., Maleva, T., & Soloviev, V. (2020). Using YOLOv3 algorithm with pre- and post-processing for apple detection in fruit-harvesting robot. *Agronomy (Basel)*, *10*(7), 1016. doi:10.3390/agronomy10071016

Kuznetsova, A., Maleva, T., & Soloviev, V. (2021). YOLOv5 versus YOLOv3 for Apple Detection. In A. G. Kravets, A. A. Bolshakov, & M. Shcherbakov (Eds.), *Cyber-Physical Systems: Modelling and Intelligent Control. Studies in Systems, Decision and Control* (Vol. 338). Springer. doi:10.1007/978-3-030-66077-2_28

Li, J., Su, Z., Geng, J., & Yin, Y. (2018). Real-time detection of steel strip surface defects based on improved YOLO detection network. *IFAC-PapersOnLine*, *51*(21), 76–81. doi:10.1016/j.ifacol.2018.09.412

Make sense. (n.d.). *Makesense.Ai*. Retrieved February 20, 2022, from https://www.makesense.ai

Mathew, M. P., & Mahesh, T. Y. (2022). Leaf-based disease detection in bell pepper plant using YOLO v5. *SIViP*, *16*(3), 841–847. doi:10.100711760-021-02024-y

Mendoza, F., Dejmek, P., & Aguilera, J. M. (2010). Gloss measurements of raw agricultural products using image analysis. *Food Research International (Ottawa, Ont.)*, *43*(1), 18–25. doi:10.1016/j.foodres.2009.08.003

Millan Escriva, D., & Laganiere, R. (2019). *OpenCV 4 Computer Vision Application Programming Cookbook: Build complex computer vision applications with OpenCV and C* (4th ed.). Packt Publishing.

Mitcham, B., Cantwell, M., & Kader, A. (1996). Methods for Determining Quality of Fresh Commodities. *Perishables Handling Newsletter*, 1–5.

Nepal, U., & Eslamiat, H. (2022). Comparing YOLOv3, YOLOv4 and YOLOv5 for Autonomous Landing Spot Detection in Faulty UAVs. *Sensors (Basel)*, *22*(2), 464. doi:10.339022020464 PMID:35062425

Nunes, M. C. N., Emond, J. P., Rauth, M., Dea, S., & Chau, K. V. (2009). Environmental conditions encountered during typical consumer retail display affect fruit and vegetable quality and waste. *Postharvest Biology and Technology*, *51*(2), 232–241. doi:10.1016/j.postharvbio.2008.07.016

PehlivanlıR.KolukırıkS.GörgülüE. (2021). Global Transformation and Differentiation III. figshare.

Péneau, S. (2005). *Freshness of Fruits and Vegetables: Concept and Perception*. Swiss Federal Institute of Technology Zurich.

Ramjan, M. D., & Ansari, M. T. (2018). Factors affecting of fruits, vegetables and its quality. *Journal of Medicinal Plants Studies*, *6*(6), 16–18.

Redmon, J. (2015). *YOLO: Real-time object detection*. Retrieved February 20, 2022, from https://pjreddie.com/darknet/yolo/

Redmon, J. (2016). *Darknet: Convolutional neural networks*. Academic Press.

Redmon, J., & Farhadi, A. (2018). *YOLOv3: An Incremental Improvement*. https://arxiv.org/abs/1804.02767

Sharma, M., Carole, S., Paliyath, G., & Subramanian, J. (2008). Changes in Nutritional Quality of Fruits and Vegetables During Storage. In G. Paliyath, D. P. Murr, A. K. Handa, & S. Lurie (Eds.), *Postharvest Biology and Technology of Fruits Vegetables and Flowers* (pp. 443–466). Wiley-Blackwell.

Valentino, F., Cenggoro, T. W., & Pardamean, B. (2021). A design of deep learning experimentation for fruit freshness detection. *IOP Conference Series. Earth and Environmental Science, 794*(1), 012110. doi:10.1088/1755-1315/794/1/012110

Verma, M. (2015). *Energy Use in Global Food Production*. Springer International Publishing.

Welcome to icrawler — icrawler 0.6.6 documentation. (n.d.). Retrieved February 20, 2022, from https://icrawler.readthedocs.io/en/latest/

Xu, C., He, H., Hauser, S. C., & Gerling, G. J. (2019). Measurement of Touch Interaction Cues in Discriminating Soft Fruit. *Proceedings of the IEEE World Haptics Conference*.

Yao, J., Qi, J., Zhang, J., Shao, H., Yang, J., & Li, X. (2021). A Real-Time Detection Algorithm for Kiwifruit Defects Based on YOLOv5. *Electronics (Basel), 10*(14), 1711. doi:10.3390/electronics10141711

Yohanandan, S. (2020). *mAP (mean Average Precision) might confuse you!* Towards Data Science. https://towardsdatascience.com/map-mean-average-precision-might-confuse-you-5956f1bfa9e2

## KEY TERMS AND DEFINITIONS

**Agriculture:** Agriculture science is a sub-discipline of the biology of processing food and fiber from the soil.

**Artificial Intelligence:** Artificial intelligence (AI) is the act of humanly intelligent act of a computer-controlled machine. Artificial intelligence is the superset of deep learning and machine learning.

**Computer Vision:** Processing of the meaningful information from digital visions within the terms of artificial intelligence.

**Deep Learning:** Deep learning is the subset of artificial intelligence and machine learning. Deep learning is used for processing meaningful data with various algorithms to teach computers actions of human intelligence.

**Freshness:** Freshness is a term that is used for indicating the condition of a product. In this chapter, it is used for indicating and categorizing the ripeness status of agricultural products.

**Machine Learning:** Machine learning is the subset of artificial intelligence and superset of deep learning. Machine learning is used for imitating the acts of humans by teaching computer-based machines and robots.

**Ripeness:** Ripeness is a property of fruits and vegetables used for explaining the quality of the product.

**YOLO:** YOLO (you only look once) is a computer vision-based deep learning algorithm used for object detection.

## APPENDIX: YOLOV5 MODEL CONFUSION MATRICES

*Table 4. Object detection confusion matrix*

| | | | | |
|---|---|---|---|---|
| Predicted | Strawberry | 1.00 | 0.00 | 0.00 |
| | Tomato | 0.00 | 1.00 | 0.01 |
| | Lemon | 0.00 | 0.00 | 0.99 |
| | | Strawberry | Tomato | Lemon |
| | | True | | |

*Table 5. Lemon freshness level detection confusion matrix*

| | | | | | | |
|---|---|---|---|---|---|---|
| Predicted | Unripe | 0.99 | 0.33 | 0.00 | 0.00 | 0.00 |
| | Slightly Ripe | 0.01 | 0.56 | 0.01 | 0.00 | 0.00 |
| | Ripe | 0.00 | 0.11 | 0.99 | 0.12 | 0.02 |
| | Overripe | 0.00 | 0.00 | 0.00 | 0.12 | 0.00 |
| | Rotten | 0.00 | 0.00 | 0.00 | 0.75 | 0.98 |
| | | Unripe | Slightly Ripe | Ripe | Overripe | Rotten |
| | | True | | | | |

*Table 6. Strawberry freshness level detection confusion matrix*

| | | | | | | |
|---|---|---|---|---|---|---|
| Predicted | Unripe | 0.99 | 0.08 | 0.00 | 0.00 | 0.00 |
| | Slightly Ripe | 0.01 | 0.83 | 0.01 | 0.00 | 0.02 |
| | Ripe | 0.00 | 0.08 | 0.98 | 0.00 | 0.03 |
| | Overripe | 0.00 | 0.00 | 0.00 | 0.91 | 0.04 |
| | Rotten | 0.00 | 0.00 | 0.00 | 0.09 | 0.91 |
| | | Unripe | Slightly Ripe | Ripe | Overripe | Rotten |
| | | True | | | | |

*Table 7. Tomato freshness level detection confusion matrix*

| | | | | | | |
|---|---|---|---|---|---|---|
| Predicted | Unripe | 0.99 | 0.09 | 0.00 | 0.00 | 0.02 |
| | Slightly Ripe | 0.01 | 0.68 | 0.01 | 0.00 | 0.00 |
| | Ripe | 0.00 | 0.23 | 0.99 | 0.17 | 0.00 |
| | Overripe | 0.00 | 0.00 | 0.00 | 0.78 | 0.02 |
| | Rotten | 0.00 | 0.00 | 0.00 | 0.06 | 0.95 |
| | | Unripe | Slightly Ripe | Ripe | Overripe | Rotten |
| | | True | | | | |

# Chapter 4
# Precision Farming Using Image Processing and Machine Learning

**Ayesha Naaz**
https://orcid.org/0000-0003-3064-4926
*Muffakham Jah College of Engineering and Technology, India*

**Sucharitha B.**
*Muffakhjam Jah College of Engineering and Technology, India*

## ABSTRACT

*Agriculture in general is plagued by numerous problems that can be solved using modernization techniques and help the farmers making them aware of problems related to crop yield, nutrient status, crop disease, etc. Precision agriculture is a promising approach to address these issues. The objective of the chapter is design and deployment of farmer-friendly precision farming using multispectral imaging techniques for providing advanced, smart, and connected agricultural management solutions by integrating high-precision hardware, software, and data. Remote sensing is a useful tool for monitoring spatio-temporal variations of crop morphological and physiological status and supporting practices in precision agriculture. Of the multiple technologies used for remote sensing, multispectral and hyperspectral remote sensing are widely in demand. The data is processed employing deep learning to allow multispectral image classification based on spectral-spatial features. Applying data analytics on the mapped data can provide suggestions/notifications/alerts to the farmers.*

## INTRODUCTION

Indian agriculture and agriculture in general is plagued by numerous problems. Also, agricultural practices in India are still largely traditional and are dependent on human involvement in a major way and these issues hamper the course of evolution. One of the answers to these types of problems is to help

DOI: 10.4018/978-1-6684-5141-0.ch004

the farmers using modernization techniques and make them aware of problems related to crop yield, nutrient status, crop disease etc.

Precision agriculture provides strategies and tools which allow farmers to improve soil quality and productivity. The chapter aims to describe the design and deployment of farmer friendly precision farming using multispectral imaging techniques for providing advanced, smart and connected agricultural management solutions by integrating high-precision hardware, software, and data. Remote sensing is a useful tool for monitoring spatio-temporal variations of crop morphological and physiological status and supporting practices in precision agriculture. Of the multiple technologies used for remote sensing multispectral and hyperspectral remote sensing are widely in demand. The data is processed employing deep learning to allow multispectral image classification based on spectral-spatial features. Applying data analytics on the mapped data this can provide Suggestions / Notifications / Alerts to the farmers.

This will add value to all farm sizes to all aspects of their operations and more importantly, it will help in crop monitoring and farm management. Thus, resulting in an environment which improves operational efficiency and productivity in farming.

Remote sensing is capable of identifying within-field variability of soils and crops and providing useful information for site-specific management practices. Multispectral imaging is facilitated by collecting spectral signals in a few discrete bands, each spanning a broad spectral range from tens to hundreds of nanometers. Multispectral imaging has been used in agriculture for a wide range of purposes, including estimating crop biochemical properties (e.g., chlorophyll, carotenoids, and water contents) and biophysical properties (e.g., LAI, biomass) for understanding vegetation physiological status and predicting yield, evaluating crop nutrient status (e.g., nitrogen deficiency), monitoring crop disease, and investigating soil properties.

Previous studies have focused on discussing one or two of the many factors impacting crop growth performance and productivity, and thus cannot evaluate crop status and growth-limiting factors comprehensively. It is important to integrate these factors to achieve a better understanding of their inter-relationships for optimal crop production and environmental protection. The first stage starts by acquiring multispectral images of the field. With this dense spectral sampling, imaging spectrometers provide images with an extra dimension to conventional imaging. Multispectral sensors can be mounted on different platforms, such as satellites, airplanes, UAVs, and close-range platforms, to acquire images with different spatial and temporal resolutions. This is followed by data preprocessing stage, issues like atmospheric noise, radiometric noise that is the prominent concern mainly due to the light conditions, needs to be considered and removed as it forms a major source of distortion. It is very important to overcome such issues to obtain the proper conditions that will allow to process and obtain an intended outcome.

The multispectral images typically have 3-10 bands, and many of them are highly correlated, therefore, dimension reduction is also an essential procedure to consider. Image compression is a technique through which the size of the image can be reduced without loss in image quality beyond the desired level. It is one of the essential steps of Image processing technique, which is included in every space mission, as it reduces the cost of bandwidth and storage equipment. In lossless mode, compression reduces the size by storing the same information with a small number of bits, by two methods using different representations, and removing existing redundancy.

Finally, Discriminant algorithms for spectral features and convolutional neural networks (CNN) to deal with spatial features are commonly used methods for multispectral image classification. The spectral-spatial residual network consisting of a supervised deep learning framework that is capable of discriminating features from abundant spectral signatures and spatial contexts will be used for processing. Machine learning algorithms, including Support Vector Machine regression (SVM) and Random Forest (RF), are powerful tools for analyzing multispectral information since they can process a large number of variables (e.g., spectral reflectance and vegetation indices) efficiently. Machine learning has been widely used in the remote sensing field for estimating properties of ground features or classifying different ground covers. Using Machine learning a model is created using training data that can determine a particular feature and after having the appropriate model, the evaluation will be performed using Precision, Recall, and F-Score. Testing data is feed-forwarded into final model to obtain the prediction of that feature/ parameter. Applying data analytics on the mapped data this can provide Suggestions/ Notifications/Alerts to farmers regarding:

a. The type of crop to be harvested for the upcoming season, based on their soil characteristics and the extent of crop already harvested
b. Their crop health and possible Pest Management techniques
c. Better fertilizer and nutrient management procedures for better productivity, maximum economic return and time to market which results in profitable farming.

## IMPACT OF IMAGE PROCESSING AND MACHINE LEARNING IN AGRICULTURE

Precision agriculture is a promising approach to address the various challenges in farming to fulfil the food demands through improving farming practices, e.g., adaptive inputs (e.g., water and fertilizer), ensured outputs (e.g., crop yield and biomass), and reduced environmental impacts. It provides the tools and technologies to identify in-field soil and crop variability, offering a means to improve sub-field level farming practices and optimizing agronomic inputs. The food demand increased three times to cater to the needs of population that doubled since the 1960s and universal agriculture has not been able to meet the demands completely. Because of the limited availability of land, the demand for food and agricultural products which is projected to further increase by more than 70% by 2050 will be met through agricultural strengthening, i.e., increased use of fertilizers, pesticides, water, and other inputs.

Precision Agriculture a technology-enabled approach to farming management that uses a set of advanced information, communication, and data analysis techniques in the decision-making process helps in enhancing crop production and reducing water and nutrient losses and negative environmental impacts. Precision agriculture uses intensive data and information collection and processing in time and space to make more efficient use of farm inputs, leading to improved crop production and environmental quality (Figure.1).

New technologies have been developed to collect high amounts of data from field experiments. Emerging technologies, such as Remote Sensing (RS), Global Positioning Systems (GPS), Geographic Information Systems (GIS), Internet of Things (IoT), Big Data analysis, and Artificial Intelligence (AI) are promising tools being utilized to optimize agricultural operations and inputs aimed to enhance production and reduce inputs and yield losses. Remote sensing is widely used in many applications as it provides high information content of images, fast data collection possibility for large territories, availability of different sensors both airborne and space-borne, and so on.

Image processing captures images through the techniques of remote sensing, involving aircraft or satellites, which are then processed and analyzed using machine learning algorithms. Image processing provides information about crops without even having a physical contact. This technique is helpful in agriculture in many ways, few of which are presented in Figure.2

Crop production plays a vital role in the food industries and Artificial Intelligence and Machine Learning paves way for the farmers to improve crop yields and reduce food production costs.

Machine Learning is a section of Computer Science where new developments evolve and help in automating and processing the crop data for agricultural deficiencies providing an economical approach in automation of agriculture. Machine learning is a type of Artificial Intelligence that provides devices with the ability to learn without being separately programmed. It gives focus on the development of computer programs that can change when exposed to new data. Finding out the suitable crops based on the soil's appearance becomes tedious for novice farmers. There also exists a need to prevent the agricultural decay.

In all, machine learning (ML), Image processing provide real-time data for increasing agricultural efficiencies,

## METHODOLOGY

The key steps followed in precision agriculture/farming are illustrated in Figure.3
The following sections present the details of the stages involved in precision agriculture.

### Data Collection

Monitoring the health status of crops requires a large effort, especially when they extend over vast areas. Productivity in an area depends on the quality of soil, the presence of parasites and fungi or irrigation problems: it is vital to be able to identify these causes in a timely manner, in order to remedy conditions that reduce productivity.

Maps of crop growth, crop diseases, weeds, crop nutrient deficiencies, and other crop and soil conditions are required for evaluating the capabilities of soil and crop. As a result, maps depicting crop and soil variability through remote sensed images acquired by sensors mounted on satellites, aircraft or ground-based equipment have become an integral part of precision agriculture.

Numerous satellites have been launched carrying sensors, instruments for capturing electromagnetic energy emitted and reflected by objects on the Earth's surface. While early remote sensing was based on photographs, most of today's remote sensing uses such sensors.

To avoid influences of spatial inconsistency as errors, the field research has applied replication, blocking and randomization in experimental design. However, conventional experimental designs are characterized by limitations and consequently may not represent a realistic cropping system. In field experiments the effects spatial inconsistency is measured through sampling, that depends on several factors (objectives, field variability, costs), and can range from one sample for several hectares to a more detail coverage of the field. Conventionally, samples are obtained for whole fields or parts of fields to provide average values.

There are several commonly used sampling methods characterized by destructive sampling (Figure.4a-4e):

- Simple random method wherein locations are randomly selected, and may not capture the variation structure of the attributes of interest.
- Stratified random method where the field is divided into several areas according to its characteristics (e.g., topography), and sampling locations are selected randomly and then composite, reducing the influence of local heterogeneity.
- Systematic (grid sampling) method is the method in which the field is divided in grids and samples are collected randomly within each cell and then composite.
- Stratified-systematic method is the systematic method where each cell is further divided into smaller cells to try to overcome the bias introduced by systematic sampling.
- Judgmental method chooses sampling locations based on observation of a specific problem (e.g., low yield) and is not statistically accurate.

To study spatial inconsistency a large amount of data is required, for which, automatic mobile soil samplers have been developed. Continuous-sampling is another emerging solution, where every location in the field is measured by non-invasive techniques (e.g., electromagnetic induction, remote sensing).

## Remote Sensing

Remote sensing applications in agriculture are based on the interaction of electromagnetic radiation with soil or plant material. Typically, remote sensing involves the measurement of reflected radiation, rather than transmitted or absorbed radiation. Remote sensing refers to non-contact measurements of radiation reflected or emitted from agricultural fields (Figure.5). The platforms for making these measurements include satellites, aircraft, tractors and hand-held sensors. Measurements made with tractors and hand-held sensors are also known as proximal sensing, especially if they do not involve measurements of reflected radiation (Sami Khanal, 2020).

Remote sensing is applicable to both soil and crop data collection. Remote sensed imagery can be used for mapping soil properties, classification of crop species, detection of crop water stress, monitoring of weeds, detection of crop diseases and mapping of crop yield.

Use of remote sensing in precision agriculture is influenced by various factors, as presented in Figure.6.

The satellite, aerial, and ground-based platforms and their associated imaging systems can be differentiated based on the altitude of the platform, the spatial resolution of the image, and the minimum return frequency for sequential imaging. Spatial resolution affects the area of the smallest pixel that can be identified. As spatial resolution improves, the area of the smallest pixel decreases, and the homogeneity of soil or crop characteristics within that pixel increases. Poor spatial resolution implies large pixels with increased heterogeneity in soil or plant characteristics. Return frequency is important for assessment of temporal patterns in soil or plant characteristics. The availability of remote sensing images from satellite and aerial platforms is often severely limited by cloud cover, whereas ground based remote sensing is less affected by this limitation (Deepak Katkani, 2022).

### *Satellite Remote Sensing*

Satellites have been used for remote sensing imagery in agriculture since the early 1970's. The first application of remote sensing in precision agriculture used Landsat imagery (1991) of bare soil to estimate spatial patterns in soil organic matter content, which were then used as auxiliary data along with ground based measurements. The spatial resolution of Landsat, SPOT and IRS satellites is fairly coarse for current applications in precision agriculture (Jonathan R. B. Fisher, 2017).

While using remote sensed images for agricultural decision-making, several issues must be carefully evaluated, including:

- How accurately the image matches the ground location (also called geometric precision);
- To what extent the image depicts features in the ground (i.e., spatial and spectral resolutions); and
- The quality of spectral information represented in acquired images.

Efforts were subsequently started to design satellite imaging systems that had the higher spatial resolution and quicker revisit cycles required for precision agriculture (Figure.7). Today, sensing technologies—both ground based and remote—continue to evolve and have become cheaper for capturing field level data.

Modern remote sensing tends to improve the spatial resolution of sensors and to make them multichannel, for example, hyperspectral and multispectral.

## Airborne Remote Sensing

Airborne platforms were the sole non-ground-based platforms for early remote sensing work (Figure.8). The first aerial images were acquired with a camera carried aloft by a balloon which are rarely used today because they are not very stable and the course of flight is not always predictable. At present, airplanes are the most common airborne platform. Nearly the whole spectrum of civilian and military aircraft is used for remote sensing applications.

Low altitude aircraft typically fly below altitudes where supplemental oxygen or pressurization are needed which are good for acquiring high spatial resolution data limited to a relatively small area. Helicopters are usually used for low altitude applications where the ability to hover is required. Ultralight aircraft are a class of aircraft that is gaining popularity which is a single seat powered flying machine. These small, often portable, aircraft are inexpensive and are able to take off and land where larger aircraft cannot. They are limited to flying at lower elevations and at slow speeds. If the demands of the remote sensing requirement are not too strict, ultralight aircraft may be a reasonable alternative to larger aircraft.

## Proximal Remote Sensing

Remote sensing allows the mapping of the Earth's surface from satellite or airborne systems, while proximal sensing systems collect detailed information near the surface. Proximal remote sensing involves the use of sensors in close proximity to the plants, such as on a tractor or harvester or spreaders or sprayers or irrigation booms or ground-based robot (Figure.9). There has been significant interest in proximal remote sensing techniques to assess crop growth and crop stress owing to the limitations of satellite remote sensing for precision agriculture.

Proximal sensing allows real-time site-specific management of fertilizer, pesticides or irrigation. This offers a great advantage over previous data sources, as repetitive wide area coverage can be performed at a low cost, and estimations are performed in a non-destructive way. Remote sensing is widely tied to the use of satellite, airborne or UAV platforms using multi- or hyperspectral imagery. In terms of proximal sensing, the sensor is close to the object and is installed on platforms ranging from handheld, fixed installations, or robotics and tractor-embedded sensors. The types of sensors range from simple RGB or grey-level-cameras to multispectral and hyperspectral high resolute imaging systems or even thermographic camera (Bosoon Park, [. L., 2015).

## Ground Based Remote Sensing

Ground-based remote sensing uses a variety of geophysical survey techniques to "see" beneath the surface of the soil, providing a map of the underlying archaeological, alluvial and geological features (Figure.10). A wide variety of ground based platforms are used in remote sensing. Some of the more common ones are hand held devices, tripods, towers and cranes. Instruments that are ground-based are often used to measure the quantity and quality of light coming from the sun or for close range characterization of objects. For example, to study properties of a single plant or a small patch of grass, it would make sense to use a ground-based instrument.

Laboratory instruments are used almost exclusively for research, sensor calibration, and quality control. Much of what is learned from laboratory work is used to understand how remote sensing can be better utilized to identify different materials. This contributes to the development of new sensors that improve on existing technologies.

Permanent ground platforms are typically used for monitoring atmospheric phenomenon although they are also used for long-term monitoring of terrestrial features. Towers and cranes are often used to support research projects where a reasonably stable, long-term platform is necessary. Towers can be built on site and can be tall enough to project through a forest canopy so that a range of measurements can be taken from the forest floor, through the canopy and from above the canopy.

From satellite remote sensing to low-flying unmanned aerial vehicles (UAVs) and ground based platform, a large amount of data is collected to help farmers and agricultural policy makers to take informed decisions. Remote and proximal sensing systems can provide a multitude of information for agricultural applications, with the main objective being to map, monitor and model agricultural resources and the environmental impacts of agriculture.

Remote sensing images, collected multiple times during a growing season, are used to determine various indicators of crop water demand such as ET, soil moisture, and crop water stress. These indicators are used to estimate crop water requirement and schedule irrigation precisely (Yichun Xie, 2008).

## Data Analysis

Multispectral data are recognized due to its unique characteristics of spatial, spectral, and temporal features. The data formation is directly associated to various characteristics (Figure.11). The most appropriate spatial and spectral resolution for precision agriculture applications depends on factors such as crop management objectives, capacity of farm equipment to vary farm inputs, and farm unit area. The multispectral images integrate with specialized agriculture software which output the information into meaningful data. As the spatial and spectral resolution of satellite imagery has improved, the suitability of using reflectance data for precision agriculture applications has also increased (Gamal El Masry, 2019).

Every surface reflects back some of the light that it receives. Objects having different surface features reflect or absorb the sun's radiation in different ways. The ratio of reflected light to incident light is known as reflectance and is expressed as a percentage.

Water, pigments, nutrients, and other parameters are each expressed in the reflected optical spectrum from 400 nm to 2500 nm, with often overlapping, but spectrally distinct, reflectance behaviors (Figure.12). These known signatures allow us to combine reflectance measurements at different wavelengths to enhance specific vegetation characteristics. By taking the ratio of red and near infrared bands from a remotely sensed image, an index of vegetation "greenness" can be defined.

## Vegetation Indexes:

Vegetation indices are mathematical expressions that combine measured reflectance in many spectral bands to produce a value that helps assess crop growth, vigor, and several other vegetation properties such as biomass and chlorophyll content. Mapping of these indices can help understand spatio-temporal variability in crop conditions, which is crucial for PA applications.

Vegetation reflectance properties are used to derive vegetation indices (VIs). The indices are used to analyze various ecologies. Vegetation Indices are constructed from reflectance measurements in two or more wavelengths to analyze specific characteristics of vegetation, such as total leaf area and water content. Vegetation interacts with solar radiation differently from other natural materials, such as soils and water bodies. The absorption and reflection of solar radiation is the result of numerous interactions with different plant materials, which varies considerably by wavelength (Su, 2017).

## Normalized Difference Vegetation Index

The Normalized Difference Vegetation Index (NDVI) is an index of plant "greenness" or photosynthetic activity, and is one of the most commonly used vegetation indices. Vegetation indices are based on the observation that different surfaces reflect different types of light differently. Photosynthetically active vegetation, in particular, absorbs most of the red light which hits it while reflecting much of the near infrared light. Vegetation which is dead or non-vegetated reflects more red light and less near infrared light (Boris Boiarskii, H. H. 2019).

NDVI can be calculated on a per-pixel basis for any sensor that contains both a red and infrared band; NDVI scores are calculated pixel-by-pixel using the following algorithm:

$$NDVI = (NIR - R) / (NIR + R)$$

Where R stands for the visible red band, while NIR represents the near-infrared band. NDVI is calculated using reflectance ratios in the NIR and red portion of the spectrum (Figure.13).

By taking this ratio, an index of vegetation "greenness" can be defined. NDVI is calculated as the normalized difference between the red and near infrared bands from an image. Also, because it is a ratio of two bands, NDVI helps compensate for differences both in illumination within an image due to slope and aspect, and differences between images due things like time of day or season when the images were acquired. Thus, vegetation indices like NDVI make it possible to compare images over time to look for agricultural and ecologically significant changes.

Various factors can affect NDVI values like plant photosynthetic activity, total plant cover, biomass, plant and soil moisture, and plant stress. Because of this, NDVI is correlated with both agricultural and ecosystem attributes, which are of interest to researchers and managers (e.g., net primary productivity, canopy cover, bare ground cover).

The NDVI provides useful information relevant to at geographical scales ranging from local to global (Figure.14).

### Normalized Difference Red Edge

Normalized difference red edge (NDRE), is a metric which can be used to analyse whether images obtained from multi-spectral image sensors contain healthy vegetation or not. It is similar to Normalized Difference Vegetation Index (NDVI) but uses the ratio of Near-Infrared and the edge of Red. NDRE uses a red edge filter to view the reflectance from the canopy of the crop (Figure.15). The red edge is a region in the red-NIR transition zone of vegetation reflectance spectrum and marks the boundary between absorption by chlorophyll in the red visible region and scattering due to leaf internal structure in the NIR region.

Understanding the levels of chlorophyll can provide the farmer with the ability to monitor photosynthesis activity. With NDRE information the farmer can optimize harvest times based on transitions of photosynthesis activity. During crop harvest events like: hull split in almonds, or max sugar content in grapes, a noticeable change in NDRE values occur. This information in invaluable as a crop management tool for harvest scheduling allowing the farmer to have the highest quality produce.

Other factors that can change chlorophyll levels and cause crop stress are insect infestations. By utilizing NDRE you can determine how severe a bug outbreak is for an almond field and then use a precise way to terminate the infestation. This not only allows one to monitor outbreaks, but also reduce costs associated with pest control.

## Data (Image) Processing

Remote sensing techniques employ sensors or camera which capture and process an image at high number of wavelengths, referred to as spectral imaging. Spectral imaging allows for extraction of additional information which the human eye fails to capture. Spectral imaging, in agriculture, refers to the detection of light reflected by the crop with the use of specialized sensors. Spectral imaging is widely used now in agriculture and precision farming. Data imagery visualizes different wavelengths that are absorbed and radiated from soils, helping farmers monitor different variables affecting their crops (Figure.16). These include soil moisture, surface temperature, photosynthetic activity and weed or pest infestations (José Paulo Molin, 2019).

Currently satellites or aerial or ground-based equipment is fitted with cameras that captures the scene with a higher spectral resolution. A traditional digital camera captures the light that falls onto the sensor in a fashion that resembles the human perception of color. For this, wideband filters are used to obtain red (R), green (G), and blue (B) channels. In contrary, higher spectral resolution cameras enables us to capture information that is neither available to the human observer, nor to the RGB camera. The images are captured using reflectance of light in several narrow frequencies called spectral bands. In this, each pixel holds a vector of intensity values (instead of a R, G, B triplet), where each value corresponds to the incoming light over a small wavelet range. The spectral bands give more comprehensive details and measures of the object or scene. The higher the number of bands the higher the accuracy, the flexibility and information content. Moreover, the individual spectral band has different kinds and levels of details. This may be achieved by collecting each band separately with narrow color filters or by scanning the image with a spectrometer.

Multispectral and Hyperspectral imaging are two primary methods for capturing images with higher spectral resolution. While multispectral cameras provide only a few (typically <10) relatively broad spectral bands, hyperspectral cameras are characterised by a higher number (often >100) of narrow spectral bands, resulting in near-continuous spectral curves, which allow even minor material differences to be distinguished (Figure.17).

Hyperspectral imaging involves narrow, usually contiguous spectral bands about hundreds or thousands of spectra, while multispectral imaging involves spectral bands of varying bandwidths that may or may not be contiguous. Multispectral imaging can be thought of as a reduced subclass of hyperspectral imaging. These are complementary technologies and are application dependent.

Although the hyperspectral images offer additional opportunities to capture variability in crop and soil conditions in agricultural sectors, there are limitations associated with their use including the high cost of the sensor, high image storage requirement and the complexities associated with image processing. Thus, when selecting a spectral sensor for agricultural decision-making, the benefits of using hyperspectral images should be weighed against the expected cost. However, their widespread adoption has been hindered because they are very expensive.

Hyperspectral imaging will always be the go-to for exploratory work, where speed, cost, and complexity matter less, and multispectral imaging systems are the natural derivative of that exploration once a production system needs to be built. Multispectral imaging can be considered for precision agriculture keeping in view its cost and higher-speed inspection.

## MULTISPECTRAL IMAGING

The primary goal of multispectral imaging in agriculture is to detect subtle variation in plant health before visible symptoms appear. For instance, a grower could spot a small reduction in a plant's chlorophyll content before the leaves start to turn yellow. Such early detection is possible because the amount of sunlight plants reflects in different wavelengths vary as their health changes. Multispectral sensors then capture and record this variation. However, some sensors capture more precise data than others. If a sensor measures too broad a region of the light spectrum, any subtle variation will be lost.

A multispectral image sensor captures image data at specific frequencies across the electromagnetic spectrum. The wavelengths may be separated by filters or by the use of instruments which are sensitive to particular wavelengths, including light from frequencies beyond our visible sight, such as infrared. A series of bandpass filters centered at regular intervals of the visible and near-infrared spectrum can be used in combination with reflective and trans-missive lighting. Each bandpass filter blocks all light except for a short range around a specific wavelength. Each pixel of the image captured by the camera is then a measurement of light as it interacted with the materials present in the parchment at one coordinate. The combination of those images, photographed under each filter, then provides a map of spectral responses for each pixel.

Multispectral imaging consists of spectral bands which are discretely positioned from each other. Multispectral technology (5–7 bands) can offer a good overview of crop such as overall growth.
Data from multispectral imaging has the following benefits;

- Identify pests, disease and weeds and hence optimize pesticide usage and crop sprays through early detection.
- Provide data on soil fertility and refine fertilization by detecting nutrient deficiencies which will help with land management and whether to take agriculture land in or out of production or rotate crops etc.

- Count plants and determine population or spacing issues.
- Estimate crop yield.
- Measure irrigation which can further control crop irrigation by identifying areas where water stress is suspected. Then, make improvement to land areas such as install drainage systems and waterways based on the multispectral data.
- View damage to crops from farm machinery and make necessary repairs or replace problematic machinery.
- Survey fencing and farm buildings.
- Monitor livestock. Now, drones with thermal cameras can be used to locate livestock at night time along with plenty of other terrific uses.

Multispectral sensors can highlight small changes in the health of crops. This is because multispectral imagery captures a critical part of the light spectrum for studying plants (712–722 nm), called the red edge band. It is in this section of the spectrum that the first signs of stress start to show. Sometimes this stress is related to disease. Using analytics generated with the red edge band, farmers can identify, monitor and track disease-related stress. This is how multispectral sensors can help farmers catch disease sooner and act faster to stop the spread.

Multispectral remotely sensed images composed information over a large range of variation on frequencies (information) and these frequencies change over different regions (irregular or frequency variant behavior of the signal) which need to be estimated properly for an improved classification. Multispectral remote sensing image data are basically complex in nature, which have both spectral features with correlated bands and spatial features correlated within the same band (also known as spatial correlation). An efficient method for utilization of these spectral and spatial (contextual) information can improve the classification performance significantly compared to the conventional non-contextual information-based methods.

In general, the spectral component of a remotely sensed image pixel may be a mixture of more than one spectral information that usually comes from the different regions of the study area. The conventional multispectral remotely sensed image classification systems detect object classes only according to the spectral information of the individual pixel/pattern in a particular frequency band, while a large amount of spatial and spectral information of neighboring pixels of different regions at other frequency bands are neglected. Hence the pixels are classified based on its spectral intensities of a specific band and does not give attention to its spatial and spectral dependencies and thus the spectral intensities of the neighbors at different frequency bands are assumed to be independent. Such approaches may be reasonable if spatial resolution is high or when the spectral intensities are well separated for different classes, which is rarely found in any real-life data sets.

One of the main advantages of digital data is that they can be readily processed and can be easily used to correct, enhance, and classify digital, remotely sensed image data (Figure.18).

## Image Acquisition

This is the fundamental step of digital image processing. In fact, this image acquisition can be as simple as being given an image which is already in digital form. This step involves pre-processing such as scaling.

## Image Enhancement

To make image data easier to interpret. These so-called image enhancement techniques include contrast stretching, edge enhancement, and deriving new data by calculating differences, ratios, or other quantities from reflectance values in two or more bands, among many others. It is the process of filtering image (removing noise, increasing contrast, etc) to improve the quality.

Image enhancement is among the simplest and most appealing areas of digital image processing. Basically, the idea behind enhancement techniques is to bring out detail that is obscured, or simply to highlight certain features of interest in an image. Such as, changing brightness & contrast etc.

To determine, image enhancement is the most straightforward and most appealing areas of image processing. The idea behind an enhancement technique is to deliver the information that is obscure or simply to highlight the particular features of interest in an image. Such as changing brightness and contrast etc.

## Image Restoration

It is the process of improving appearance (e.g,reducing blurring) of an image by mathematical or probabilistic models. Image restoration is an area that also deals with improving the appearance of an image. However, unlike enhancement, which is subjective, image restoration is objective, in the sense that restoration techniques tend to be based on mathematical or probabilistic models of image degradation.

## Image Compression

In fact, the compression deals with techniques are useful for reducing the storage required to save an image or the bandwidth from transmitting it. Particularly, in the uses of Internet, it is very much mandatory to compress data.

## Image Segmentation

Segmentation procedures partition an image into its constituent parts or objects. In general, autonomous segmentation is one of the most difficult tasks in digital image processing. A rugged segmentation procedure brings the process a long way toward successful solution of imaging problems that require objects to be identified individually.

Description and representation follow the output of a segmentation stage, that is usually is raw pixel data. At the same time, choosing an observation is only part of the solution for converting raw data into the form that is useful for subsequent digital signal processing projects. It deals with extracting attributes which result in some quantitative details of interest or is fundamental for differentiating one class of objects from another.

## Image Correction

The reflectance at a given wavelength of an object measured by a remote sensing instrument varies in response to several factors, including the illumination of the object, its reflectivity, and the transmissivity of the atmosphere. Furthermore, the response of a given sensor may degrade over time. With these factors in mind, it should not be surprising that an object scanned at different times of the day or year will exhibit different radiometric characteristics.

Analysts have developed numerous radiometric correction techniques, including Earth-sun distance corrections, sun elevation corrections, and corrections for atmospheric haze. In addition to radiometric correction, there is a need for images to be geometrically corrected. Geometric correction and orthorectification are two methods for converting imagery into geographically-accurate information. Geometric correction is applied to satellite imagery to remove terrain related distortion and earth movement based on a limited set of information. In contrast, orthorectification uses precise sensor information, orbital parameters, ground control points, and elevation to precisely align the image to a surface model or datum.

### Image Classification

A multispectral image covers enormous areas of land cover and is inherently difficult to process on this entire multispectral image, hence image analysts have developed a family of image classification techniques that automatically sort pixels with similar multispectral reflectance values into clusters that, ideally, correspond to functional land use and land cover categories.

## MACHINE LEARNING

Machine Learning is the subfield of computer science that gives computers the ability to learn without being explicitly programmed. A machine or intelligent computer program learns and extract knowledge from the data, builds a framework for making predictions or intelligent decisions. Thus, the ML process is divided into three key parts, i.e. data input, model building, and generalization (Figure.19). Generalization is the process for predicting the output for the inputs with which the algorithm has not been trained before. ML algorithms are mainly used to solve complex problems where human expertise fails such as weather prediction, spam filtering, disease identification in plants, pattern recognition.

At present Machine learning, deep learning, computer vision, image processing, robotics and IoT technologies are very supportive to farmers. They are very helpful in agriculture field to minimize the manpower, improve the quality of crops, water and soil management and detect crop disease in early stages. These technologies are very helpful for farming because it makes it easier to monitor, scan and analyze the crops by providing high quality images. This technology is useful to identify the progress of the crops. In addition, farmers can decide whether the crops are ready for harvest or not (RavesaAkhter, S. A., 2021).

ML algorithms as presented in Figure.20 are broadly categorized as:

1. supervised learning,
2. unsupervised learning, and
3. reinforcement learning

A brief discussion about these algorithms follows.

## Supervised Learning

In a supervised learning model, the algorithm learns on a labeled dataset, to generate expected predictions for the response to new data. This set of algorithms works with labeled data-set which means corresponding to each input there are outputs. The algorithm builds an input-output relationship with this labeled data set and thereafter generalize or predicts outputs for unseen inputs.

Supervised learning algorithms are used for predicting the categorical value are known as classification algorithms and the algorithms that are used for predicting the numerical value are known as regression algorithms. In Classification, a computer program is trained on a training dataset, and based on the training it categorizes the data in different class labels. This algorithm is used to predict the discrete values such as male or female, true or false, spam or not spam, etc.

K-Nearest Neighbor can be used for both classification and regression. It is a non-complex algorithm which stores all the available cases and classifies new cases based on some similarity measure. The sample set is classified based upon the "closeness" that is the distance measure such as Euclidean distance or Manhattan distance.

Naive Bayes classifier is a simple probabilistic classifier which works based on applying Bayes' theorem (from Bayesian statistics) with strong naive independence assumptions. Naive Bayes is a technique for constructing classifier models which assign class labels to problem instances which are represented as vectors of feature values, where the class labels are drawn from some finite set. It is not just a single algorithm for training such classifiers, but a family of algorithms based on a common principle. All naive Bayes classifiers assumes that the value of a particular feature is independent of the value of any other feature, given the class variable. These Learners predict the class label for each of the training data set. The class label that is predicted by the majority of the models is voted through the majority voting technique and the class label of the training data set is dec

The task of the regression algorithm is to find the mapping function to map input variables(x) to the continuous output variable(y). Regression algorithms are used to predict continuous values such as price, salary, age, marks, etc.

## Unsupervised Learning

Unsupervised learning algorithms works with unlabelled data and discovers unknown objects by grouping similar objects. The goal of an unsupervised learning algorithm is to extract hidden knowledge from the training data set thus this approach is difficult to implement than supervised learning algorithms.

## Reinforcement Learning

Reinforcement learning (RL) is a type of machine learning where the model learns to behave in an environment by performing some actions and analysing the reactions. RL takes appropriate action in order to maximize the positive response in the particular situation. The reinforcement model decides what actions to take in order to perform a given task that's why it is bound to learn from the experience itself.

In data mining algorithms were applied on a data set containing different crops, rainfall, production, area and pH values. The dataset includes rainfall, soil pH, soil attributes. The raw dataset is collected and data is preprocessed to remove noise. On the processed dataset data mining algorithms are applied and accuracy of the dataset is calculated using F-measure. Comparative study is done to analyze and understand various other algorithms. And finally best algorithm is found based on accuracy.

## MACHINE LEARNING FOR PRECISION AGRICULTURE

Machine learning enables computer to learn from experience and here experience is in the form of dataset. More the accuracy in dataset more accurate the results are. Accuracy depends on:

1. Duration of the dataset captured,
2. Features considered,
3. Frequency of the data recording,
4. Duration of the data considered for analysis

A dataset can be prepared from historical data available and the monitoring device.

The necessary steps involved in crop cultivation are Land suitability analysis, appropriate crop selection, crop production, crop protection, nutrient supply, water supply, crop health monitoring (pest and weed control), human and animal attack detection, yield management, and post-harvesting.

Although these steps are common for all types of crops, soil nourishment value and chemical composition determine the techniques adopted in each level. Also, this paves a significant consideration of fertiliser supply when multi-cropping is selected. This multi-cropping technique has been in evolution decades back and done explicitly in the hill areas with meagre farming areas yielding better productivity.

In the conventional machine learning the primary step is to select the data on the problem under investigation and to select the parameters for the examination. The model is trained by a sample set of data (termed as training data) to gain experience in the environment and make the model fit. Later, the model evaluated using a sample set of data (termed as test data). So this is the primary step involved in any machine learning model, i.e., Train-Test-Predict. Usually, the data set was divided into two viz., training (70%) and testing (30%). Testing data is kept separate and not used in the preparation.

The dataset with many alternatives is collected and pre-processed using any normalisation or standardisation methods. The pre-processed data set was divided as train and test data set. The machine algorithms take the train data as input to train the model or to learn for the historical information. The trained model is evaluated with test data. The data visualisation tools are used for visualising the prediction or classification results.

Machine learning utilises a secondary dataset (termed as validation data) for training the model further to avoid the overfitting of the model by the trained data. If the model generates more error on validation data, that means the model overfitted with the prepared data so that training stopped. Now the data split can be done like 60, 10, and 30 per cent of training, validation, and testing, respectively.

### Machine Learning in Land Suitability Analysis

To ensure proper analysis, it is to be tested if it is agricultural land or not, This is referred to as suitability analysis, which is very important for crop cultivation which otherwise may lead to an enormous waste of time, more fertiliser supply, abnormal and water requirements. Few of the factors considered for land

suitability analysis are soil quality parameters (pH, organic carbon content, salinity, texture, slope), topography, water availability, essential nutrients, socio-economic factors.
The key characteristics that determine the suitability analysis are:

- Soil quality
- Moisture content in the soil
- Soil fertility levels

Various Machine learning methods that can be used for the purpose of analysis are:

- Support Vector Machines (SVM) preferred for classifying the suitable area for agriculture of rain-fed wheat based on thirteen factors relating to property, topography, climate, and soil.
- Parallel Random Forest (PRF), SVM, Linear Regression (L.R.), K- Nearest Neighbour (KNN) algorithm, Linear Discriminant Analysis (LDA), and Gaussian-Naïve Bayesian also proved to ensure the land suitability for crop cultivation.
- KNN algorithm was found to be more accurate for farmer's assessments with regard to soil moisture.

A better understanding of the land suitability of the agricultural field under consideration will assist in selecting suitable crops as well as supplying fertiliser to make it better nourished for growing the required plants. It followed by crop production, water supply, and Nutrient management.

## Machine Learning in Crop Production

Crop Production and management include crop selection, soil preparation based on suitability analysis, sowing seeds, application of manure & fertiliser, water management through proper irrigation mechanisms, and harvesting. Machine learning in agriculture crop production links various participants in the food chain or agricultural chain. Machine learning helps the agriculturalists in making better decisions in crop quality determination, yield prediction, plant species determination, crop disease prediction, and harvesting techniques.

The machine learning algorithms data acquired from IoT sensors in the agricultural field. Once the data feed, ML algorithms train the model using history and can make predictions at any stage of production to determine the different features required to predict the yield. It will help to improve the nutritional value (if deficient in the current return predicted) in the next production. Consequently, the crop production price will show a dramatic improvement in the upcoming yield. Application of A.I. in agriculture will enable the farmers to get up to minute information about current production, suggestions on next production, plant species identification, and quality improvement.

Once Land suitability analysis for cultivation is done, crop species selection has to be done based on suitability. Based on the nourishing factor in the soil and nutrient capability, a crop can be selected appropriately. Multi-criteria decision-making models used to get land suitability analysis. Image processing techniques integrated with machine learning suggested for plant species identification for the given crop image.

The various Machine Learning techniques that can be employed for crop selection based on environmental parameters are:

- K- Nearest Neighbour (KNN) algorithm for the data obtained through multiple IoT sensors and specific yield prediction by considering Crop yield prediction using topological algorithms like ANN, back propagation, and Multi-layered perceptron through the implementation of a neural network.
- Support vector regression (SVR.) a variant of SVM used for crop yield prediction.

Machine learning plays an important role in nitrogen management also. As nitrogen is an essential component for photosynthesis, it is one of the important parameters in prediction of yield. The various Machine Learning approaches and signal processing methods used for crop yield prediction and optimised techniques for nitrogen management is Least Squares SVM.

One or more stages of crop cultivation will give information to other steps and vice versa. Depending on soil test results done during land suitability and crop health monitoring, the fertilisers will be recommended. Consequently, water and nutrient management carried out. The ML approaches work best for fertilizer recommendation, the relevant algorithms are:

- Water management is ML neural network with Backpropagation algorithm based on soil nutrient content,
- Gradient boosting and Random Forest for soil nutrient assessment and
- Multivariate Relevance Vector Machine and Multilayer Perceptron for estimating the water requirement based on evapotranspiration and climatic data.

Periodic Drought assessment is essential for crop maintenance and water management. Machine learning approaches used for drought assessment are:

- Random Forest,
- Cubist,
- Boosted regression trees,
- Support vector regression,
- Coupled wavelet ANNs, and
- ANN.

## Machine Learning in Crop Protection

Crop protection implies the protection of crops from unwanted plants that grow in the land, pests (insects, bugs), and intruders (an animal which intends to graze the crops and human for theft). Accurate detection of weeds is more significant and done using Machine learning algorithms integrated with sensor data. The machine learning techniques employed in Precision Agriculture for crop protection from weeds include:

- Counter Propagation ANN,
- XY-Fusion Network and
- Supervised Kohonen Network (SFN)

The machine learning techniques used for crop protection from pests are:

- CNN (though images)
- Decision trees, SVM, ANN (best for prediction and classification of pests),
- Bayesian Networks and KNN (for training data set).

Animal intrusion detection is one of the threats to the agricultural crop. These intrusions identified and detected to avoid loss of crop production. IoT sensors provide periodic alerts on the detection of an animal object like rats, cow, sheep, elephant, and other wild animals. It can be detected effectively and prevented through wireless sensors alerts to farmers mobile and machine learning algorithms can be used for object classification. Also, Machine learning algorithms used to predict the animal or human object entry prior by training the model with past data from IoT sensors.

## CONCLUSION

Several machine learning approaches have become popular for achieving superior and precision agriculture. The following sub-section discusses certain machine learning approaches that have been deployed for achieving enhanced agricultural benefits. In the perspective of machine learning, supervised learning is a phenomenon that encompasses both the input and the sought-after target values. Besides, both the input and target data are in labelled form, which offers a learning platform for processing data in the future. Further, when this model is offered a new test dataset (with a similar background) since the model is already trained, it generates the accurate output for the test data.

## REFERENCES


Boris Boiarskii, H. H. (2019). Comparison of NDVI and NDRE indices to detect differences in vegetation and chlorophyll content. *J. Mech. Cont.& Math. Sci*, *1*(Special Issue), 20–29. doi:10.26782/jmcms.spl.4/2019.11.00003

Bosoon Park. (2015). *Hyperspectral Imaging Technology in Food and Agriculture*. Springer.

Deepak Katkani, A. B. (2022). A Review on Applications and Utility of Remote Sensing and Geographic Information Systems in Agriculture and Natural Resource Management. *International Journal of Environment and Climate Change*.

Gamal ElMasry, N. M.-R. (2019). Recent Applications of Multispectral Imaging in Seed Phenotyping and Quality Monitoring—An Overview. *Sensors*.

Jonathan, R. B., & Fisher, E. A.-F. (2017). Impact of satellite imagery spatial resolution on land use classification accuracy and modeled water quality. *Remote Sensing in Ecology and Conservation*.

José Paulo Molin, T. R. (2019). Sensor Systems For Mapping Soil Fertility Attributes: Challenges. *Engenharia Agrícola*.

RavesaAkhter, S. A. (2021). Precision agriculture using IoT data analytics and machine learning. *Journal of King Saud University - Computer and Information Sciences*.

Sami Khanal, K. K. (2020). Remote Sensing in Agriculture—Accomplishments. *Remote Sensing*.

Su, J. X. (2017). Significant Remote Sensing Vegetation Indices: A Review of Developments and Applications. *Journal of Sensors*.

Yichun Xie, Z. S. (2008). Remote sensing imagery in vegetation mapping: A review. *Journal of Plant Ecology*, 9–23.

# Chapter 5
# The Use of Pesticide Management Using Artificial Intelligence

**Sapna Katiyar**
*ABES Institute of Technology, Ghaziabad, India*

## ABSTRACT

*The population of the world may reach almost 10 billion by 2050, and currently, approximately 37.7% of land is used for the production of crops. Agriculture is a major source of revenue for any country. Globally, automation in agriculture is in demand. Innovation and integration of technologies contributes for challenges faced by farmers with enlarged revenue and employment opportunities. Artificial intelligence has brought a revolution in agriculture. Crop wellbeing is important as it is a crucial factor that relates all parameters directly; therefore, crop health examination is mandatory. Premature detection of pests also reduces the quantity and frequent use of pesticides, but human intervention in process makes it time consuming and expensive. Time and techniques to use the pesticides in large farmland using AI along with computer vision and IoT converts traditional processes into smart agriculture. This chapter presents the assessment and implementation of an intelligent system for pesticide management.*

## INTRODUCTION

Agriculture is India's foundation. As per the Economic survey, modernization and automation is required in farming. Pests is one of the biggest issues as they can soil crops, with which every farmer has to struggle. Every year billions of USD loss are observed in crop due to disease and pests (Abhilash & Singh, 2009; Popp et al. 2013). To protect crop against pests, farmers used to spray pesticides regularly in farmland. Pesticides have numerous drawbacks also like it may harm farmers during spray sometimes may cause cancer, impact on consumer health due to supply chain and damage to non-infected crop (Grewal et al. 2017; Zang 2018). Excessive use of pesticides in crops also increases pesticides residue in food products and by controlling it, farmers pesticides expenditure can be reduced by around 90%.

DOI: 10.4018/978-1-6684-5141-0.ch005

Technology is contributing a vital role in agrobusiness. Involvement of many technologies in traditional agriculture has transformed it into digital agriculture. Artificial Intelligence is one of the promising technologies which can be used for pesticides management (Talaviya et al. 2020; Bannergee et al. 2018). If it is used accurately, it increases efficiency and reduces the adverse impact of pesticides on environment. Presently AI is costly and sometimes limit its applications into agriculture domain. These days many industries are working for it to reduce the initial cost, creating awareness and making eco-friendly devices so that wider population can be beneficial. Internet of Things (IoT) technology can be integrated for developing connected applications for farmers.

AI can support farmers to analyze multiple issues related with pest, climate, soil, water etc. Massive data handling with supreme precision is the strength of AI, which is used for decision making in harvesting (Panpatte 2018; Katiyar & Farhana 2021). Many tasks handled by human labor in agricultural lank such as pesticides spraying, weed removal, seed sowing. These tasks can be handled by AI through Robots, Drones, UAV (Schor et al., 2018; Ahirwar et al. 2019). These days farmers are using innovative approaches like AI, IoT and Computer Vision to protect their crops, increasing profits and sustainable farming (Arakeri et al. 2017; Elijah et al. 2018). These emerging technologies identify pests very precisely and pesticides spraying can be done at designated planes only.

Few challenges are there to implement such technologies like internet availability and awareness issues in rural areas (Abate et al. 2000). Once these challenges are overcome, pesticide management using AI will be observed as novel and feasible solution for crop protection from pests. The objective of this chapter to discuss about pesticides management i.e. partial and sufficient amount of pesticides must be applied on specified target area. Technologies intervention predict the possibility of disease and pest in future so action can be taken accordingly (Oluwole & Cheke, 2009; Bon et al., 2014). Soil and crops can be protected and hance farmers annual revenue must be increased.

## ARTIFICIAL INTELLIGENCE IN AGRICULTURE

Agriculture lands everyday creates massive data related to water usage, soil, temperature and weather. Artificial Intelligence has proven as a boon for agriculture. Basically, AI is an intelligence demonstrated by machines. It is a simulation of intelligence, knowledgebase and problem-solving capabilities built on data. Latest technologies like Artificial Intelligence, Machine Learning, Deep Learning acquire,

*Figure 1. Usage of AI in agriculture*

monitor and analyze real time data to get valuable insights such as pesticides requirements, irrigation requirements, sow seeding, weed removal, crop selection (Bannergee et al., 2018; Arakeri et al., 2017). AI technology is one among many, which is popular these days for pest detection, disease detection, plants nourishment (Subeesh & Mehta, 2021; Food, 2017), shown in Figure 1.

Many industries are working with AI technology and Computer Vision, for developing Robots to monitors weeds, pest and spraying pesticides. AI enabled spraying mechanism reduces the excess spray of chemicals on crops and hence improve the quality and quantity of agricultural products (Katiyar & Farhana, 2021; Sharma, 2021).

Artificial Intelligence offers many advantages in agriculture domain to resolve farming challenges:

- Cost effective
- Remote monitoring
- Better decision making
- Resolve human labor shortage issues
- Reduce huma error
- Availability 24×7
- Digital assistance
- Manages repetitive jobs effectively

Besides many advantages of AI for precise farming, some of the disadvantages/limitations in agriculture faced due to the adoption of AI are:

- Technology adoption issues
- Implementation cost is high
- Unemployment of labor in agricultural land
- Missing out of box thinking
- Security and privacy issues

Integration of Artificial Intelligence with Agriculture domain offers many advantages in numerous processes. Some of them are:

- Analysis of market demand
- Risk handling
- Seeds breeding
- Crop protection
- Crop nourishment
- Soil strength monitoring
- Irrigation
- Harvesting

Agriculture consists of a number of phases and procedures. By integrating approved technologies, AI is capable to execute more complex and day to day tasks. Sometimes it can acquire and process Bigdata on digital platform, and resulted into optimal action plan which can be integrated with other budding technologies. Figure 2 depicts information management cycle for AI in agriculture. Artificial Intelligence can be integrated with other technologies (Liu et al. 2019; Gondchawar et al., 2019). These days industries require more experts as Bigdata engineer or Data analyst. Some use cases of AI in agriculture are:

- **IoT sensors for data capturing and analysis:** Sensors can be placed in farmland which senses and send data as per their nature. Secondary technologies like Drones, Robots, GIS can be used to monitor, store and analyze data. Combining these technologies with AI, farmers get real time precise data to take appropriate decisions. Remote monitoring can also be done. These days few mobile apps are available and through them farmers get notifications on their mobile phones.

*Figure 2. IM cycle for AI in agriculture*

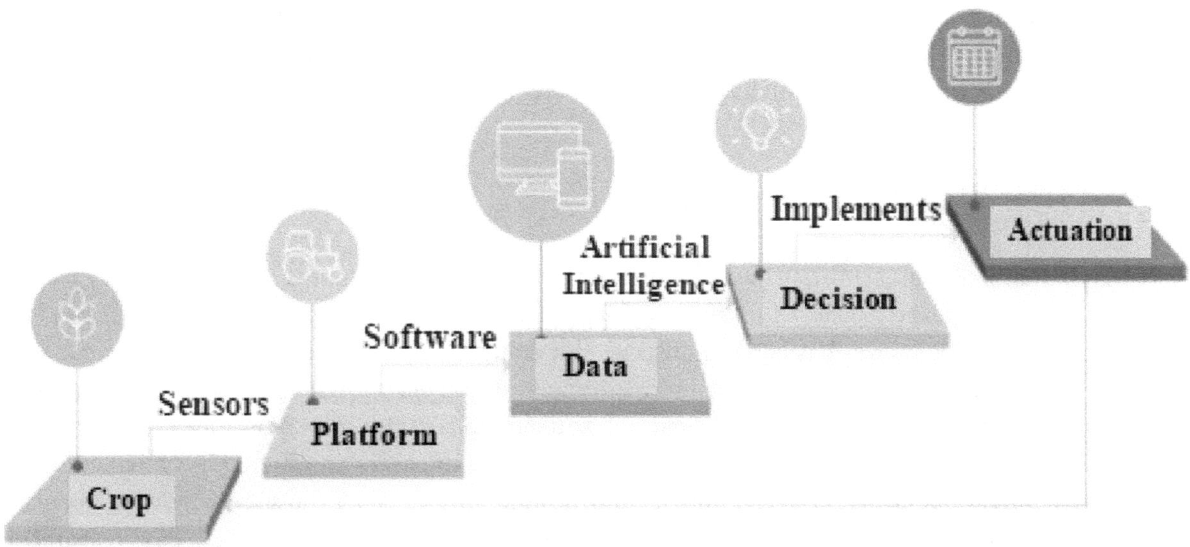


- **Robotics & Automation for reduced manual work:** AI can be integrated with Unmanned Aerial Vehicle and IoT sensors for spraying pesticides, crop monitoring and weed detection. AI can also be combined with autonomous tractor to fulfill farming task. In this way efforts and requirements of human labor can be minimized. These technologies are very efficient, precise with minimum error. Agricultural Robots are already used to perform many tasks in land such as crop picking, spraying, weed removal. Robots can work for longer duration and perform complex task in comparison to humans. They are less prone to error, immune to environmental variations and very precise.
- **Bigdata for decision making:** Massive data is produced every day from farmland, data analyst makes use of this Bigdata for better decision making. Based on real time environment, farmers can receive effective recommendations related to crop or farming. In turn which facilities farmers for precision farming and increases productivity.

## AI IN AGRICULTURE MOBILE APP

Demand for food production increases as population increases. Artificial Intelligence is more than a buzzword in modern technology. As per the survey report it has been listed that by year 2025, revenue from AI will reach around USD 36 billion. UN Food and Agricultural Organization mentioned that worldwide every year 20-40% of crop is destroyed due to unmanaged pests and diseases. Crop loss can be avoided and quality can be maintained with quickly and precise solutions through artificial intelligence.

There are some android mobile apps where information and photos need to be uploaded for expert guidance. Experts use image classifier and algorithms to recognize pest and rodents. Once any bug is identified, app offers some solutions to help experts to treat problems. Plantix is a mobile app widely used in India, North Africa, Brazil. It helps farmers for quick diagnosis, solutions, feedback and preventive measures. Artificial Intelligence, Machine Learning, Deep Learning technologies in mobile apps develops agriculture data more robust and richer. Some new solutions of old problems from this data may be generated when agricultural efficiency and food quality is more important.

## PESTICIDES AND ITS CLASSIFICATION

Human population is increasing day by day and even Green Revolution is not sufficient to supply enough food. In earlier days farmers used to apply pesticides to keep safe their crops (Obopile et al., 2008). The very first known pesticide was basic sulfur dusting. In modern agriculture, yields have been increased to a certain extent primarily by using agrochemicals or agricultural chemicals and enhanced management practices. Agrochemicals comprises of pesticides, fertilizers, soil conditioners, hormones, acidifying and liming agents etc. In recent agriculture, use of agricultural chemicals is an efficient promising feature. Insects are the major concern for agriculture as they destroy crops. As per the survey there is an annual crop production loss of 30% due to the plant disease and pests (Khan et al., 2010). Now a days researchers are trying to reduce this loss to solve the problem of food in country. Pest is characterized by an organism which can cause financial loss or injured the wellness of human being, some types is shown in Figure 3.

Pests have a tendency to destroy crops and sometimes creates disease in humans or animals. Pesticides are the category of chemicals used to destroy insect, fungus, weed, microbes etc. Figure 4 depicts that 76% of the pesticides used by India is insecticides while this is 44% globally (Agnihotri, 2000). On the other hand, use of fungicides and herbicides is very less in India. Pesticides are mainly used for cotton, wheat etc. in India. Some key benefits of using pesticides are: increased productivity, protecting crop loss, caring yield generated, food quality and quantity, food storage, transportation etc.

### Types of Pesticides

Agricultural pesticides are preferred by farmers to manage crop quality and quantity against pests' impact. It has been identified that several pesticides are injurious to human, animals or for environments (Sankararamakrishnan et al., 2005). Figure 5 depicts the various types of pesticides.

a) **Insecticides:** These substances are used to kill insects. Insecticides have extensive applications in the field of medicine, agriculture and industries. They are toxic to humans and animals. They also

Figure 3. Types of pests

Figure 4. Consumption pattern of pesticides

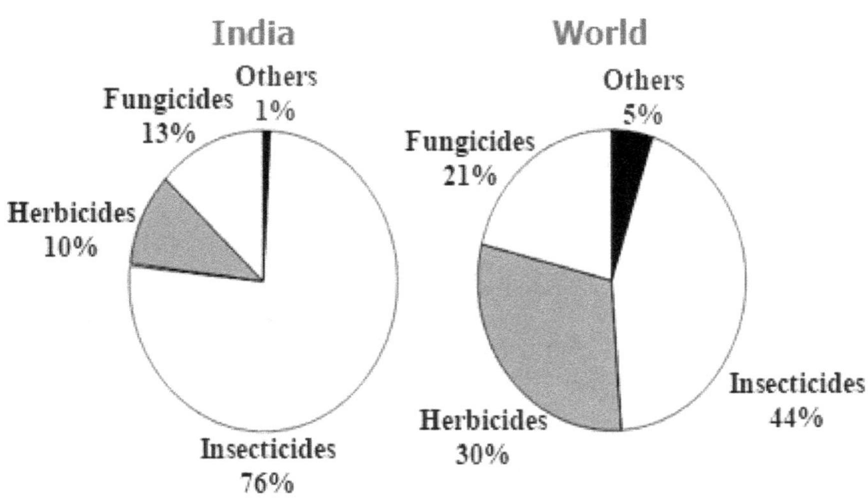

*Figure 5. Types of pesticides*

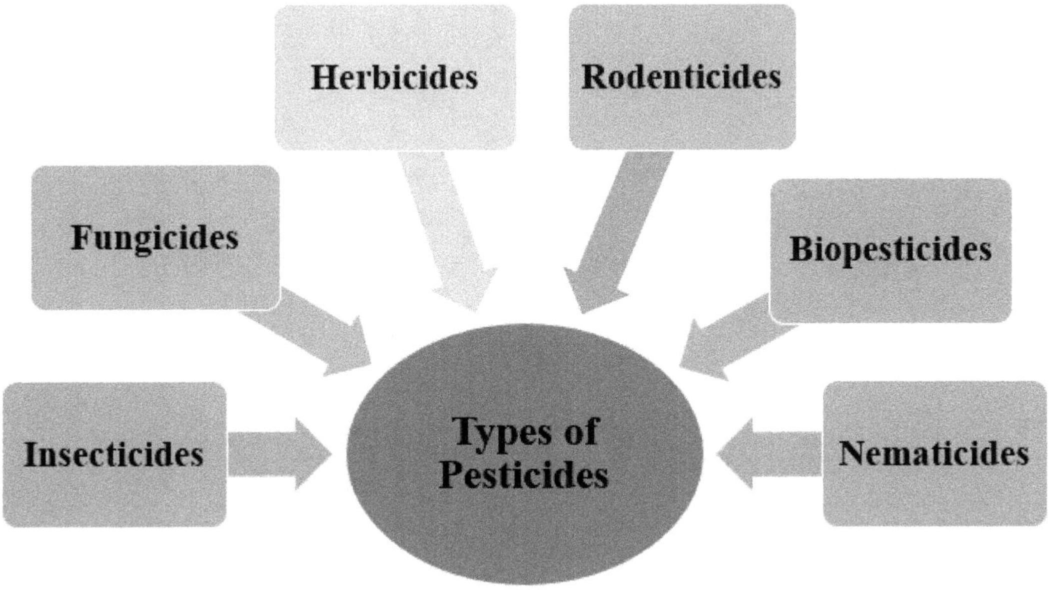


*Figure 6. Detailed insecticides classification*

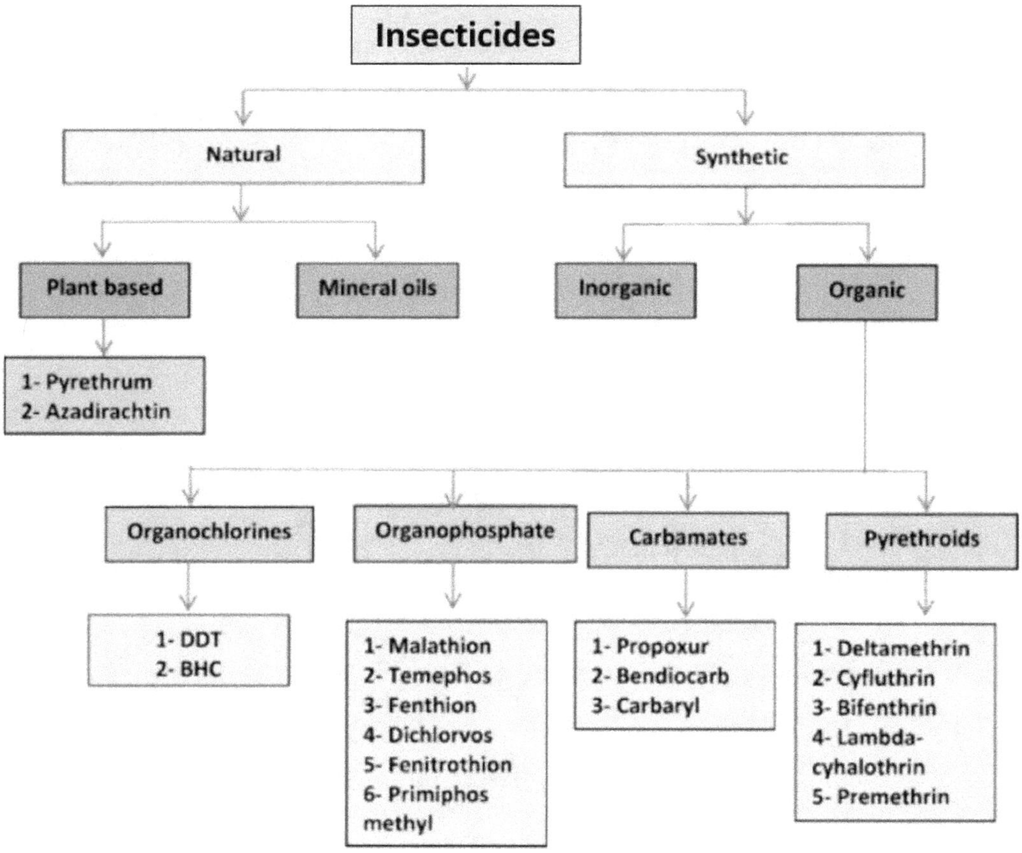

have the probability to modify the environment components mostly. Figure 6 shows the popular insecticides (Samada & Tambunan, 2020). Few insecticides turn into concentrated because they multiply in food chain. Broadly there are three types of insecticides.

- **Contact Insecticides:** It acts as a shot which is used kill some specific targeted insect upon application. Generally domestic insect sprays are a type of contact insecticides as they directly hit and kill the insect.
- **Systemic Insecticides:** These substances usually introduced into soil so that plants root can absorb them. When plants root absorbs the insecticides, it tends to move towards external portion like leaves, twigs, branches and fruits. There it forms a coating on the plant surface area. This coating acts as a toxic for any insect which is coming to bite the plants.
- **Ingested Insecticides:** These types of substances can be given to rats and roaches.

## Classification of Insecticides

Depending upon various characteristics, classification (Tudi et al., 2021; Berg et al. 2017) has been depicted in Figure 7.

*Figure 7. Insecticides classification on various attributes*

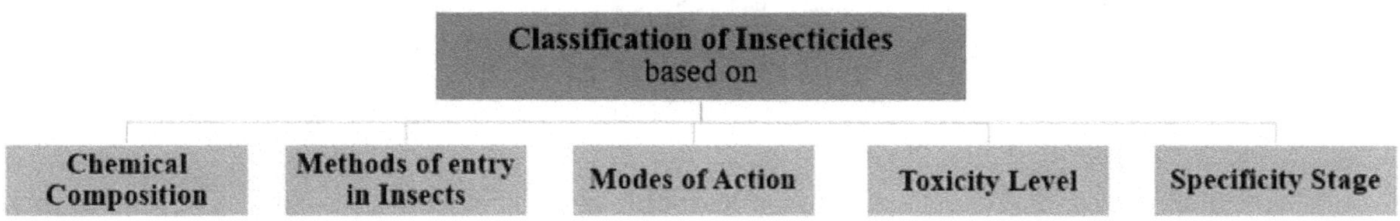


1) **Insecticides classification depending on chemical composition:** Chemical composition of insecticides is extremely diverse in nature. They can be classified into four groups.
   - **Organochlorines:** Generally, these are organic compounds which have been chlorinated. They have a tendency to combine and reflect much empathy for animal's fatty tissues. Organochlorines have small biodegradation tendency and easily accumulated in environment which may cause severe troubles. Some examples are: DDT (Dichlorodiphenyl trichloroethane), Aldrin, BHC, Endosulphan, Mirex etc. Their features and description are in given Figure 8.
   - **Organophosphates:** Schrader has discovered the insecticidal features of organophosphates. It slows down the cholinesterase which is an essential enzyme for nerve impulse transmission across synapse. These are very dangerous to verterbrates. Asian Countries mainly use three organophosphates: Parathion, Malathion and Fenitrothion, shown in Figure 9.
   - **Synthetic Pyrethroids:** These are the variety of man-made pesticides like Pyrethrum which is natural pesticide. Till now more than 1,000 Pyrethroids have been manufactured. Few known names are: Resmethrin, Allethrin, Cyfluthrin, Permethrin. These substances are less toxic in comparison to organophosphates and carbamates. This group compounds are mainly applied for household pests.

*Figure 8. Few popular organochlorides*

| | |
|---|---|
| **DDT** | ▪ First synthesized by a German chemist Othnar Zeidler in 1874<br>▪ Insecticidal value was discovered by Paul Muller in 1939<br>▪ Most famous pesticide of the world and is a non-biodegradable pollutant<br>▪ Spraying of DDT on crops produces pollution of air, soil and water<br>▪ It has become ineffective for killing mosquitoes because of the development of adaptive resistance |
| **Lindane** | ▪ γ-hexachlorocyclohexane/Gammaxene/Lindane was 1st synthesized by Michael Faraday in 1825<br>▪ Its insecticidal value was discovered by Dupire (1941) in France and Leicester (1942) in England<br>▪ It is most common pesticide used in India, represents about 50% of total volume of pesticides used in India<br>▪ It is used in shampoos and lotion |
| **Aldrin** | ▪ Aldrin is an insecticide named after German chemist Kurt Alder<br>▪ It is applied to foundations of buildings to prevent termites.<br>▪ It has been successfully used in control of locusts and grasshoppers in Asian countries.<br>▪ Aldrin, Dieldrin and Endrin are very poisonous pesticides. |
| **Endosulphan** | ▪ It is a pesticide and is useful used in agriculture in the control of insect pests including whiteflies, aphids, leafhoppers, Colorado potato beetles and cabbage worms.<br>▪ It is also endocrine disruptor and carcinogenic to humans. |
| **Mirex** | ▪ it is insecticide used to kill fire ants in agricultural lands<br>▪ It was banned in USA because of biomagnifications to the turtles, coyotes, and other animals<br>▪ It is potent endocrine disruptor to animals including human being |

- **Carbamates:** They are derived from carbonic acid. They contains –OCON=group in their molecule. Carbamates are very helpful in controlling snails and nematodes, and their derivatives are also used as fungicides and herbicides. Their action is very much similar to organophosphates. Few popular carbamates are Propoxur, Carbofuran and Aldicarb. On 3rd Dec, 1984 in Bhopal Gas Tragedy India, Methyl isocyanate gas was used as raw material for synthesizing Carbaryl.
- **Triazines:** These substances are primarily used to control weeds in cotton, tea and tobacco. They fall under the category of herbicides and extracted from urea. Few examples are: Atrazine and Simazine.

2) **Insecticides classification depending on methods of entry in insects:** Further it can be categorized into four categories:
   - Contact Poisons
   - Systemic Poisons
   - Stomach Poisons
   - Fumigants Poisons
3) **Insecticides classification depending on methods of action:** It can be categorized further as:
   - Physical Poisons
   - Respiratory Poisons
   - Protoplasmic Poisons
   - Nerve Poisons

*Figure 9. Organophosphates used by Asian countries*

**Malathion**

- One of the two active ingredients in Flit

**Pyrathion**

- It is derived from Chrysanthemum cinerariifolium
- Examples of pyrethroids are Allethrin, Cyclethrin and Barthrin
- Mosquito-repelling coils contain pyrethrin
- Naled is insecticide used to control for the spread of Zika virus in USA during 2015

**Fanthion**

- It is used to control mosquito in India, but its manufacturing is banned in 2017 due to environmental impact

  - General Poisons
  - Chitin Poisons

4) **Insecticides classification depending on Toxicity Level:** It can be categorized further into four types, listed as:
   - Less Toxic: Color: Green; Symbol: Caution; Oral LD50:>5000
   - Moderate Toxic: Color: Blue; Symbol: Danger; Oral LD50: 501-5000
   - Highly Toxic: Color: Yellow; Symbol: Poison; Oral LD50: 51-500
   - Extreme Toxic: Color: Red; Symbol: Poison and Skull; Oral LD50: 1-50
5) **Insecticides classification depending on Specificity Stage:** Here it can be further subdivided into four categorized, listed as:
   - Ovicides
   - Pupicides
   - Adulticides
   - Larvicides

b) **Fungicides:** These pesticides are used to get rid of fungal infection in crop and their spores. They precisely manage and control fungus or rust or mildews which can damage plants. Majority of fungicides are in liquid or solid form but major compound found is sulphur with diverse concentration. Few well known fungicides are: Organic fungicides and Inorganic fungicides. Organic fungicides comprise mercury composite, oxanthiins, benzimidiazole derivatives, dithane M-22, dithane Z-78, dithane S-21. Inorganic fungicides include burgundy fusion, mercuric chloride, sulphur, bordeus mixture etc. Figure 10 shows the three categories of fungicides.
   - **Systemic fungicides** are most efficient because they restructure themselves according to the plant vessel. Some fungicides can be observed as locally systemic, extended all through the plant tissue and on the surface itself.

*Figure 10. Types of fungicides*

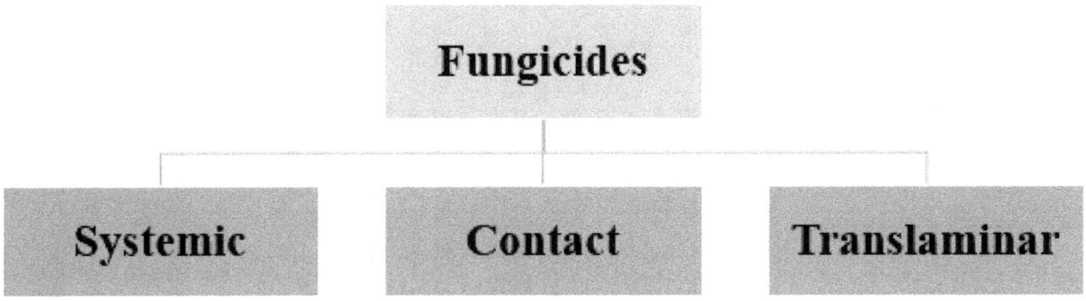


- **Contact fungicides** protect that part of plant only where it is deposited during spraying. It is not scattered all throughout the plant tissue.
- **Translaminar fungicides** restructure themselves from upper sprayed part of plant to lower unsprayed part.

c) **Herbicides:** These substances are applied in agricultural land to destroy unwanted vegetation or weeds. Herbicides contribute for maximizing crop productivity while reducing weeds. Mostly they are applied before planting or during planting, commonly in row crop farming. Depending upon the modes of action, various types of herbicides are depicted in Figure 11.

- **Contact herbicides**: They have restricted movement inside the plant therefore overall coverage of target is difficult. These herbicides have a tendency to show the symptoms quickly, mostly within 24 hours duration.
- **Selective herbicides:** They don't kill plants. Only target weeds are destroyed when it is applied at precise application rate.
- **Non-selective herbicides:** They are also known as knockdown herbicides. These types of herbicides mostly damage the plants also. Some known names are: paraquat and glyphosate.
- **Translocated herbicides:** They shift within plant towards the action location through transport mechanism. Nutrition form soil to developed sites is transported by xylem. The phloem transports photosynthesis goods towards developed and storage sites. Symptoms on target can be observed sometimes in two weeks or more depending upon the species, circumstance and pace of herbicides.
- **Residual herbicides:** These substances remain active for longer duration mostly for months long. Therefore, they can be active on following weed germinations.
- **Non-residual herbicides:** They have lesser or no activity in soil and hence rapidly deactivated in soil. Sometime they are bound to soil particles or broken down, and becomes less noticeable to developing plants. They have fewer tendencies to be immersed by rots.
- **Pre-emergent and post-emergent herbicides:** These are related to the timing and goal of herbicides application. Pre-emergent application means applying herbicides to soil earlier than weeds have emerged. Post-emergent application means applying herbicides when weeds have emerged from soil.
- **Herbicides Mixture:** To control the large variety of weed species, amalgamation of two or more than two herbicides is applied. Sometimes the herbicides are aggressive so could not mixed collectively and hence they must be applied sequentially in any application.

*Figure 11. Types of herbicides*

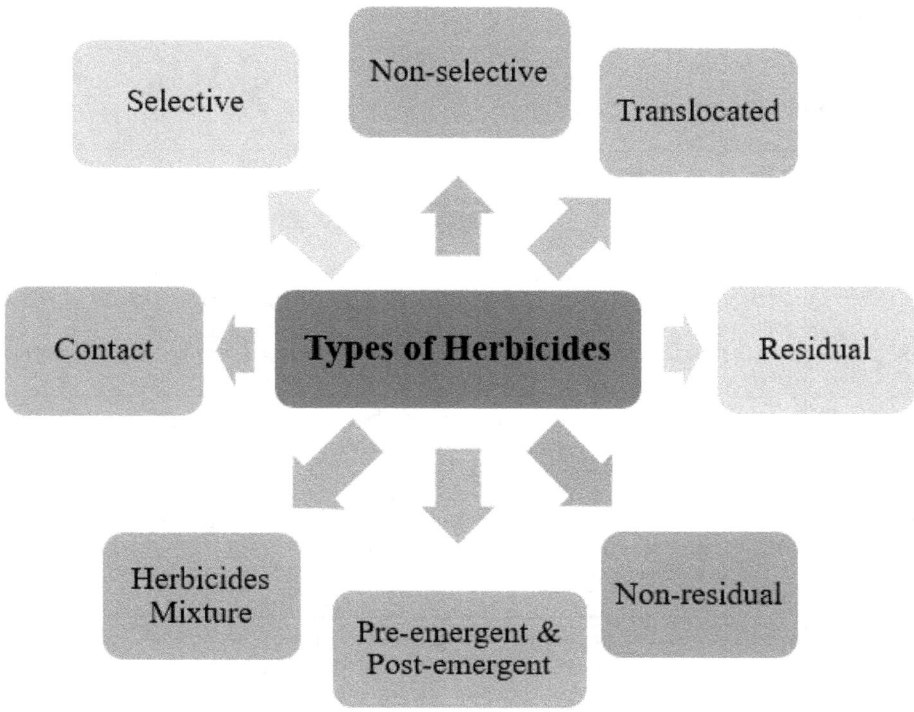


d) **Rodenticides:** This substance is used to kill mice, rodents and rats. Mostly they are formulated with striking substances such as molasses or peanut butter. Rodenticides are very poisonous for humans. Pests are killed through preventing blood clotting and causing internal hemorrhaging. Red squill, sodium fluroacetate, zinc phosphide are frequently used rodenticides.

e) **Biopesticides:** These are naturally derives compounds or agents. They are derived from plants, microorganism, raw minerals or animals. These agents are eco-friendly which causes smaller amount harm to environment and humans in comparison to chemical based pesticides. Some known and used canola oil and baking soda commonly used to control pests.

f) **Nematicides:** It is sometimes known as nematocides. It is a type of chemical substance used worldwide to manage tiny bloodsucking worms like threadworms and roundworms. One commonly used commercial nematode is Aldicard. It is very important for production of potatoes or tobacco, which kill soil borne pests.

## Hazards of Pesticides

Pesticides can have many types of effects on health after the exposure of days, months or years (Akashe et al., 2018; Hassaan & Nemr, 2020). Table 1 demonstrates various pesticides hazards.

- Short term health effects are known as acute effects
- Long term health effects are known as chronic effects

*Table 1. Various pesticides hazards*

| | |
|---|---|
| **Impact on humans** | ▪ Many evidence shows that some of these chemicals have a potential risk to humans and other life along with unwanted side effects to the environment<br>▪ No segment of the population is entirely protected against exposure to pesticides and the possibly serious health effects<br>▪ The world-wide deaths and chronic diseases due to pesticide poisoning number is about 1 million per year<br>▪ The high-risk groups exposed to pesticides include production workers, formulators, sprayers, mixers, loaders and agricultural farm workers<br>▪ During manufacture and formulation, the probability of hazards may be higher because the processes involved are not risk free. |
| **Impact through food commodities** | ▪ In year 1996, seven pesticides names are: acephate, chlopyriphos, chlopyriphos-methyl, methamidophos, iprodione, procy- midone and chlorothalonil were analyzed in apples, tomatoes, lettuce, strawberries and grapes<br>▪ An average of about 9,700 samples has been analyzed for each pesticide<br>▪ For each pesticide, 5.2% of the samples were found to contain residues and 0.31% had residues higher than the respective MRL for that specific pesticide<br>▪ In India the first report of poisoning due to pesticides was from Kerala in 1958, where over 100 people died after consuming wheat flour contaminated with parathion |
| **Impact on environment** | ▪ Pesticides can contaminate soil, water, and other vegetation<br>▪ Pesticides can be toxic to a host of other organisms including birds, fish, beneficial insects, and non-target plants<br>▪ Insecticides are generally the most acutely toxic class of pesticides, but herbicides can also pose risks to non-target organisms |
| **Surface water contamination** | ▪ Pesticides can reach surface water through runoff from treated plants and soil<br>▪ Contamination of water by pesticides is widespread. |
| **Ground water contamination** | ▪ Groundwater pollution due to pesticides is a worldwide problem<br>▪ As per the survey at least 143 different pesticides and 21 transformation products have been found in ground water |
| **Soil contamination** | ▪ Not many of all possible pesticide transformation products have been monitored in soil<br>▪ Persistency and movement of these pesticides and their TPs are determined by some parameters like water solubility, soil-sorption constant, the octanol/water partition coefficient and half-life in soil |
| **Effect on soil fertility** | ▪ Heavy treatment of soil with pesticides can cause populations of beneficial soil microorganisms to decline<br>▪ Overuse of chemical fertilizers and pesticides have effects on the soil organisms that are similar to human overuse of antibiotics |
| **Contamination of air, soil, and non-target vegetation** | ▪ Pesticide sprays can directly hit non-target vegetation or can drift or volatilize from the treated area and contaminate air, soil, and non-target plants<br>▪ Some pesticide drift occurs during every application, even from ground equipment<br>▪ Drift can account for a loss of 2 to 25% of the chemical being applied, which can spread over a distance of a few yards to several hundred miles |

## SPRAYING PESTICIDES IN AGRICULTURE

Plant protection is the most important task throughout the crop production. In agriculture, pesticides application is repetitive and substantial task which require types of spraying controllers and monitors. Conventional spraying techniques have multiple inconsistencies, distribution error, time taking and hence intelligence in spraying is required (Ahmad et al., 2021). Some important parameters must be taken into consideration for effective pesticides spray are: crop height, crop volume, farm area, plant leaf size etc.

Innovative technologies have been used to make process intelligent and adaptive in nature. Researchers are conducting regularly experiments are optimizing various require parameters for precision farming. In las few years use of Robotics, Drones, UAV, automation along with Artificial Intelligence have been preferred. AI helps to protect environment, reduced cost, minimizing human labor, path optimization, less pesticide residue in real time environment.

**a) Path Optimization**

These days many industries are working for automatic route planning for pesticides application. Manual process of applying pesticides in agricultural land is very tedious and time consuming. Path optimization algorithms or tools are not replacing humans but inculcating their expertise and domain knowledge (Srivastava et al., 2020; Fang et al. 2021).

Chen et al. (Chen & Jhao, 2013) proposed multi-objective constraints optimization problem. Optimal and non-optimal tool path have been generated and compared. Optimal tool path takes less time and gives satisfactory performance. Proposed algorithm can be applied for surface cleaning robot or grinding robot. Muthukumaran et al. demonstrated the optimized collision free path in agricultural farmland. New hybrid algorithm of Cuckoo Search and Dragonfly is proposed. This hybrid algorithm generates sequential path for pesticides spraying in farmland. Agriculture path optimization is related with Travelling Salesman Problem (TSP) and result reflects their computational efficiency and quality solution.

**b) Drones**

It is one of the popular technologies used by researchers and industrialist for executing various task. Pesticides application can be done efficiently by using Drones as it provides very clear views of every corner of farmland, shown in Figure 12. As per the survey, drone-based solutions and applications are appreciated globally and its market worth is USD 127.3 billion. Among all applications, worth for drones in agriculture is USD 32.4 billion. Use of drones in agriculture is significant these days especially in use of pesticides management, precision farming, increased productivity, yield management etc. High resolution cameras in drones creates 3D field image (Zangina et al., 2021; Kurihara et al., 2020). This image helps to recognize the areas where weed is grown, irrigation status, infected plants. Airborne spraying is one most significant task handled by drones in association with computer vision and artificial intelligence. Multispectral images of plants help in identification of infected plant and their locations in large land.

Srivastava et al. (Srivastava et al., 2020) describes an algorithm for automatic spraying through Unmanned Aerial Vehicles. Most favorable optimal route for pesticide application in agricultural land is identified. Best location of spray points for a particular worried region is calculated by novel techniques. This technique works by dividing worried region into many circular segments by considering center as a spray point. Determination of best path is done via that algorithm, and this path connects all spray points once. Travelling Salesman Problem (TSP) based path planning is done here. Uniform spraying is administered in spite of variations in deficiencies. Therefore, simple approach for automatic pesticides application without human intervention is implemented via UAV. The proposed approach directly impacts the quantity and quality of crop, accuracy for spraying requirement calculation, path planning for spraying, minimum time required, efficiency and place coverage.

**c) Spot spraying on Targets**

Integration of Artificial Intelligence into existing systems of spraying on targets provides ultra-high precision and localized spray. Such AI enabled systems incorporates very high-resolution embedded camera along with sensors. Industries developed systems are capable for pesticides spray in ground for a day or night. These modern technologies are efficient for weed detection, disease detection, pest identification and their treatment for growing crop in real time environment.

Fang et al. explains the intelligent adaptive algorithms for pesticides spray planning. When helicopter is used for executing spraying task, pilot expertise matters a lot for path planning (Faical et al., 2014). To increase efficiency and land coverage, two algorithms: Ant Colony Optimization (ACO) is combined with Genetic Algorithm (GA). Performance evaluation and testing have been done self-developed area, which is further compared with other algorithms. Optimal route calculation reduces fuel consumption and pesticides consumptions also, which intends the smart forestry management (Muthukumaran et al., 2021).

**d) Self-driven Robots**

In last two decades the nutrition content in vegetables is reduced approximately 40% and soil wellbeing is also compromised by using harsh pesticides in farmland. It has drawn the attention of researchers and farmers because in long run it is going to impact the health of individuals. Industries have developed self-driven robots, capable to execute massive task in simplified manner. One example is 10,000 weeds can be destroyed by robot in an hour by using laser technique.

Zangina et al. demonstrates that mobile robot performs very accurately and efficiently in agricultural domain for spraying pesticides. In a structured environment, self-government of mobile robot can be improved to determine best route in adaptive environment. Robotics has proven an appreciable improvement for performing agricultural task automatic in manner like planting, spraying, harvesting etc. Navigation strategies must be incorporated for increasing crop quality and quantity with reduced cost. Selection of type of pesticides and quantity of pesticides required for plant can also be controlled. In this view Vehicle Routing Problem (VRP) is implemented for robot navigation to take quick decisions for satisfying pesticides requirements for distinct plants. Researchers have tested and identified optimal route for agricultural robot with increased efficiency. Every plant may require different dosage of pesticides but pesticides carrying capacity of robot is fixed. Here the novel approach is the resolution of path planning in greenhouse surrounding by adaptive spraying procedure.

**e) Smart Pest Monitoring**

Whichever the crop is grown in farmland is impacted by at least one insect pest. Usually, farmers can control these pests through some traditional methods or by some amalgamations of interventions. But they must understand initially the nature of insect and their growth. These days sensors are placed in agricultural land so when insects came, their picture is taken by sensors and their nature can be figure out. AI enabled systems can accordingly take necessary action for amount and time of pesticides application (Katiyar et al., 2021).

*Figure 12. Pesticides spraying using drones & robots*

**f) Sustainable Agriculture**

Varying climatic conditions, using pesticides and deficiency of skilled farmers have raised multiple concerned which finally impact the worth and quality of agricultural procedures. These days the objective of researchers is to make agricultural process much more supportable and sustainable by using cloud-based technology and artificial intelligence. Agricultural sector is facing numerous challenges and hence the developed technology must be adaptive in nature.

## PESTICIDES IN FOOD PRODUCTION

Pesticides are the substance commonly used in making food to protect it from fungi, pests, weeds, bacteria. Simultaneously pesticides ensure that enough food products are available for growing population as they protect crops. Now a days people are concerned about pesticides use because they can be very poisonous to humans and animals. There are some benefits of pesticides also such as: antimicrobics which controls bacteria, gems fungal problems, unwanted plant growth and weed growth. The role of pesticides governs the extent to which they are harmful. Pesticide's quantity and concentration directly relates it effect. Artificial Intelligence contributes for assessment of pesticides at various stages (Sahni et al., 2021). Pesticides requirements as per the distinct crops can be evaluated in accordance with environmental conditions.

As per World Health Organization (WHO), herbicides are less toxic than insecticides. Large use of pesticides in longer duration may cause cancer or sometimes impact the reproduction also. Introducing enhanced quantity of pesticides may cause poisoning and sometimes it may appear after couple of days, weeks or months after use. AI along with IoT is capable to predict the level of symptoms in advance so that appropriate action can be taken (Ahmed et al., 2018; Hirsch et al. 2019). Figure 13 depicts some observed mild, moderate and severe poisoning symptoms. In India out of 234 pesticides, 4 pesticides are WHO class la, 15 pesticides are WHO class lb, 76 pesticides are WHO class ll; are grouped as 40% of registered pesticides.

*Figure 13. Categorization of poisoning symptoms*

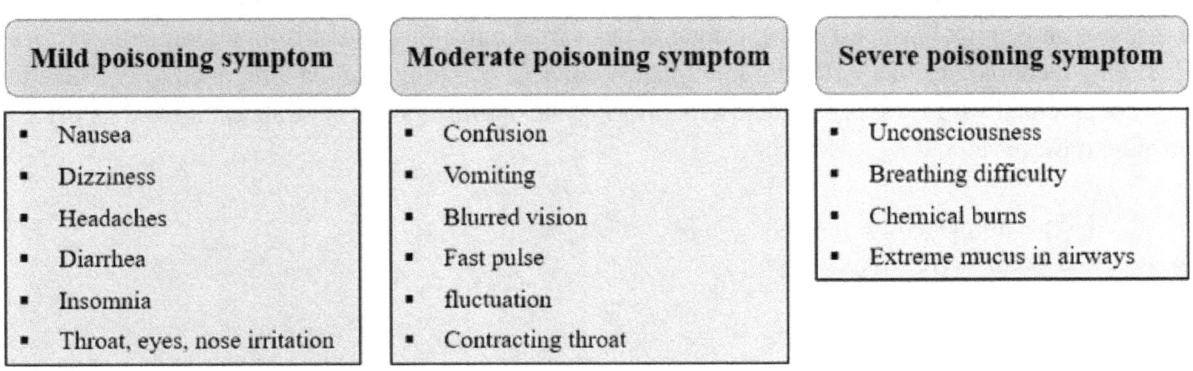

People are exposed to pesticides because their residue exist in water and food. Spraying pesticides on crop leaves some residue on harvest. Workers are in direct contact working in warehouses and in agricultural land. Few pesticides are absorbed by ground and enter water channels. Though people consume pesticides residue and it impacts on their health, but Environmental Protection Agency (EPA) has fixed a benchmark for safe level of pesticides which can be used in food. It is the responsibility of EPA to ensure that pesticides used in foods and vegetables must meet safety standards. EPAs keep evaluating existing and new registered pesticides under the Food Quality Protection Act (FQPA) standards (Carvalho et al., 2009).

Presence of pesticides residue on vegetables and fruits does not mean that they are unsafe to consume. This residue level decreases from the stage of crop till transportation, exposed to sun light, washing, preparation and finally cooking stage (Luchini et al., 1995; Lostil et al., 2014). As per the report presented by USDAs Pesticide Data Program (PDP), residue levels detected are much lower than the levels set for health risk. Table 2 show some of commonly used pesticides in India and U.S., which have very less probability to impact human health.

*Table 2. Some pesticides used in India and U.S*

| Pesticides used in India | Pesticides used in U.S. |
|---|---|
| Acephate | Altrazine |
| Benomyl | Metolachlor-S |
| Chlorothalonil | Glyposate |
| Dicofol | Dichloropropene |
| Propoxur | 2,4-D (type of herbicides) |

The U.S. Department of Agriculture (USDA) tested more than ten thousand samples and then Environment Working Group (EWG) developed a guide for consumers. This guide contains list of vegetables and fruits having lowest and highest levels of pesticides, mentioned in Figure 14.

It is impossible to avoid pesticides in food entirely. These days organic foods are in much demand and farmers are putting their efforts for organic crops as per government guidelines. Organic foods are very expensive and do not have pesticides residue. It has varied nutrition content from nonorganic products. Emerging technologies like Artificial Intelligence, Internet of Things can be integrated to organize the concept practically for various parameters monitoring and optimization. The strategy adopted for organic farming must have:

- Avoid use of pesticides
- Avoid synthetic fertilizers
- Encourage animal wellbeing
- Conservation of wildlife
- Improving water and soil quality
- No use of genetic enhanced organism

*Figure 14. Pesticides levels in vegetable and fruits*

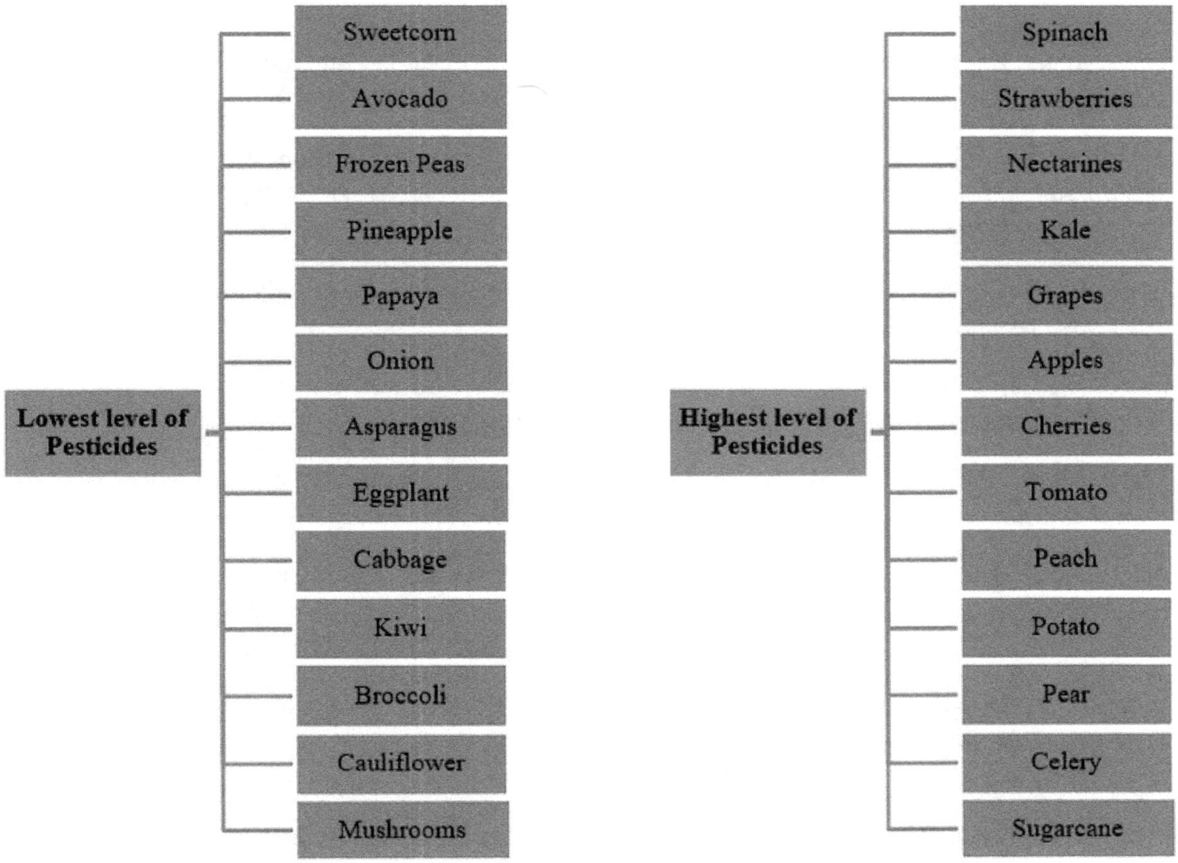


## CONCLUSION

Artificial intelligence enabled solutions in agriculture are very intelligent and precise for problems faced by farmers worldwide. AI provides solutions for every kind of problems starting from crop monitoring to spraying till harvesting and supply chain. Farmers are using pesticides from more than 60 years as it is a low cost and quick substance for disease, pests and weed management. These days India has become a major contributor for pesticides manufacturing. Inclusion of various other technologies like machine learning, deep learning, computer vision, internet of things has aided various advantages for farmers. Efficient use of pesticides for agriculture, food processing and food storage can be achieved with emerging technological solutions. Ultimately it is helpful in conservation of environment.

## REFERENCES

Abate, T., van Huis, A., & Ampofo, J. K. O. (2000). Pest Management Strategies in Traditional Agriculture: An African Perspective. *Annual Review of Entomology*, *45*(1), 631–659. doi:10.1146/annurev.ento.45.1.631 PMID:10761592

Abhilash, P. C., & Singh, N. (2009). Pesticide use and application: An Indian scenario. *Journal of Hazardous Materials*, *165*(1–3), 1–12. doi:10.1016/j.jhazmat.2008.10.061 PMID:19081675

Agnihotri, N. (2000). Pesticide Consumption in Agriculture in India - an Update. *Pesticide Research Journal*, *12*, 150–155.

Ahirwar, S., Swarnkar, R., Bhukya, S., & Namwade, G. (2019). Application of drone in agriculture. *International Journal of Current Microbiology and Applied Sciences*, *8*(1), 2500–2505. doi:10.20546/ijcmas.2019.801.264

Ahmad, F., Khaliq, A., Qiu, B., Sultan, M., & Ma, J. (2021). *Advancements of Spraying Technology in Agriculture*. Technology In Agriculture.

Ahmed, N., De, D., & Hussain, I. (2018). Internet of things (IoT) for smart precision agriculture and farming in rural areas. *IEEE Internet of Things Journal*, *5*(6), 4890–4899.

Akashe, M. M., Pawade, U. V., & Nikam, A. V. (2018). Classification of Pesticides: A Review. *International Journal of Research in Ayurveda and Pharmacy*, *9*, 144–150.

Arakeri, M. P., Kumar, B. V., Barsaiya, S., & Sairam, H. V. (2017). Computer vision based robotic weed control system for precision agriculture. *Proceedings of International Conference on Advances in Computing, Communications and Informatics*, 1201–1205. 10.1109/ICACCI.2017.8126005

Bannerjee, G., Sarkar, U., Das, S., & Ghosh, L. (2018). Artificial intelligence in agriculture: A literature survey. *International Journal of Scientific Research in Computer Science Applications and Management Studies*, *7*(3), 1–6.

Berg, H., Manuweera, G., & Konradsen, F. (2017). *Global trends in the production and use of DDT for control of malaria and other vector-borne diseases* (Vol. 16). Malaria Journal.

Bon, H. D., Huat, J., Parrot, L., & Sinzogan, A. (2014). Pesticide risks from fruit and vegetable pest management by small farmers in sub-Saharan Africa. A review. *Agronomy for Sustainable Development*, *34*(4), 723–736. doi:10.100713593-014-0216-7

Carvalho, F. P., Villeneuve, J. P., Cattini, C., Rendón, J., & Oliveira, J. M. (2009). Pesticide and PCB residues in the aquatic ecosystems of Laguna de Terminos, a protected area of the coast of Campeche, Mexico. *Chemosphere*, *74*(7), 988–995.

Chen, W., & Zhao, D. (2013). *Path Planning for Spray Painting Robot of Workpiece Surfaces*. Robust Control with Engineering Applications.

Elijah, O., Rahman, T. A., Orikumhi, I., Leow, C. Y., & Hindia, M. N. (2018). An overview of Internet of Things (IoT) and data analytics in agriculture: Benefits and challenges. *IEEE Internet of Things Journal*, *5*(5), 3758–3773. doi:10.1109/JIOT.2018.2844296

Faiçal, B. S., Costa, F. G., Pessin, G., Ueyama, J., Freitas, H., Colombo, A., Fini, P., Villas, H. L., Osório, F. S., Vargas, P. A., & Braun, T. (2014). The use of unmanned aerial vehicles and wireless sensor networks for spraying pesticides. *Journal of Systems Architecture*, *60*(4), 393–404.

Fang, S., Ru, Y., Liu, Y., Hu, C., Chen, X., & Liu, B. (2021). *Route Planning of Helicopters Spraying Operations in Multiple Forest Areas* (Vol. 12). Forests.

F. A. O. Food. (2017). *The future of food and agriculture–Trends and challenges*. Annual Report, 296.

Gondchawar, N., & Kawitkar, R. S. (2016). IoT based smart agriculture. *International Journal of Advanced Research in Computer and Communication Engineering*, *5*(6), 838–842.

Grewal, A. S., Singla, A., Kamboj, P., & Dua, J. S. (2017). Pesticide Residues in Food Grains, Vegetables and Fruits: A Hazard to Human Health. *Journal of Medicinal Chemistry and Toxicology*, *2*(1), 40–46. doi:10.15436/2575-808X.17.1355

Hassaan, M. A., & Nemr, A. E. (2020). Pesticide's pollution: Classifications, human health impact, extraction and treatment techniques. *The Egyptian Journal of Aquatic Research*, *46*, 207–220.

Hirsch, C., Bartocci, E., & Grosu, R. (2019). Capacitive soil moisture sensor node for IoT in agriculture and home. *Proc. IEEE 23rd International Symposium on Consumer Technologies*, 97–102.

Katiyar, S., & Farhana, A. (2021). Smart Agriculture: The Future of Agriculture using AI and IoT. *Journal of Computational Science*, *17*(10), 984–999. doi:10.3844/jcssp.2021.984.999

Katiyar, S., Khan, R., & Kumar, S. (2021). Artificial Bee Colony Algorithm for Fresh Food Distribution without Quality Loss by Delivery Route Optimization. *Journal of Food Quality*.

Khan, M. J., Zia, M. S., & Qasim, M. (2010). Use of pesticides and their role in environmental pollution. *World Academy of Science, Engineering and Technology*, *72*, 122–128.

Kostik, V., Angelovska, B., Petreska, E. K., & Bauer, B. (2014). Determination of pesticide residues in plant-based foods from the Republic of Macedonia. *Journal of Food Nutrition Science*, *2*, 124–129.

Kurihara, J., Ishida, T., & Takahashi, Y. (2020). *Unmanned aerial vehicle (UAV)-based hyperspectral imaging system for precision agriculture and forest management. In Unmanned Aerial Vehicle: Applications in Agriculture and Environment*. Springer.

Liu, S., Guo, L., Webb, H., Ya, X., & Chang, X. (2019). Internet of Things Monitoring System of Modern Eco-Agriculture Based on Cloud Computing. *IEEE Access: Practical Innovations, Open Solutions*, *7*, 37050–3705. doi:10.1109/ACCESS.2019.2903720

Luchini, L. C., Peres, T. B., & deAndrea, M. M. (1999). Monitoring of Pesticides Residue in a cotton crop soil. *Journal of Environmental Science and Health*, *35*, 51–59.

Muthukumaran, S., Ganesan, M., Dhanasekar, J., & Loganathan, G. B. (2021). Path Planning Optimization for Agricultural Spraying Robots Using Hybrid Dragonfly – Cuckoo Search Algorithm. *Journal of Agriculture Science*, *36*(1).

Obopile, M., Munthali, D., & Matilo, B. (2008). Farmers' knowledge, perceptions and management of vegetable pests and diseases in Botswana. *Crop Protection (Guildford, Surrey)*, *27*(8), 1220–1224. doi:10.1016/j.cropro.2008.03.003

Oluwole, O., & Cheke, R. A. (2009). Health and environmental impacts of pesticide use practices: A case study of farmers in Ekiti State, Nigeria. *International Journal of Agricultural Sustainability*, *7*(3), 153–163. doi:10.3763/ijas.2009.0431

Panpatte, D. G. (2018). *Artificial Intelligence in Agriculture: An Emerging Era of Research. Intuitional Science.*

Popp, J., Pető, K., & Nagy, J. (2013). Pesticide productivity and food security. A review. *Agronomy for Sustainable Development*, *33*(1), 243–255. doi:10.100713593-012-0105-x

Sahni, V., Srivastava, S., & Khan, R. (2021). Modelling Techniques to Improve the Quality of Food using Artificial Intelligence. *Journal of Food Quality*.

Samada, L. H., & Tambunan, U. S. F. (2020). Biopesticides as Promising Alternatives to Chemical Pesticides: A Review of Their Current and Future Status. *The Journal of Biological Sciences*, *20*(2), 66–76. doi:10.3844/ojbsci.2020.66.76

Sankararamakrishnan, N., Sharma, A. K., & Sanghi, R. (2005). Organochlorine and organophosphorous pesticide residues in ground water and surface waters of Kanpur. *Environment International*, *31*(1), 113–120. doi:10.1016/j.envint.2004.08.001 PMID:15607785

Schor, N., Bechar, A., Ignat, T., Dombrovsky, A., Elad, Y., & Berman, S. (2016). Robotic disease detection in greenhouses: Combined detection of powdery mildew and tomato spotted wilt virus. *IEEE Robotics and Automation Letters*, *1*(1), 354–360. doi:10.1109/LRA.2016.2518214

Sharma, R. (2021). Artificial Intelligence in Agriculture: A Review. *International Conference on Intelligent Computing and Control Systems*.

Srivastava, K., Pandey, P. C., & Sharma, J. K. (2020). An Approach for Route Optimization in Applications of Precision Agriculture Using UAVs (Vol. 4). Academic Press.

Subeesh, A., & Mehta, C. R. (2021). Automation and digitization of agriculture using artificial intelligence and internet of things. *Artificial Intelligence in Agriculture*, *5*, 278–291. doi:10.1016/j.aiia.2021.11.004

Talaviya, T., Shah, D., Patel, N., Yagnik, H., & Shah, M. (2020). Implementation of artificial intelligence in agriculture for optimization of irrigation and application of pesticides and herbicides. *Artificial Intelligence in Agriculture*, *4*, 58–73. doi:10.1016/j.aiia.2020.04.002

Tudi, M., Ruan, H. D., Wang, L., Lyu, J., Sadler, R., Connell, D., Chu, C., & Phung, D. T. (2021). Agriculture Development, Pesticide Application and Its Impact on the Environment. *International Journal of Environmental Research and Public Health*, *18*(3), 1112–1112. doi:10.3390/ijerph18031112 PMID:33513796

Zangina, U., Buyamin, S., Abidin, M. S. Z., & Mahmud, M. S. A. (2021). Agricultural rout planning with variable rate pesticide application in a greenhouse environment. *Alexandria Engineering Journal*, *60*, 3007–3020.

Zhang, W. (2018). Global pesticide use: Profile, trend, cost/benefit and more. *Proc Int Acad Ecol Environ Scince*, *8*, 1–27.

# Chapter 6
# Role of Deep Learning in Weed Detection

**Kavita Srivastava**
*Institute of Information Technology and Management, GGSIP University, India*

## ABSTRACT

*Deep learning, robotics, AI, and automation have lots of applications that are beneficial to society at large. In fact, nearly every sector, such as transportation, industries, manufacturing, healthcare, education, retail, and home automation, are adopting AI, machine learning, IoT, and robotics to their advantage. Of course, agriculture is no exception. The chapter starts with an introduction to the applications of deep learning in agriculture. Next, a comprehensive survey of the research work done in recent years is provided. It is followed by the description of various techniques of deep learning (DL). The next section briefly describes the traditional ways of weed detection and removal. Next, the architecture of deep learning for weed detection and removal is presented along with the associated code. Further, the chapter goes on to discuss the pros and cons of this approach. Finally, the chapter concludes by citing the important points discussed in this study.*

## INTRODUCTION

A weed is an undesirable plant that grows along with the crop. Weeds usually grow aggressively and compete with desirable crops for resources. The resources for which the weeds compete with desirable crop plant include space, water, soil nutrients and sunlight. Therefore they hamper the growth of the crops. Although weeds can coexist with crops for few week but ultimately they interfere in the growth of desired flora and fauna and hamper the overall crop yield.

Weed detection and removal is important to agriculture. When done manually, it is a time-consuming and tedious task. Therefore, a number of automatic and semi-automatic weed detection methods are considered for this purpose.

There are several traditional methods of weed detection and removal in practice. Yet these traditional methods are either highly inefficient or costly or they cause much damage to the soil if the chemicals are applied.

DOI: 10.4018/978-1-6684-5141-0.ch006

In fact, identification of weeds and to distinguish it from the crops is absolutely necessary for effectiveness of the weed control. Firstly, it would help us in targeted weed removal and secondly, we can precisely spray pesticides and herbicides. Hence, it will result in reduction of cost incurred in the spray of pesticides as well as the reduction in soil damage.

Hence automating the task of weed detection and their subsequent removal can greatly benefit the farmers and save their valuable time which they can spend in taking care of their crops. However, this task is quite challenging since the weeds and the crop are very similar and it is hard to distinguish the crop from the weed. It is much more difficult in the growth phase since crop and the weeds have similar shape and texture.

The task of weed detection can be automated by using the Computer Vision methods. The Computer Vision methods for weed detection can be categorised as the methods based on Image Processing and the methods based on the Machine Learning.

Weed detection methods based on Image Processing collect images from the field using high resolution cameras. The images contain both the crop and the weeds. Images are resized and pre-processing is performed. The pre-processing involves adjusting the brightness and noise reduction. Then feature extraction is performed. After that it applies a classification method to classify the weeds. However traditional image processing based weed detection methods are far from good mainly because they are too slow and inaccurate. Furthermore, for new unseen images these methods don't generalize well and don't yield acceptable results. In contrast, Deep Learning methods for weed detection can effectively improve the efficiency of weed detection. The emergence of weed removing robots can significantly reduce the cost of farming as well as the use of pesticides. A weed removing robots makes use of AI, Machine Learning, and Deep Learning to detect the presence of weed.

This chapter focuses on the Deep Learning methods for weed detection. The chapter starts with a literature survey on weed detection methods. After that, the role of UAV is discussed. Next, a number of Deep Learning based object detection methods are described. Applications of some of these methods in weed detection is given next. The chapter goes on by discussing the advantages, drawbacks, and current challenges of the approaches discussed. Finally, the Future Scope and Conclusion are presented.

## LITERATURE SURVEY

This section presents a survey of state-of-the-art papers on weed detection. The survey comprises of papers on Image Processing, UAV, and Deep Learning.

In image processing, the line segmentation is most basic technique for distinguishing two or more types of objects in an image. Mao et al., (2007) describes a method for weed detection that is based on the optimized segmentation line of weed and crop. In this study, the spectral information in the visible light captured by the CCD camera is used. Basically, the use of both G-R and H-S Optimized Segmentation line helps in detecting the flixweed (tansy mustard which is a broadleaf weed and can be highly toxic to cattle) and separates it from the wheat crop. Cheng et al., (2015) presented a method that automatically discriminates rice and weed using a feature-based Machine Learning (ML) agent. Basically, the intelligent agent used in this study makes use of sensors in order to automate a process that separates rice plants from weeds. The algorithm used for this purpose is Harris Corner Detection algorithm that determines the points of interest. Further, the proposed approach extracts multiple features surrounding each point. Then a Machine Learning algorithm is used for training. Finally, a clustering algorithm removes the noise.

Unmanned Aerial Vehicles (UAV) have clear advantage of monitoring the field in shorter duration of time. Pérez-Ortiz et. al., (2016) proposed an approach for monitoring the weeds using Unmanned Aerial Vehicles (UAV). In their approach the UAV takes the images of crops and then the image analysis is performed. The images taken by the UAV are divided and the rows of crops are detected. Then, the classification model is trained. Bah et. al, (2019) discussed a weed detection technique that collects the image data of weeds and crops using UAV (Unmanned Aerial Vehicle). The proposed method uses deep features and one-class classification on unsupervised data for weed detection in UAV images. Dasgupta et al., (2020) proposed an approach that makes use of wireless technologies for weed detection. Their approach combine Wireless Sensor Network (WSN) and AI techniques. In this study, a frame-capturing drone has been used that takes images of the field. Further, Deep Learning (DL) techniques have been used for classification that distinguish weeds from the crops.

Several research papers have proposed the use of Deep Learning (DL) technique in identification of weeds in the fields by investigating the images captured. Many Deep Learning algorithms work well with image dataset. The most widely used algorithm is the Convolution Neural Networks (CNN) that has been used in weed detection. Dos et. al, (2017) proposed a technique for weed detection using ConvNets or Convolution Neural Networks (CNN). In this study Soyabean crops have been used. Besides, a collection of shape, colour, texture, and feature extraction techniques are also used in this work. Evidently, an accuracy of 98% is achieved in detecting the broadleaf and grass weeds. Tang et. al, (2017) proposed a method of weed detection that combines both K-Means Clustering and Convolution Neural Network (CNN). This approach makes use of K-Means clustering which is an unsupervised machine learning technique for feature learning. After that a CNN model is trained on the given dataset. It leads to higher weed identification accuracy. Andrea et al., (2017) proposed a weed detection method using Convolution Neural Network (CNN). The proposed method distinguishes the weeds from the maize crop. The proposed method is based on image segmentation and classification. In fact, image segmentation retrieves the target plant from the crop image. While, the classification determines the class label of the image and CNN distinguishes Maize plant from the weed in real-time.

Yu et al., (2019) proposed an approach for weed detection using Deep Convolution Neural Network (CNN). In this study, this method is used on the Prennial Ryegrass. This method is basically uses the machine vision for autonomous weed control. Also, several number of DCNN has been used in this study that classify the weeds. The model has been trained using both positive and negative images. Mioa et al., (2019) proposed a weed detection technique that is also based on Convolution Neural Network (CNN). CNN is basically a Deep Learning technique that is used extensively for image classification. In this study, the image dataset has been prepared during the carrot seeding period. Further, a CNN model has been trained using this dataset that identifies the weeds.

Moazzam et al., (2019) presented a review of various kinds of Deep Learning (DL) applications that have been used for the classification of weeds and crops. The survey highlighted the use of Deep Convolution Neural Network (CNN) architecture in separating the weeds from the crops. It is found in the survey that the Deep CNNs used for this purpose have different architectures in terms of the number and types of the layers. Asad et al., (2019) proposed a method based on Deep Convolution Neural Network (CNN) for weed detection. The image data from the canola field is used here. First the maximum likelihood classification is used to segment the background and foreground of the image. Then, manual labelling of weed pixels is performed. With this labelled data semantic segmentation model is trained. The trained model is used for classification. The crop image and the background pixels are considered as one class. While, all other vegetation are considered as the second class. Further, in this paper, the performance of different Deep Learning architectures such as SegNet, UNET, VGG16, and ResNet-50 are compared.

Ngo et. al, (2019) describes a weed detection technique using Convolution Neural Network (CNN). Basically, CNN is a Deep Learning technique and works well with the image data. CNN is kind of Artificial Neural Network (ANN) that can perform the classification function. In this study, a Lego Mindstorm EV3 is used which is connected to a computer. The camera fitted in the robot captures the images of weeds and crops. Then a classification system is applied that distinguishes the weeds from the crops. Further, the herbicides are sprayed selectively on the weeds. Gothai et al., (2020) used the Convolution Neural Network (CNN) for the identification of the weeds. However, they have used different CNN architectures differing in the number of convolution layers. Precisely, the CNN architectures with four convolution layers, six convolution layers, eight convolution layers, and thirteen convolution layers have been used.

Champ et al., (2020) proposed a technique for detection of weed plants by using precision agricultural robots. In this method, an instance segmentation Convolution Neural Network is trained. The trained CNN model segments and identifies each plant specimen visible in the images captured by the field robots.

Apart from CNN, several other Machine Learning (ML) methods have been used for weed detection. Alam et al., (2020) presented a method for crop and weed detection and classification that makes use of a Machine Learning (ML) model in real-time. It is basically a real-time computer vision based method. A Random Forest Classifier has been used in this study. The model has been trained offline and used in variable-rate spraying.

Many research studies have used other Deep Learning and Machine Learning methods using both supervised and unsupervised machine learning techniques. These techniques include Long Short Term Memory (LSTM), Transfer Learning (TL), Support Vector Machines (SVM), and K-Means Clustering. Espejo-Garcia et. al., (2020) proposed a weed identification method using Transfer Learning (TL). The proposed approach takes a fine-tuned pre-trained model for classification of weeds and apply it to a different dataset. The experiments have shown that there is very little difference in the performance of training and test datasets.

Gao et al., (2020) discussed a technique based on Deep Convolution Neural Network (CNN) for detecting Convolvulus Sepium or hedge bindweed in sugar beet fields. This approach uses a tiny YOLOv3 architecture. A total of 2271 synthetic images are generated and combined with 452 actual images from the field. The box size is calculated using K-Means clustering. Osorio et al., (2020) presented three different methods of weed estimation. The first of these method uses a Support Vector Machines (SVM) classifier. While, the second method uses YOLOv3 for object detection. The third method uses a mask R-CNN (Region Based Convolution Neural Network) for the purpose of getting an instant segmentation of individuals. The three methods achieved an F1-Score of 88%, 94%, and 94% respectively.

Arif et. al, (2021) proposed a classification method of several kinds of weeds using Long Short Term Memory (LSTM). In their experiment they have used nine different kinds of weeds like vine weeds, three-leaf weeds, spiky weeds, and invasive creeping weeds. The LSTM based classifier achieved 99.36% accuracy in classifying the weeds. Jin et al., (2021) proposed a method for weed identification which is based on Deep Learning (DL), and image processing. The proposed method aims to identify weeds in vegetable plantation. Evidently, the weed detection in vegetable plantation is more difficult as compared to weed detection in crop plantation. Here, a trained CenterNet model is used to detect vegetables and then it draws boundary boxes around the detected vegetable images. Remaining green areas are considered as weeds.

Sarvini et al., (2019) presented a comparison of performance achieved by different weed detection algorithms. In this study three different classifiers are used for classification of weeds and crops. The three classifiers used for weed detection are Support Vector Machines (SVM), Artificial Neural Network (ANN), and Convolution Neural Network (CNN). The images of four different commercial crops and two types of weeds are used as dataset.

Many research methods in weed detection have used several image processing methods along with Machine Learning and Deep Learning techniques. Umamaheshwari et al., (2018) proposed a weed detection method using parallel image processing. In the proposed system real-time images of the farm have been taken. These images are served as the input data for classification. A Deep Learning (DL) model is trained using the input images of crops and weeds. The model performs feature extraction and classification. The results of automated weed detection are used in the precision agriculture. Khan et. al, (2020) proposed a method called CED-Net which is a semantic segmentation method based on cascaded encoder-decoder network. The proposed method differentiates weeds from the crops and precisely identifies the location of the weeds in an agricultural field.

The next section describes the role of UAV in weed detection.

## USING UAV FOR WEED DETECTION

### Data Collection

Data collection is the first step towards weed detection. Basically, we need to collect data from a field which has a variety of weeds along with crops.

As can be seen from the literature survey most of the weed detection techniques are based on Deep Learning (DL). In particular, Convolution Neural Network (CNN) has been used widely in many studies. Since, Deep Learning algorithms require labelled data for training the neural network model, it is necessary to collect the suitable data and perform labelling. Moreover, Deep Learning techniques are data hungry, therefore training requires the use of a large dataset. Hence, automated methods of data collection should be used in order to gather large volume of data in short duration of time.

Further, most weed detection techniques work by segregating weed image from the crop image, the dataset comprises of sufficient number of images from the field. The image dataset can be prepared by either an Unmanned Aerial Vehicle (UAV) or an Unmanned Ground Vehicle (UGV). Equipped with sensors and cameras, a UAV or a UGV captures the images at specific interval of time. Basically, UAV or Drones collect data in the form of RGB images. The drones perform an important task in weed detection called as the Weed Mapping.

### Weed Mapping

As drones fly over the field to capture the images they enable weed identification and classification using aerial imagery. After the acquisition of aerial data, the images can be processed further. Image processing require classification of weed cover, crop cover, and the soil cover.

Drones are specifically helpful when it is difficult to cover the field area by navigating on the ground. For instance, suppose the field comprises of dense forest and the area which is harder to reach or there are rivers and lakes in-between. One such example is the aquatic weed management, where the weeds float on the water. The UAVs equipped with high-resolution cameras can capture the images more efficiently as compared to manual collection by travelling through boats.

The use of drones for remote imaging analysis help in studying the variety of weeds. Not only it helps in selective spraying of herbicides but also allows us to further study the behaviour of weeds. It has been noted that not all weeds are detrimental to the growth of the crop.

## Applications of UAV in Detecting Weeds

Unmanned Aerial Vehicles (UAVs) or drones have been widely used in automating the task of weed detection. The benefit of covering large agricultural area using drones cannot be underestimated.

Drones perform the task of weed detection by applying a number of approaches which are discussed below.

1. Drones capture RGB images which are used as the training data for a machine learning model. Particularly, the RGB images are used to train a Convolution Neural Network (CNN). This is how an identification model is built (Liang et al., 2019). The advantage of weed detection using drones is that as soon as the model detects an image, a push message is generated containing the location of the weed and sent to the user so that the weeds can be dealt with at the earliest in order to stop further damage to crops.
2. In addition to applying a machine learning model, a drone can also apply several image processing techniques for identifying and segregating weed image part. For instance, first of all the soil and residue parts are removed from the image. Next, the crop lines are determined by calculating superpixels. Interline weeds can be detected by intersection of detected lines with objects created by an algorithm of blob colouring (Bah et al., 2017). Therefore superpixel relationships and crop lines can tell us the presence of weeds. This method is particularly useful when the characteristics of the field are not known. The method is an example of image segmentation in determining the presence of weeds with crops along the crop lines.
3. A UAV can also apply a combination of both machine learning techniques and image segmentation to detect weeds in the crop lines. For instance, a CNN based Hough Transform along with SegNet can be used to create a model (Bah et al., 2019). A UAV captures the images from the field and feeds the captured images in the trained model. This method results in a high efficiency of the weed detection.
4. UAVs and field robots can monitor the farm on a per-plant basis which is very difficult manually. Especially, when the field is very large, the manual monitoring of each plant for detecting the weeds becomes impractical. As a result without the information about exact location of weeds, the herbicides and pesticides are sprayed all over the field irrespective of the presence of weeds. Hence, the soil is unnecessary polluted and plant nutrients also get damaged. UAVs have excellent survey capabilities. Equipped with feature extraction techniques, the UAVs and field robots can quickly and accurately detect the weeds and help in saving cost, time, as well as manual efforts.

## Deep Learning Techniques for Weed Detection

Almost all weed detection methods rely on identifying weeds in an image that also contains the crop part. There are many Machine Learning (ML) techniques that we can use for image classification. Image classification is basically a supervised learning method. In case of supervised learning, we build a model that classifies the data belonging to one of the predefined class. In order to prepare the model that classifies with high accuracy, we first need to train the model. Therefore, as the first step to build a Machine Learning model, we first prepare a dataset for training. The dataset need to be sufficiently large so that the model is trained well and performs classification accurately.

Likewise, for image classification, an image dataset is prepared. Along with the image dataset, we also define a set of classes. These classes are also called as labels. For instance, suppose we have a set of images taken from the agriculture field that contain both crops and weeds, we need to define the classes as the name of crop and weeds. Before defining the classes we first need to perform image segmentation so that the crop image and weed image are segregated. Image segmentation is also important since the image contains both soil and plant cover. Image segmentation extracts plant cover from the image.

First of all we capture a reference image that contains a weed. Subsequent images are compared with this reference image. Hence we perform an XOR operation of the reference image with the new image. The image part that is missing in the reference image is considered as the crop. However, this method has a drawback. It doesn't detect the weeds in between the rows.

Another technique used for detecting a weed in an image is called as Inter Plant Weed Detection. This method first takes an RGB image and detects the green colour in it. This process is called as colour segmentation and it separates all visually distinguishable colours from one another. Next RGB to grey colour conversion is performed and then the edge detection is done. The output image has only two colours – the green colour for crop and weeds and the remaining part of the image is represented by the black colour. After edge detection we get an image that consists of image of both crop and weed in white colour.

The next step in weed detection is filtering. Filtering is done on the basis of edge frequency of weed and crop. A threshold value is defined and filtering process separated weed edge from the crop edge on the basis of whether the edge frequency exceeds the threshold value. Hence, filtering process detects the weed blocks.

Other image processing techniques that we use for weed detection include background removal, histogram equalization, sharpening, and glare removal.

The process of image segregation is performed on all images in the dataset. As a result, we get two sets of images where one is classified as the weed image and other one is classified as crop image.

The next step is to create a Machine Learning (ML) model that learns from these images.

Most weed detection techniques employ supervised machine learning algorithms since they achieve high accuracy.After image acquisition and image processing is done, the next step is feature extraction followed by image detection and classification.

## Image Classification using Deep Learning

Deep Learning has emerged from Artificial Neural Network (ANN) and has remarkable applications in the field of Computer Vision (CV) and Image Classification. Basically, Deep Learning is a sub-field of Machine Learning. Still, it is a continuously developing field. There are many Deep Learning algorithms that we can use for image classification such as Convolution Neural Network (CNN), Recurrent Neural Network (RNN), Long Short Term Memory (LSTM), and Deep Neural Network (DNN). Even a pre-trained CNN can be used for classification and it is termed as Transfer Learning.

Some popular Deep Learning algorithms for image classification are described below.

## Convolution Neural Network (CNN)

CNN takes the source images as the input data and breaks them into smaller chunks called convolutions. The source data is fed into the CNN with appropriate dimensions of width×length×color channels. The actual convolution is done by conv2D layer. We can get a 1D layer from a 2D layer using flatten layer. Therefore, the typical transformation is shown in figure 1.

*Figure 1. Layers in a convolution neural network (CNN)*

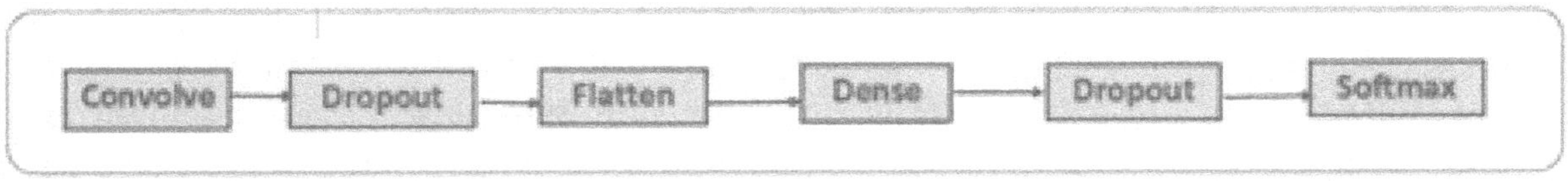


Dropout prevents over-fitting. Max Pooling can substantially reduce the training time without compromising the quality of training. We choose a group of pixels and select the highest value pixel in that group. It speeds up the processing considerably. We can add multiple layers of convolution for achieving the reasonable prediction accuracy. Between each layer of convolution, a layer of max pooling is added. Max pooling is also referred as down sampling. This technique improves the computational efficiency of CNN by reducing feature map dimensionality while preserving features.

In order to avoid over-fitting, we also add a layer of dropout. During the training process, usually neurons develop co-dependency amongst each other. Therefore some regularization technique such as Dropout is required to avoid over-fitting.

Dropout helps the network to generalize without overfitting. The final layer of output uses the Dense layer with Sigmoid activation function.

Confusion matrix is used to find out predicted class vs. true class. It will indicate how many samples are correctly predicted. Diagonal numbers show that how many elements are correctly classified. Outside diagonal the number indicates that how many samples are misclassified.

Although RNN and LSTM can also be used for image classification, CNN is the best model and has been used in a significantly large number of image classification problems. Evidently, most of the research work regarding weed detection using image datasets have also used CNN as the most preferred method for image classification.

## Region Proposal Network (RPN)

A Region Proposal Network (RPN) is also a type of Convolution Neural Network (CNN). It has been used for prediction of object bounds along with objectness scores at each position.

## R-CNN (Region Based Convolution Neural Network)

While, a Convolution Neural Network (CNN) performs the image classification efficiently, still we need a way to visualize the classification. For this purpose, we can use an R-CNN (Region Based Convolution

Neural Network) that places a bounding box around the identified object. Further, R-CNN also indicates the category name along with the bounding box. Since R-CNN tells us about the location of classified object as well as its category, it has various applications in Computer Vision. For instance, R-CNN can be used in applications of tracking objects in images captured by a drone-mounted camera.

Hence, R-CNN is an important technique in the application of weed detection since it can be used to identify the location of weeds from the images taken by camera using UAV.

In order to place a bounding box around an object in the image, R-CNN performs Selective Search to fetch the Regions of Interest (ROI). Basically, ROI is nothing but a rectangle representing the boundary of the identified object in the image. After applying the boundary boxes, it is passed to a Machine Learning classifier as input. The Machine Learning classifier determines the category or class of the object which is enclosed in an ROI. Hence, the output of classifier represents the objects and their class.

In a weed detection application the output of such machine learning classifier indicates the images which are identified as weeds and possibly the type of weeds. In any case, it helps us in determining the images which contain weeds. Since images taken by UAV also have location information, we can easily identify the location of weeds in the agriculture field. Later, this information helps us in selective spraying at the weed location only and therefore amount of chemicals sprayed in the field are substantially reduced.

R-CNN has many extensions. Some of these extensions are described below.

- **Fast-RCNN**
  Like R-CNN, the fast R-CNN also employs the feature of Selective Search in order to determine the Regions of interest (ROI). However, it runs the Neural Network model on the whole image only once in a single forward pass and uses ROI pooling that cuts each ROI from thee output. Then, fast R-CNN reshapes the ROI and finally classifies it.
- **Faster R-CNN**
  In this approach, generation of ROI is integrated within the neural network (CNN model) itself and improves accuracy, precision, and recall value.
- **Mask R-CNN**
  Mask R-CNN performs the additional task of instance segmentation along with image segmentation.

Region Proposal Network (RPN) is the backbone of an R-CNN. Basically, RPN comprises of a classifier as well as a regressor. Probability of the proposal containing the target object is determined. Coordinates of proposal are computed using regression. RPN scans every location in the image efficiently in very less duration of time. The following list indicates the working of RPN.

1. A 3 X 3 window slides over a convolution feature map
2. The 3 X 3 window is resampled to a 256 dimensional vector
3. Then it is fed into two fully connected convolution layers
4. It is also fed into a box regression layer that computes the box offset
5. It is also fed into a box classification layer that computes the confidence scores
6. These confidence scores are associated with the probability of objectness

## Graph Convolution Network (GCN)

A Graph Convolution Network (GCN) is a neural network that works directly on graphs.A GCN can be used for multi-class crop and weed identification. Further, this method works well with limited availability of labelled data where hidden layers don't share weights with each other.

## Multi-Label Image Classification

The Multi-Label image classification represents a technique that applies more than one label to an image. For instance, an image can be labelled as either a crop plant or weed. In addition, another label can be applied to the same image indicating the type of crop or weed. Some of the approaches for performing Multi-Label Image Classification are given below.

1. Convolution Neural Networks are good for single label image classification. However, another deep learning technique known as Long Short Term Memory (LSTM) can be used as decoder to generate multiple labels for an image (Li et al., 2018). Also, bridging the semantic gap between labels is also necessary so that labels representing a particular image become related. For instance, a label indicating type of the weed is in coordination with another label that indicates that the image is a weed.
2. A combination of CNN and RNN (Recurrent Neural Network) can be used for multi-label image classification (Wang et al, 2018). The CNN-RNN framework learns a joint image-label embedding to characterize the semantic label dependency as well as the image-label relevance.
3. A trained CNN classifier used for single label classification can be used for multi-label classification as well. The classifier can be trained with objectness measure and selective search (Shahriyar et al., 2018).

## Weed Identification using TensorFlow and Keras

The dataset for weed identification is available at https://www.kaggle.com/c/plant-seedlings-classification/data

The dataset contains two folders namely test and train. Further, the train folder contains subfolders representing the crops and weeds. The name of a subfolder represents the name of a crop or a weed. Also, the same is a class label for the images contained in the corresponding subfolder.

However, the test folder contain the images for which the class label is to be predicted. The hierarchy of image dataset folders is shown in figure 2.

*Figure 2. Hierarchy of image dataset folders*

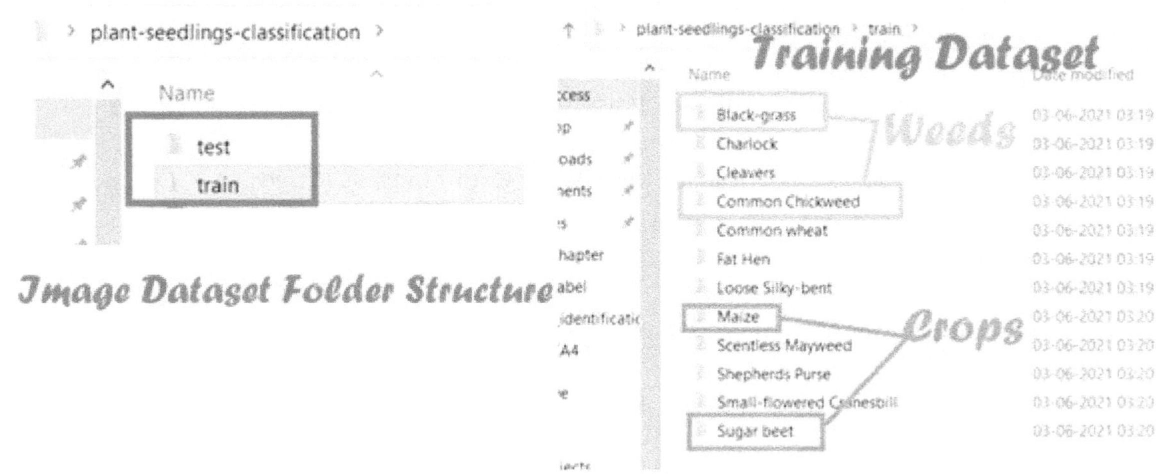


A sample of test dataset is shown in the figure 3.

*Figure 3. Sample test dataset*

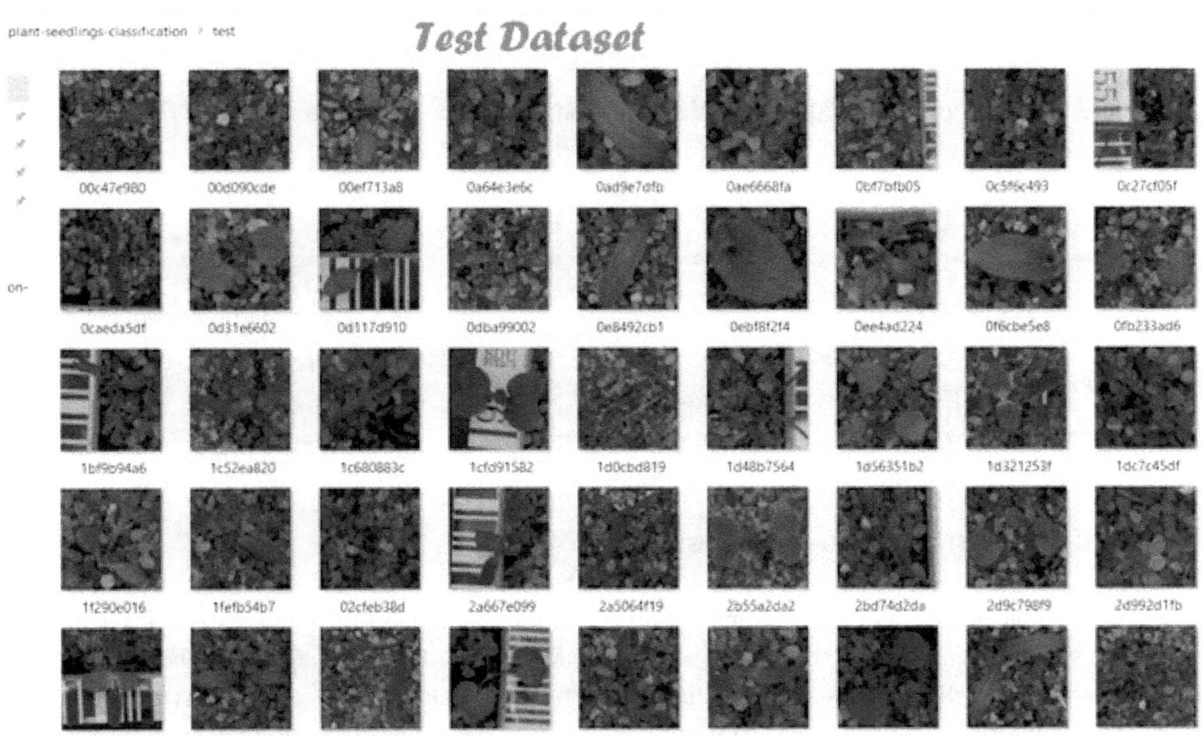


A method for weed identification is described in this section.

1. Import the libraries OS and CV2. Read all images from the training dataset and resize them to a size of 128X128. Initialize a variable called X_train from these images.
2. Import numpy library and convert all training dataset images into a numpy array.
3. Now, convert the named categorical labels to one-hot encoded format.
4. Retrieve the labels in a variable called y_train.
5. Split the dataset into training and testing dataset using the function train_test_split.
6. Now fetch the test images and create the variable X_test from these images.
7. Convert the test dataset into numpy array
8. Import layers library from tensorflow.keras
9. Define the model in sequential manner
10. Add a convolution layer using the relu activation function.
11. Add a max pool layer
12. Flatten the layers
13. Perform batch normalization
14. Add a dense layer with relu activation function
15. Perform a dropout
16. Add a dense layer with softmax activation function
17. Compile the model with adam optimizer
18. Train the model
19. Train the model with validation data
20. Evaluate the model
21. Predict the class of weeds using the test dataset

The figure 4 figure summarizes the weed identification process described above.

*Figure 4. Weed identification process*

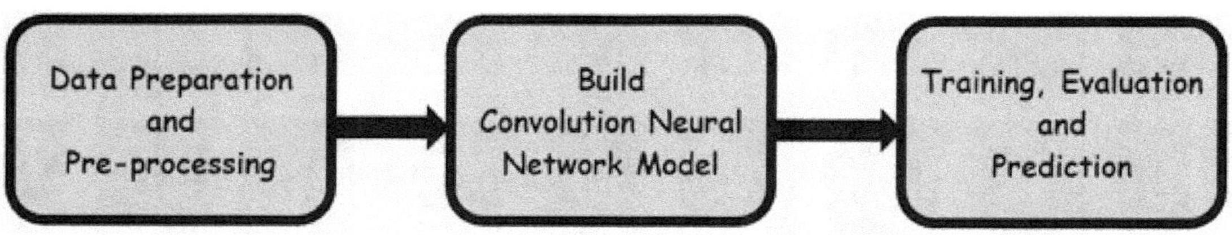


## Data Preparation and Pre-Processing

The data preparation task requires fetching the images from a specified folder and resizing all images to a uniform size. In order to feed the images as input to a CNN model, the images must be converted to a numpy array. Therefore, the images contained in both train and test folders are converted to a numpy array. The categories must be converted to numbers. Hence the categorical variables are converted to one-hot encoded format. It is essential since many machine learning algorithms can't operate on label data directly. Finally, the dataset used for training is split into train and test datasets.

## Building a CNN Model

A CNN is created with a sequential model that defines a stack of layer with each layer has one input tensor and one output tensor. First, a convolution operation is carried out that produces a feature map. Once the feature map is created, a max pooling layer is defined. The max pooling layer considers the feature map produced by convolution layer and retrieves the maximum elements from the feature map. The max pool operation is followed by a flatten operation. Basically, the flatten operation takes into consideration the pooled feature map and converts them into a single column. In other words, the flatten operation creates the data suitable to be fed to the fully connected layer.

In a sequential model the input and output of all layers should be standardized. Batch normalization is basically a layer in CNN that allows learning in each layer to be done independently and it normalizes the output of the previous layer. After the batch normalization layer, a Dense layer is defined. A fully connected layer in a CNN has neurons that receive input from each neuron from the previous layer. A Dense layer in CNN is created using a fully connected layer.

The Dropout layer is required in the training process. It is not required when inferences are made using the trained CNN model. Dropout is performed in order to prevent overfitting.

The dropout operation selects the neurons randomly and ignores them from being part of the training process. The randomly selected neuron are removed from the forward pass as well as the backward pass. After dropout operation chances are that very important features may be missed. Therefore the dropout layer is followed by another dense layer with a softmax activation function.

Next, an optimization is performed in order to minimize the cost function. Then, training and validation of the CNN model takes place.Further, model evaluation determines the accuracy of CNN model. Finally, prediction operation predicts the classes using the trained model.

Figure 5 depicts the general approach for data preparation and pre-processing for weed identification.

*Figure 5. Data preparation and pre-processing*

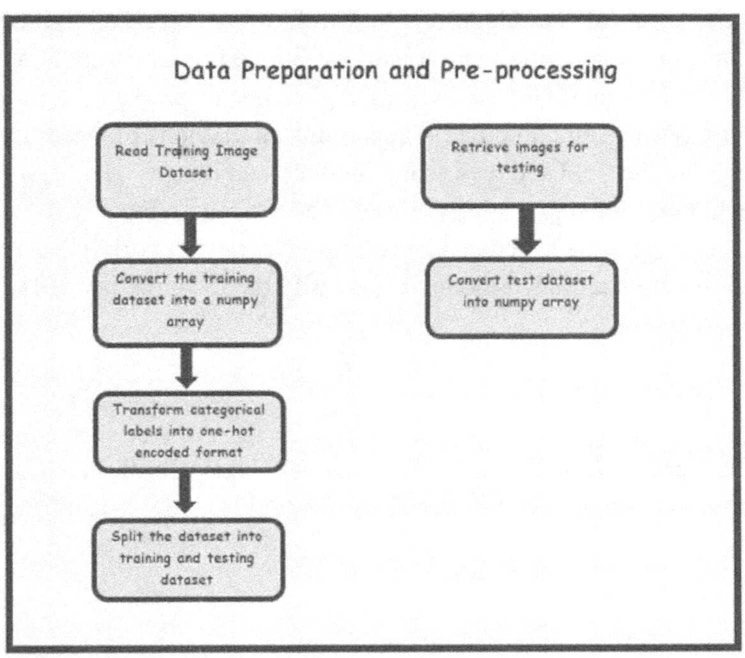

## Challenges in Weed Detection and Removal

1. UAVs can't detect all weeds in the fields. The weeds which are hidden deep under the crop plants are not captured by drones. Therefore, some of the weeds remain undetected.
2. The Deep Learning methods detect known species of weeds which have been labelled so far. However, there are many more varieties of weeds which may not be detected since they are not learned by the model.
3. Weed detection using UAV and Deep Learning doesn't necessarily result in their precise removal since it requires accurate determination of their location. Otherwise, targeted spraying of chemicals for removing weeds may harm crops.
4. The automatic weed detection method is costly and require skilled manpower.

## Advantages of Deep Learning Techniques in Weed Detection

The most obvious advantage of using Deep Learning techniques in weed detection is the accuracy and precision of detection. Moreover, the detection is faster as compared to the traditional image processing techniques.

### Drawbacks

Deep Learning techniques require a large image dataset for training. The data collection task is costly and time consuming. Furthermore, labelling is required so it is possible for some of the previously unknown weeds to remain undetected.

# CONCLUSION

The weed detection and removal is an important task in farming that takes significant amount of time and efforts. Weeds hamper the growth of crop plants by competing for essential resources.

Traditionally, the weed removal is done by spraying chemicals all over the field. This procedure is not only costly but also harms soil and crops. The automated methods of weed detection work by first capturing the images from the field and then searching for weed images. Hence, it requires a number of image processing techniques for segregating the weed part from an image.

Further, a Deep Learning model such as Convolution Neural Network (CNN) can be used to learn the weed images and therefore a trained model can identify the location of weeds.

# REFERENCES

Adeux, G., Vieren, E., Carlesi, S., Bàrberi, P., Munier-Jolain, N., & Cordeau, S. (2019). Mitigating crop yield losses through weed diversity. *Nature Sustainability*, *2*(11), 1018–1026.

Alam, M., Alam, M. S., Roman, M., Tufail, M., Khan, M. U., & Khan, M. T. (2020, April). Real-Time Machine-Learning Based Crop/Weed Detection and Classification for Variable-Rate Spraying in Precision Agriculture. In *2020 7th International Conference on Electrical and Electronics Engineering (ICEEE)* (pp. 273-280). IEEE.

Andrea, C. C., Daniel, B. B. M., & Misael, J. B. J. (2017, October). Precise weed and maize classification through convolutional neuronal networks. In *2017 IEEE Second Ecuador Technical Chapters Meeting (ETCM)* (pp. 1-6). IEEE.

Arif, S., Kumar, R., Abbasi, S., Mohammadani, K., & Dev, K. (2021). *Weeds Detection and Classification using Convolutional Long-Short-Term Memory*. Academic Press.

Asad, M. H., & Bais, A. (2019). Weed detection in canola fields using maximum likelihood classification and deep convolutional neural network. *Information Processing in Agriculture.*

Bah, M. D., Hafiane, A., & Canals, R. (2017, November). Weeds detection in UAV imagery using SLIC and the hough transform. In *2017 Seventh International Conference on Image Processing Theory, Tools and Applications (IPTA)* (pp. 1-6). IEEE.

Bah, M. D., Hafiane, A., & Canals, R. (2019). CRowNet: Deep network for crop row detection in UAV images. *IEEE Access: Practical Innovations, Open Solutions*, *8*, 5189–5200.

Bah, M. D., Hafiane, A., Canals, R., & Emile, B. (2019, November). Deep features and One-class classification with unsupervised data for weed detection in UAV images. In *2019 Ninth International Conference on Image Processing Theory, Tools and Applications (IPTA)* (pp. 1-5). IEEE.

Champ, J., Mora-Fallas, A., Goëau, H., Mata-Montero, E., Bonnet, P., & Joly, A. (2020). Instance segmentation for the fine detection of crop and weed plants by precision agricultural robots. *Applications in Plant Sciences*, *8*(7), e11373.

Cheng, B., & Matson, E. T. (2015, June). A feature-based machine learning agent for automatic rice and weed discrimination. In *International Conference on Artificial Intelligence and Soft Computing* (pp. 517-527). Springer. 10.1007/978-3-319-19324-3_46

Dasgupta, I., Saha, J., Venkatasubbu, P., & Ramasubramanian, P. (2020). AI Crop Predictor and Weed Detector Using Wireless Technologies: A Smart Application for Farmers. *Arabian Journal for Science and Engineering*, *45*(12), 11115–11127.

Dos Santos Ferreira, A., Freitas, D. M., da Silva, G. G., Pistori, H., & Folhes, M. T. (2017). Weed detection in soybean crops using ConvNets. *Computers and Electronics in Agriculture*, *143*, 314–324. doi:10.1016/j.compag.2017.10.027

Espejo-Garcia, B., Mylonas, N., Athanasakos, L., Fountas, S., & Vasilakoglou, I. (2020). Towards weeds identification assistance through transfer learning. *Computers and Electronics in Agriculture*, *171*, 105306.

Esposito, M., Crimaldi, M., Cirillo, V., Sarghini, F., & Maggio, A. (2021). Drone and sensor technology for sustainable weed management: A review. *Chemical and Biological Technologies in Agriculture*, *8*(1), 1–11.

Gao, J., French, A. P., Pound, M. P., He, Y., Pridmore, T. P., & Pieters, J. G. (2020). Deep convolutional neural networks for image-based Convolvulus sepium detection in sugar beet fields. *Plant Methods*, *16*(1), 1–12.

Gibson, D. J., Young, B. G., & Wood, A. J. (2017). Can weeds enhance profitability? Integrating ecological concepts to address crop-weed competition and yield quality. *Journal of Ecology*, *105*(4), 900–904.

Gothai, E., Natesan, P., Aishwariya, S., Aarthy, T. B., & Singh, G. B. (2020, March). Weed Identification using Convolutional Neural Network and Convolutional Neural Network Architectures. In *2020 Fourth International Conference on Computing Methodologies and Communication (ICCMC)* (pp. 958-965). IEEE.

Jiang, H., Zhang, C., Qiao, Y., Zhang, Z., Zhang, W., & Song, C. (2020). CNN feature based graph convolutional network for weed and crop recognition in smart farming. *Computers and Electronics in Agriculture*, *174*, 105450.

Jin, X., Che, J., & Chen, Y. (2021). Weed Identification Using Deep Learning and Image Processing in Vegetable Plantation. *IEEE Access: Practical Innovations, Open Solutions*, *9*, 10940–10950.

Khan, A., Ilyas, T., Umraiz, M., Mannan, Z. I., & Kim, H. (2020). Ced-net: Crops and weeds segmentation for smart farming using a small cascaded encoder-decoder architecture. *Electronics (Basel)*, *9*(10), 1602.

Li, Y., & Wang, Y. (2018, June). A multi-label image classification algorithm based on attention model. In *2018 IEEE/ACIS 17th International Conference on Computer and Information Science (ICIS)* (pp. 728-731). IEEE.

Liang, W. C., Yang, Y. J., & Chao, C. M. (2019, November). Low-Cost Weed Identification System Using Drones. In *2019 Seventh International Symposium on Computing and Networking Workshops (CANDARW)* (pp. 260-263). IEEE.

Lottes, P., Khanna, R., Pfeifer, J., Siegwart, R., & Stachniss, C. (2017, May). UAV-based crop and weed classification for smart farming. In *2017 IEEE International Conference on Robotics and Automation (ICRA)* (pp. 3024-3031). IEEE.

Mao, W., Hu, X., & Zhang, X. (2007, August). Weed detection based on the optimized segmentation line of crop and weed. In *International Conference on Computer and Computing Technologies in Agriculture* (pp. 959-967). Springer.

Miao, F., Zheng, S., & Tao, B. (2019, January). Crop Weed Identification System Based on Convolutional Neural Network. In *2019 IEEE 2nd International Conference on Electronic Information and Communication Technology (ICEICT)* (pp. 595-598). IEEE.

Millar, K., Gibson, D. J., Young, B. G., & Wood, A. J. (2007). Impact of interspecific competition on seed development and quality of five soybean cultivars. *Australian Journal of Experimental Agriculture*, *47*(12), 1455–1459. doi:10.1071/EA07034

Millar, K. D. L., Ebbs, S. D., Gibson, D. J., Wood, A. J., & Young, B. G. (2011). Evaluation of physiological parameters for the prediction of seed yield and quality for soybean (Glycine max) plants grown in the presence of weed competition. *Plant Biosystems*, *145*(1), 1–11. doi:10.1080/11263504.2010.514129

Moazzam, S. I., Khan, U. S., Tiwana, M. I., Iqbal, J., Qureshi, W. S., & Shah, S. I. (2019, October). A Review of Application of Deep Learning for Weeds and Crops Classification in Agriculture. In *2019 International Conference on Robotics and Automation in Industry (ICRAI)* (pp. 1-6). IEEE.

Nasierding, G., & Sajjanhar, A. (2013, December). Multi-label classification with clustering for image and text categorization. In *2013 6th International Congress on Image and Signal Processing (CISP)* (Vol. 2, pp. 869-874). IEEE. 10.1109/CISP.2013.6745287

Ngo, H.C., Hashim, U.R., Sek, Y.W., Kumar, Y.J., & Ke, W.S. (2019). Weeds Detection in Agricultural Fields using Convolutional Neural Network. *International Journal of Innovative Technology and Exploring Engineering.*

Osorio, K., Puerto, A., Pedraza, C., Jamaica, D., & Rodríguez, L. (2020). A Deep Learning Approach for Weed Detection in Lettuce Crops Using Multispectral Images. *AgriEngineering*, *2*(3), 471–488.

Pérez-Ortiz, M., Gutiérrez, P. A., Peña, J. M., Torres-Sánchez, J., López-Granados, F., & Hervás-Martínez, C. (2016, December). Machine learning paradigms for weed mapping via unmanned aerial vehicles. In 2016 IEEE symposium series on computational intelligence (SSCI) (pp. 1-8). IEEE. doi:10.1109/SSCI.2016.7849987

Ren, S., He, K., Girshick, R., & Sun, J. (2015). *Faster r-cnn: Towards real-time object detection with region proposal networks.* arXiv preprint arXiv:1506.01497.

Sarvini, T., Sneha, T., GS, S.G., Sushmitha, S., & Kumaraswamy, R. (2019, April). Performance Comparison of Weed Detection Algorithms. In *2019 International Conference on Communication and Signal Processing (ICCSP)* (pp. 843-847). IEEE.

Shahriyar, S. A., Alam, K. M. R., Roy, S. S., & Morimoto, Y. (2018, December). An approach for multi label image classification using single label convolutional neural network. In *2018 21st international conference of computer and information technology (ICCIT)* (pp. 1-6). IEEE.

Tang, J., Wang, D., Zhang, Z., He, L., Xin, J., & Xu, Y. (2017). Weed identification based on K-means feature learning combined with convolutional neural network. *Computers and Electronics in Agriculture, 135*, 63–70. doi:10.1016/j.compag.2017.01.001

Umamaheswari, S., Arjun, R., & Meganathan, D. (2018, October). Weed detection in farm crops using parallel image processing. In *2018 Conference on Information and Communication Technology (CICT)* (pp. 1-4). IEEE.

Wang, J., Yang, Y., Mao, J., Huang, Z., Huang, C., & Xu, W. (2016). Cnn-rnn: A unified framework for multi-label image classification. In *Proceedings of the IEEE conference on computer vision and pattern recognition* (pp. 2285-2294). 10.1109/CVPR.2016.251

Wu, Z., Chen, Y., Zhao, B., Kang, X., & Ding, Y. (2021). Review of Weed Detection Methods Based on Computer Vision. *Sensors (Basel)*, *21*(11), 3647.

Yu, J., Schumann, A.W., Cao, Z., Sharpe, S.M., & Boyd, N.S. (2019). Weed detection in perennial ryegrass with deep learning convolutional neural network. *Frontiers in Plant Science, 10*, 1422.

# Chapter 7
# Evaluation of Machine Learning Techniques for Crop Yield Prediction

**Divya Goel**
*J. C. Bose University of Science and Technology, YMCA, India*

**Payal Gulati**
*J. C. Bose University of Science and Technology, YMCA, India*

**Suman Kumar Jha**
*IIMT College of Engineering, India*

**Dr Nitendra Kumar**
https://orcid.org/0000-0001-7834-7926
*IIMT College of Engineering, India*

**Ayoub Khan**
https://orcid.org/0000-0002-0398-3722
*University of Bisha, Saudi Arabia*

**Priti Kumari**
*B. S. M. College of Engineering, India*

## ABSTRACT

*The agricultural segment is a major supporter of the Indian economy as it represents 18% of India's GDP, and it gives work to half of the nation's work power. The farming segment is required to satisfy the expanding need for food because of the increasing populace. Therefore, to cater to the ever-increasing needs of people of the nation, yield prediction is done prior. The farmers are also benefited from yield prediction as it will assist the farmers to predict the yield of crops prior to cultivating. There are multiple parameters that affect the yield of crops like rainfall, temperature, fertilizers, pH level, and other atmospheric circumstances. Thus, considering these factors, the yield of a crop is thus hard to predict and becomes a challenging task. In this chapter, the dataset of different states producing different crops in different seasons is considered; further, after preprocessing the data, the authors applied machine learning algorithms, and their results are compared.*

DOI: 10.4018/978-1-6684-5141-0.ch007

## INTRODUCTION

India is a horticultural nation, with the majority of its population reliant on the industry. It is one of the most prevalent occupations in our country and contributes significantly to the country's overall development. Due to the reality that it's far the spine of the Indian economy. Around 60% of land is devoted to agriculture so that it will meet the needs of 1.2 billion people (Khan et al., 2021). Thus, modernization in horticulture is crucial and could advantage our nation's ranchers. The level of manufacturing in India is lowering due to different factors consisting of environmental changes, choppy rainfall, water management, and immoderate pesticide use. For a whole lot of reasons, the bulk of farmers fall quick of the anticipated crop yield. To comprehend production levels, crop yield prediction is used, which incorporates crop yield predictions into the data. Previously, crop yield estimation was highly dependent on the type of crop and the farmer's cultivation understanding.

There are several methods to address improvements and growth harvest yield and quality. Data mining strategies also are useful for forecasting crop yields. Generally, Data Mining is used to take a look at records from numerous frameworks and summarise it as beneficial records. Data mining software is a valid device that permits customers to explain and summarise obvious relationships, in addition, in addition, to take taking a look at records at numerous estimations or edges. Indeed, Data Mining is the technique of figuring out institutions or examples of fields (Priya et al., 2018). Data may be used to offer context for connections, institutions, or fashions. Then, this understanding may be converted into recorded fashions and records that may be used to create destiny examples or fashions. For instance, an overview of agricultural topics implores ranchers to indicate and avoid destiny harvest incidents.

Numerous analysts have been directed to develop a robust strategy for forecasting yields. However, the centre has consistently focused on quantifiable strategies and a small amount of work that has been prepared using a machine learning approach. Crop production is dependent on a variety of different parameters (Chowdary & Venkataramana, 2018), which vary according to climatic conditions such as temperature, humidity, rainfall, and soil pH value, as well as region geography and fertilisers. Numerous prediction tools for various crops make use of subsets of these parameters. Thus, prediction models are classified into two broad categories. There are statistical models that employ a single prediction task that encompasses all possible example spaces. Another approach is machine learning, in which information for data searches is used to connect the input and output variables.

AI can acquire capability in conjunction with the machine without requiring knowledge of PC programming; as a result, it advances machine completion by recognising and depicting the stability and example of driven data. We used supervised learning techniques to forecast crop yields in this study. For precise and effective methods, learning data is used to associate desired outputs with an input-output mapping standard. It entails developing a machine learning representation that is dependent on labelled sample.

Until now, prediction has been limited to a single state or crop. However, this proposed system investigates the application of supervised machine learning approaches in determining the yield production of a variety of crops in a variety of different states and their districts. This section collects, analyses, and tactics records from diverse states and crops. For yield estimation, techniques which include Linear Regression, Random Forest Regression, Gradient Boosting Regression, Polynomial Regression, Decision Tree Regression, and Ridge Regression had been used.

This chapter is organized in the following manner: Section II discusses related work in the fields of data mining and machine learning. Segment III consists of an in-depth discussion of the proposed framework. Segment IV will refer to the communiqué and its outcome. Finally, Section V completes the examination work.

## RELATED WORK

Agribusiness is taken into consideration a smart area for Machine Learning, as a variety of work has been done with the assistance of AI in the field of horticulture. Numerous researchers in the fields of agribusiness and associated sciences have evolved and evaluated several methods of reasoning (Khan et al., 2021; Katiyar et al., 2021; Khan et al., 2021. Sahni et al., 2021 & Kumar, 2018).

Chaudhary and Venkataramana (2018) evolved an id3 calculation to enhance the incredible nature of tomato harvest yield. It is implemented in PHP and the datasets are in CSV format. Temperature, region, relative humidity, and the development of the tomato crop are all used as barriers on this study. Sujatha and Isakki (2017) forecast the usage of facts-digging strategies. This version eroded various boundaries, for example, land region, crop name, soil type, pH value, seed type, and water, and moreover predicted plant blasts and diseases, and thus engaged in plum crop selection based on climatic information and required parameters. Gandhi et al. (2016) used the SVM to forecast rice crop yields. The dataset is segmented into numerous boundaries on this approach, including spot, temperature, precipitation, and assembling. The completed classifier is incrementally improved with this dataset. Additionally, they organized the dataset using the Weka device to generate the current dataset's rule arrangement. Python results have been introduced via SVM calculation. Veenadhari et al., (2014) created an intuitive website for tracking the environmental impact and harvest creation using the c4.5 calculation. Crop Advisor has also been created, which is subject to c4.5 calculation, choice tree, and governance. It considers how various climatic boundaries affect crop development. The data on the natural boundaries between the connected years, such as precipitation and temperature, were accumulated. The decisions were made in accordance with the zone surrounding the harvested crop. Jun et al. (2009) proposed a decision tree that is capable of aggregating all possible cultivating accounts. A decision tree classifier was transformed into proposed agribusiness data. It employs new realities that can be addressed individually and throughout the record. The dataset, horse-colic, and soybean datasets are all checked using the 10-overlap cross approval system. KiranMai et al. (2006) demonstrated in their review how information mining is combined with other cultivating data, such as meteorological data and pesticide usage, to aid in pesticide mitigation. Effective data was represented in relation to the subject of farming with adjacent properties. Verheyen et al. (2001) clarified quantifiable mining methods in their review, stating that they are consistently used to observe soil properties. With GPS-based development, k-means is used to separate soils in a mix.

## PROPOSED SYSTEM

A machine learning system solves issues wherein the relationship among the enter and output variables is unknown or tough to obtain. The term "learning" refers to the programmed acquisition of auxiliary depictions from instances of the described object. Unlike traditional factual techniques, machine learning makes no assumptions about the correct structure of the model dataset that contains the data. This property is extremely useful for simulating complex non-direct practises, such as the capacity to forecast crop yields. Crop Yield Prediction is the most successful application of machine learning strategies. The techniques of Supervised Learning include a dependent/outcome variable that is to be predicted from a given arrangement of independent variables. We create a function that directs inputs to outputs using these factor arrangements. The preparation process is repeated until the model reaches the desired degree of precision with respect to the preparation data.

The proposed framework collects data on various states and crops from a variety of different sources, resulting in the creation of a dataset. Modelling is accomplished through the use of the supervised learning tool. This section discusses the predicted yield and the order in which they will be produced. This proposed framework is divided into the following stages: data collection, pre-processing of data, feature selection, and obtaining the outcome using various machine learning algorithms.

a. Collection of Datasets: Information is gathered from a variety of sources and then broken down and prepared. This data is used to conduct an illustrative investigation. The dataset used in this study spans multiple states (Maharashtra, Uttar Pradesh, West Bengal, and Gujarat), multiple harvests (sugarcane, coconut, wheat, and gramme), multiple seasons (Kharif, Rabi, Whole, summer, and so on), multiple yield years, and a variety of boundary conditions such as rainfall, temperature, pH, and humidity. Currently, either a state or a single yield is used. Nonetheless, the entire list above is used in this paper, and the dataset is subsequently acquired.
b. Data pre-processing: Pre-processing is the first and most critical step in machine learning, and it refers to the raw data. As a result, it is well-suited for developing and training Machine Learning models.
c. Highlight Extraction: Highlight extraction should minimise the amount of information required to address a massive dataset. Its objective is to strip information of valuable attributes. The characteristics include maximum, minimum, and mean temperatures, soil pH, air moisture, and precipitation.
d. Separate the Training and Testing Datasets: This stage entails preparing and testing the data. The stacked data is partitioned into groups. For instance, getting ready and checking out the facts. The training set is meant to be used together with the training set, and level facts are to be tested in the aftermath of gaining from preceding perceptions. The final piece of information is framed and handled through a artificial intelligence module.
e. Procedures for Using Machine Learning Techniques: Numerous supervised machine learning techniques are used to forecast crop yields, as illustrated in Fig. 1.

*Figure 1. Proposed system architecture*

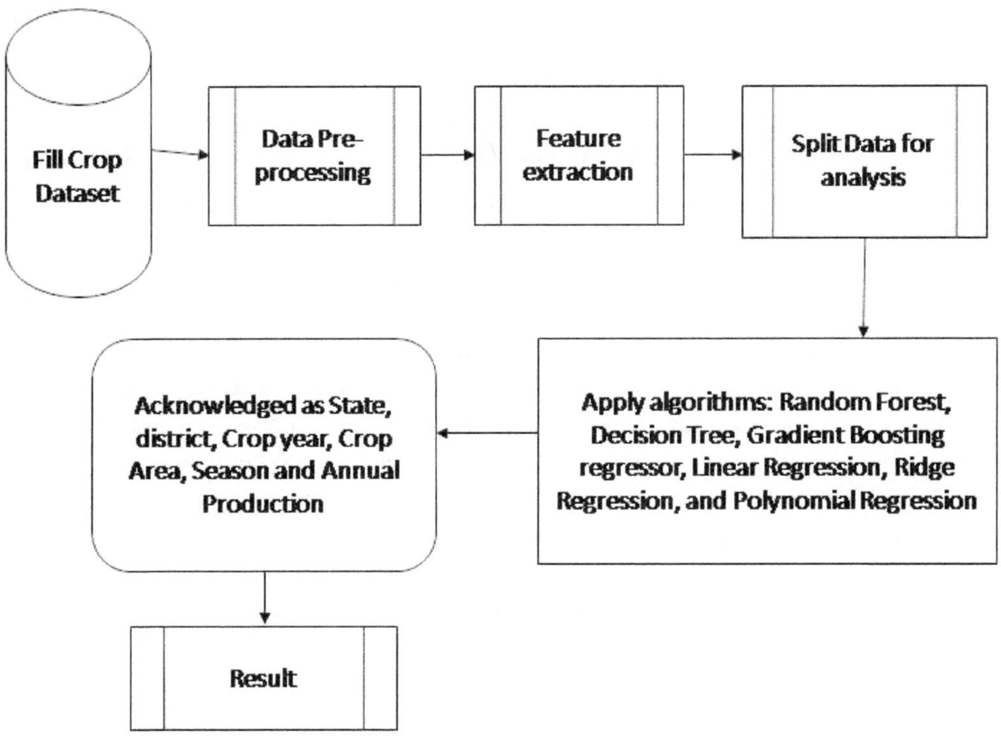


i. **Random Forest**

Random Forest (RF) is a well-known and powerful supervised machine learning tool able to perform each category and regression analysis. This is achieved through training a large number of decision trees and then outputting the class that corresponds to the mode of training as classification or mean predictions as regression for each tree. When there are more trees in the forest, the prediction becomes more accurate. Each tree is trained on a subset of the data in this case, as there are multiple trees. This algorithm is parallelizable.

ii. **Random Forest pseudocode:**

- Select the random features (i) from the entire features (n). Where i<< n
- Calculate the node (m) among (i) through the use of the first-class cut up point.
- Split the node into daughter nodes by using the first-class cut up.
- Repeat the process from 1 to 3 again, until the number of nodes (z) has been reached.
- Forest is built by repeating the steps from 1 to 4, for (y) number of times to create (y) number of trees.

iii. **Decision Tree**

It is a predictive technique that starts through checking situations at every degree of the tree and proceeds to the bottom of the tree, wherein numerous alternatives are recorded. The equipment determines the situation, and the final results can be in phrases of decisions. The numerous choice tree algorithms consist of the c4.5, CART, and ID3 algorithms.

### iv. Gradient Boosting

The purpose of the Gradient boosting algorithm for regression is to construct an additive model in a stage-wise manner while taking into account the advancement of subjective differentiable loss functions. The regression tree-based method fits the given loss function's negative gradient in each stage. Boosting arose from the question of whether a slow learner can be coached to improve. Frail speculation or a feeble learner is defined as someone whose exhibition is somewhat superior to random possibility in any case.

### v. Linear Regression

Linear regression is the most frequently used statistical technique in data analysis and predictive modelling. It is a linear approach that is used to model the relationship between two variables. It is denoted as

$Y = mx + c$; where x, Y, m and c are independent variable, dependent variable, slope of the line and intercept respectively.

### vi. Polynomial Regression

It is a type of regression analysis, which the relationship between the independent variable i and the dependent variable d is modelled as a degree polynomial. It is compatible with the non-linear relationship between the value of i and the consequent conditional mean of d, which is represented by E(d|i). Due to the linear nature of the statistical estimation problem, the regression function E(d|i) is linear for unknown parameters estimated from the data.

### vii. Ridge Regression

It is a specialised technique for analysing multiple regression data that is multi-collinear in nature. It is a fundamental regularisation technique that not many people use due to the complexity of the underlying concepts.

## DISCUSSION AND RESULTS

This section displays the results were obtain after performing machine learning calculations on the dataset. The dataset is subjected to six distinct calculations. We used Jupiter in this paper, which is a collection of programming components used in intelligent figuring. The distinct boundaries established for these strategies were mean outright error, mean squared error, root-mean-square error, $r^2$, and cross approval, which were use to determine their productivity.

MAE, MSE, RMSE, and $r^2$ are calculated using the following formulae:

$$MAE = \frac{\sum_{i=1}^{n} \left| y_i - x_i \right|}{n} = \frac{\sum_{i=1}^{n} \left| e_i \right|}{n}$$

$$MSE = \frac{1}{n} \sum_{i=1}^{n} (y_i - \tilde{y}_i)^2$$

$$RMSE = \sqrt{\frac{1}{n} \sum_{i=1}^{n} (y_i - \tilde{y}_i)^2}$$

$$r^2 = 1 - \frac{\Sigma (y_i - \tilde{y}_i)^2}{\Sigma \left( \boldsymbol{y}_i - \bar{\boldsymbol{y}}_i \right)^2}$$

In this section, we consider two cases and their associated target variables. In the first case, we use the target variable "yield," and the Gradient Boosting Regressor produces a more accurate result of 87.9% with cross-validation runs, as shown in the above Table 1. Fig. 8 compares cross-validation accuracy for the target variable "yield" using actual and predicted values.

*Figure 2. Gradient boosting graph*

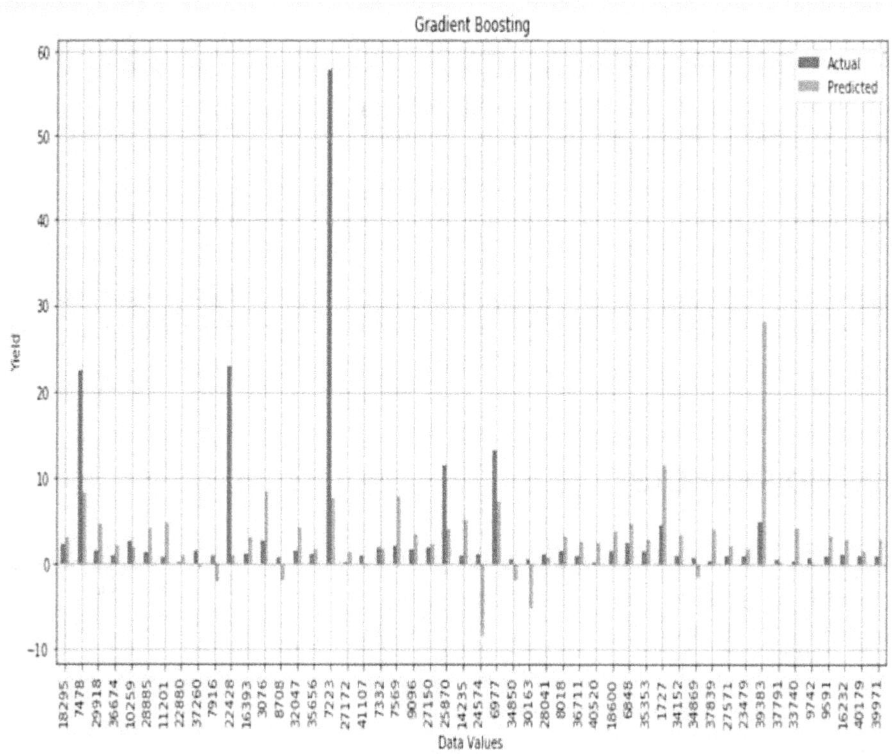

*Figure 3. Ridge regression graph*

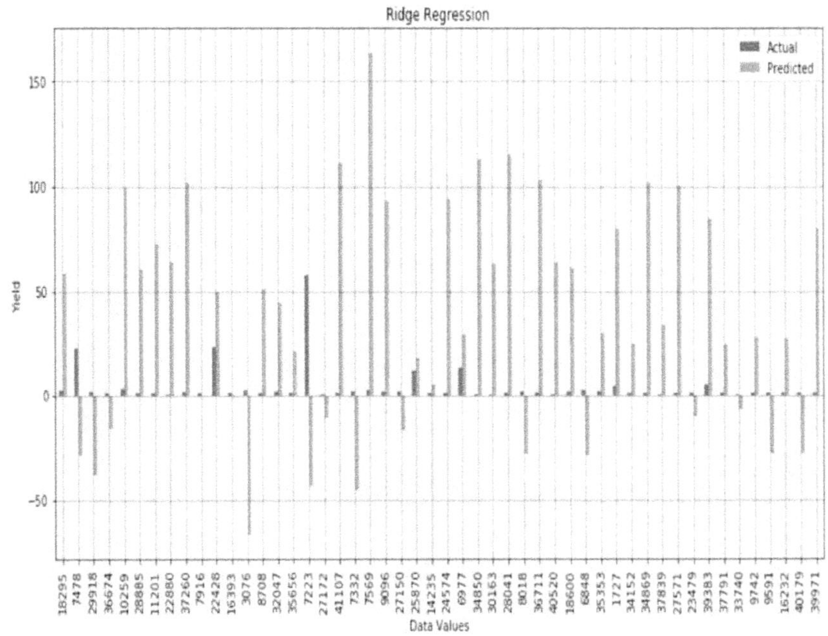


*Figure 4. Random forest graph*

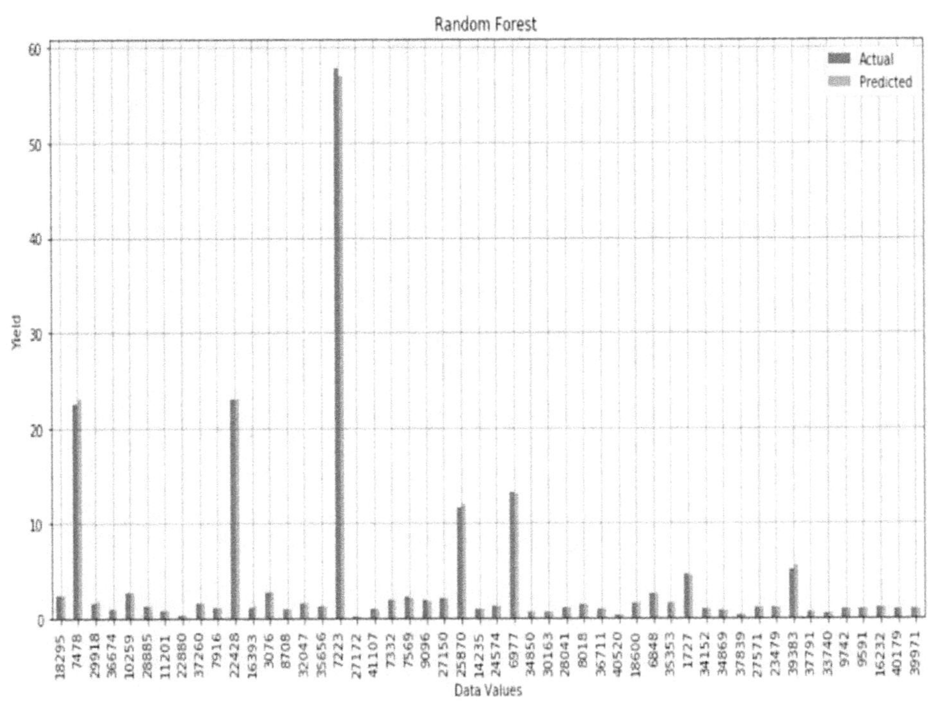

*Figure 5. Decision tree graph*

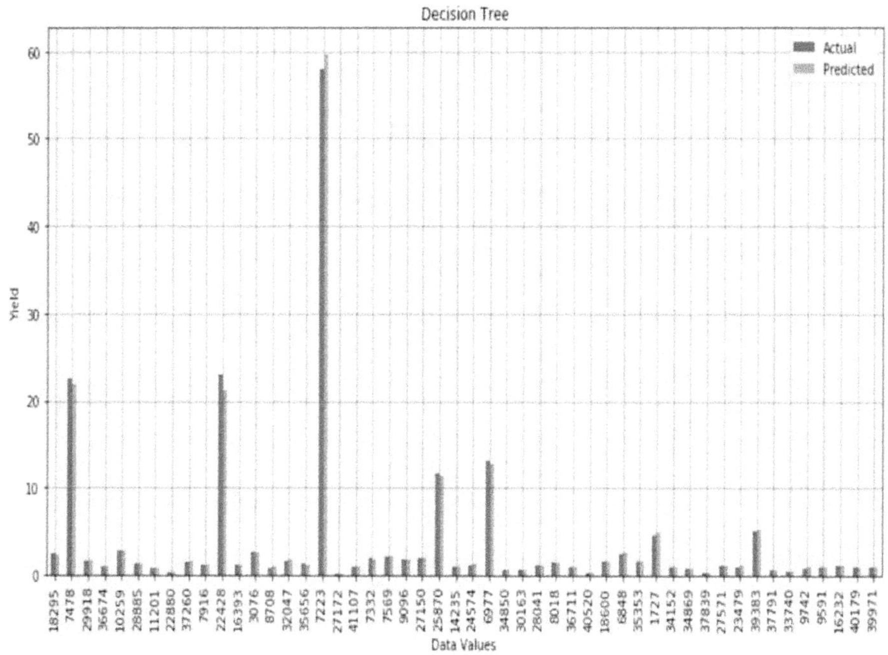


*Figure 6. Linear regression graph*

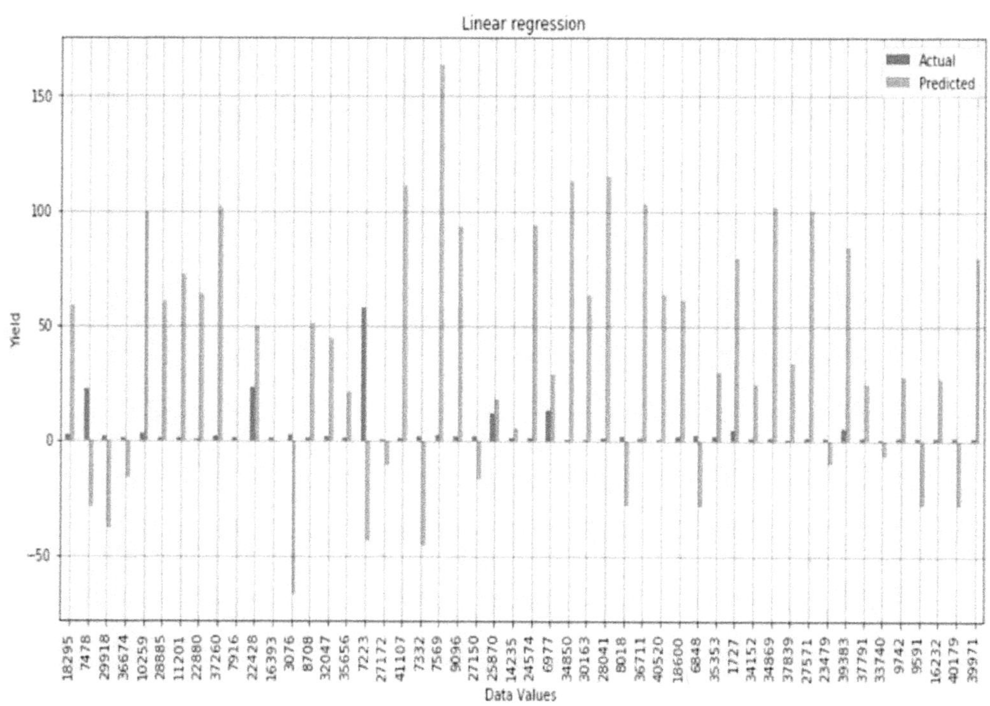

*Figure 7. Polynomial regression graph*

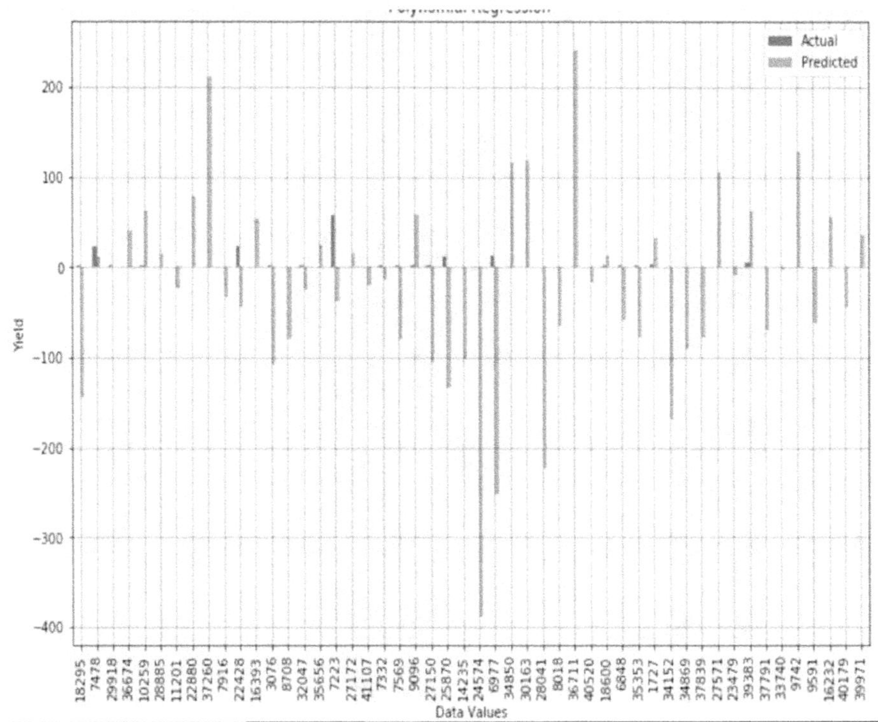


*Table 1. Different techniques accuracy for target variable as "yield"*

| | Test MAE | Test MSE | Test RMSE | Test R-Square | CrossValidation |
|---|---|---|---|---|---|
| **Gradient Boosting Regressor** | 23.744808 | 4.676468e+05 | 683.847059 | 0.796290 | 0.879121 |
| **Ridge Regression** | 83.371910 | 2.289880e+06 | 1513.235102 | 0.002513 | 0.001220 |
| **Random Forest Regressor** | 19.437881 | 7.771160e+05 | 881.541815 | 0.661483 | 0.767536 |
| **Decision Tree Regressor** | 22.049008 | 1.137754e+06 | 1066.655599 | 0.504387 | 0.730118 |
| **Linear Regression** | 83.371914 | 2.289880e+06 | 1513.235103 | 0.002513 | 0.001220 |
| **Polynomial Regression** | 146.157727 | 2.236869e+06 | 1495.616746 | 0.025605 | 0.001220 |

*Figure 8. Comparison graph for cross validation accuracy for yield*

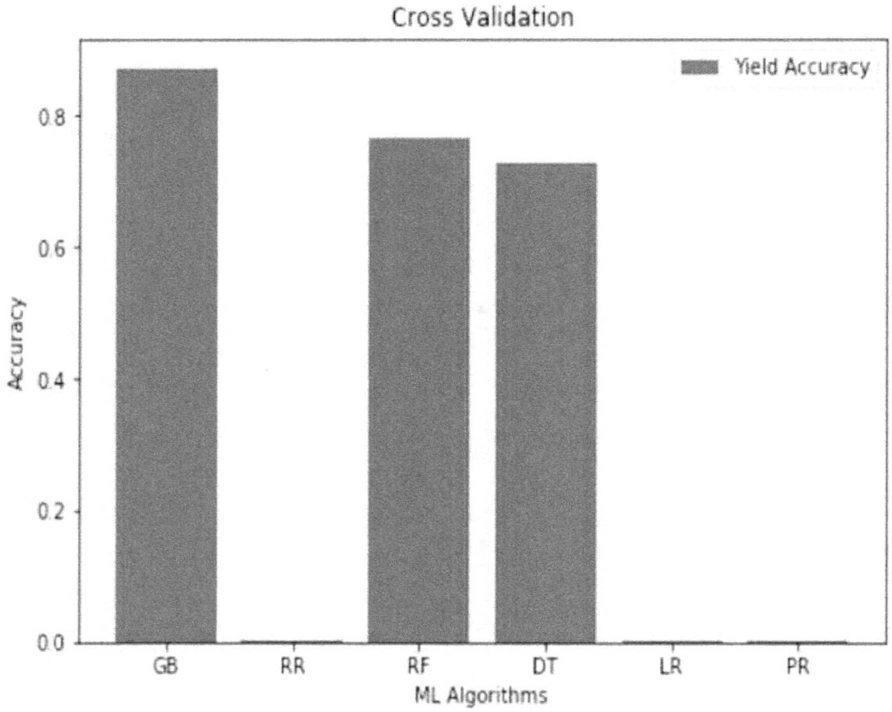


*Figure 9. Gradient boosting graph*

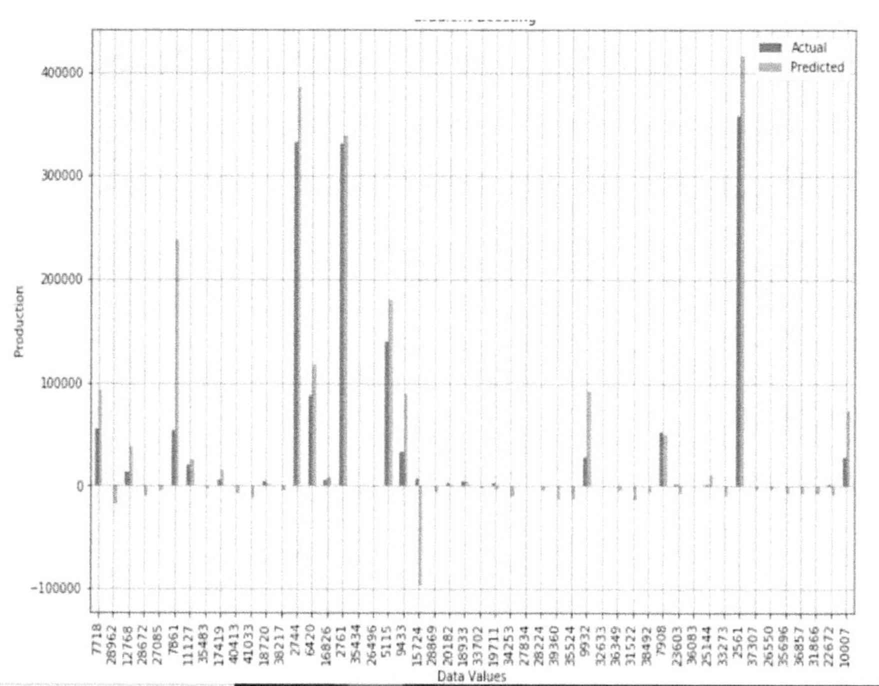

In another case, we are taking the target variable as "Production", then the Random Forest Regressor is providing more accuracy with cross-validation runs as 98.9% as shown in Table 2. A graph of the highest accuracy with their production actual and predicted values with their data values are shown below. Figure 15 shows the comparison for cross-validation accuracy when the target variable is "production".

*Figure 10. Ridge regression graph*

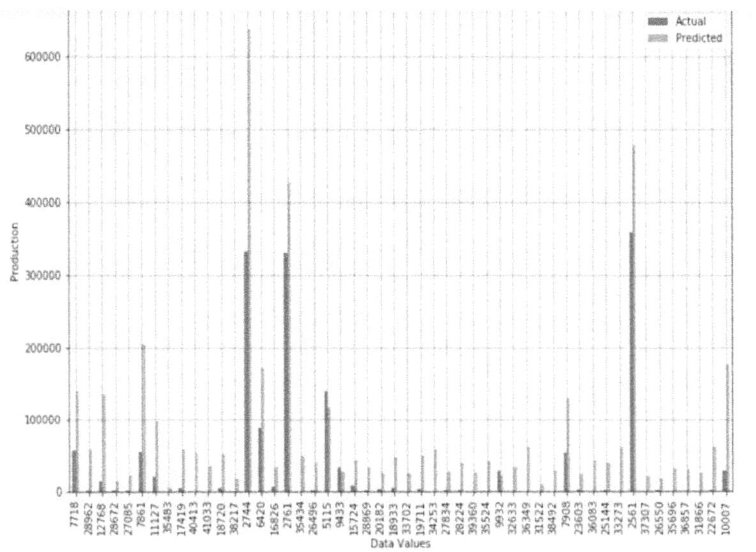


*Figure 11. Random forest graph*

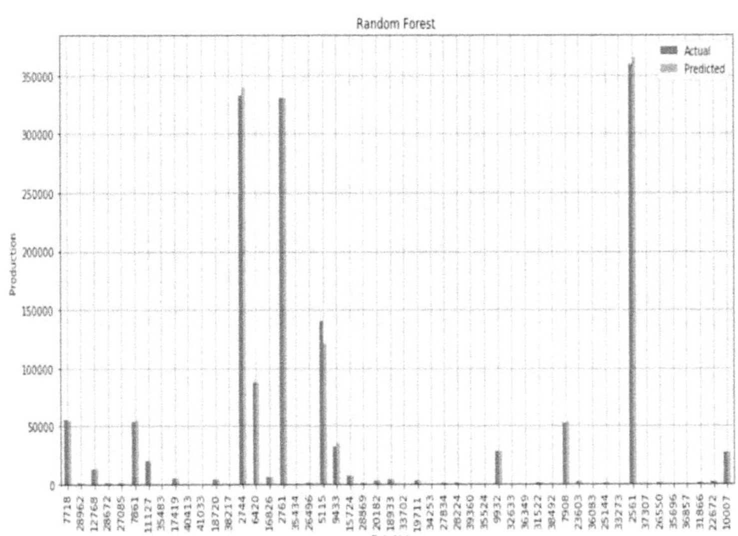

*Figure 12. Decision tree graph*

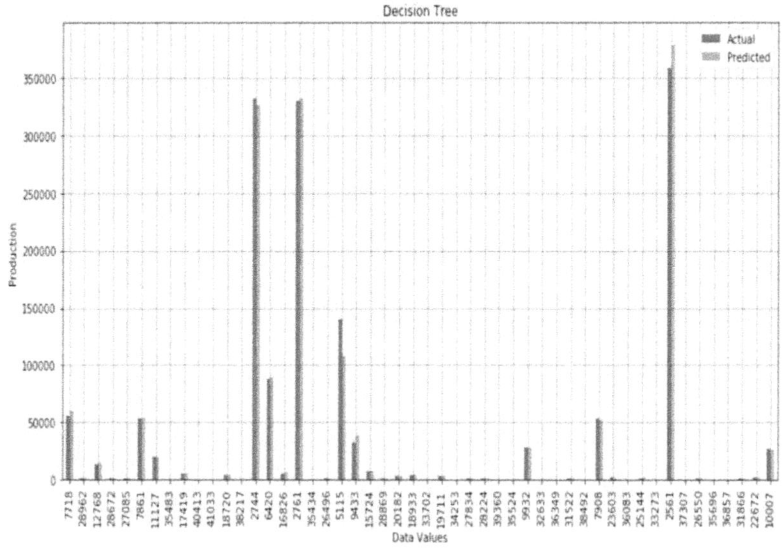


*Figure 13. Linear regression graph*

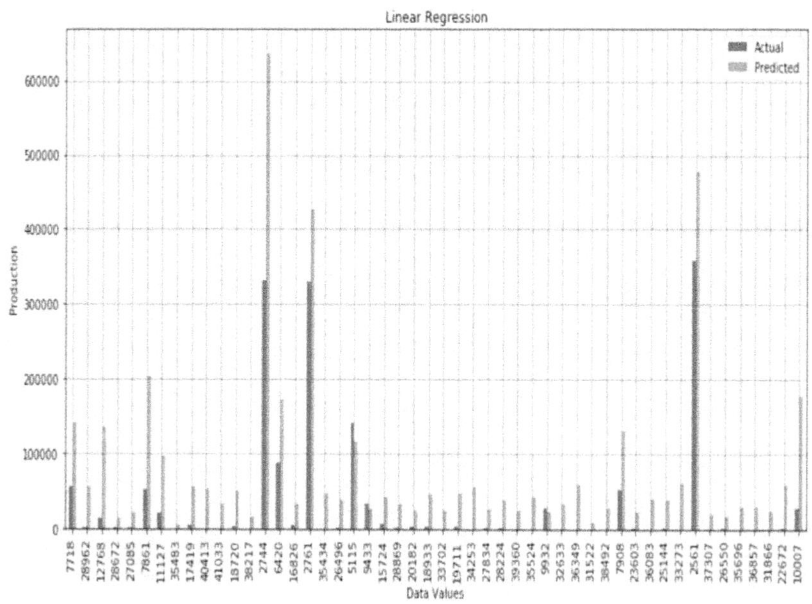

*Figure 14. Polynomial regression graph*

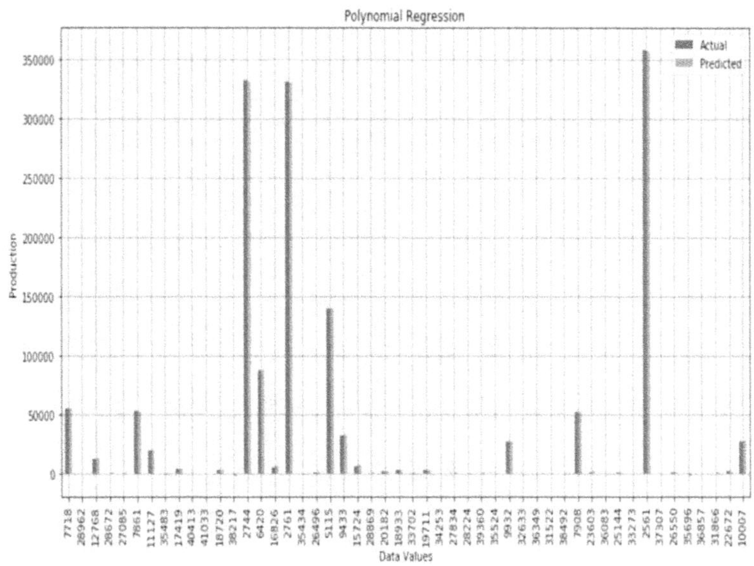


*Figure 15. Comparison graph for cross validation accuracy for production*

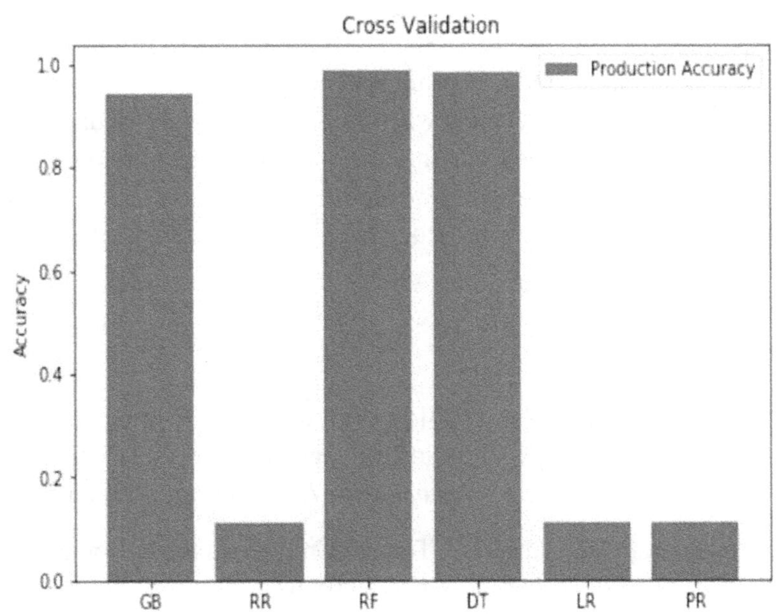

*Table 2. Different techniques accuracy for target variable as "production"*

| | Unnamed: 0 | Test MAE | Test MSE | Test RMSE | Test R-Square | CrossValidation |
|---|---|---|---|---|---|---|
| 0 | Gradient Boosting Regressor | 33085.368684 | 1.378599e+10 | 117413.755279 | 0.937892 | 0.940246 |
| 1 | Ridge Regression | 110728.210855 | 1.961227e+11 | 442857.411142 | 0.116443 | 0.109717 |
| 2 | Random Forest Regressor | 3560.027445 | 1.076494e+09 | 32809.969701 | 0.995150 | 0.989678 |
| 3 | Decision Tree Regressor | 5857.325847 | 1.929709e+09 | 43928.455944 | 0.991306 | 0.983545 |
| 4 | Linear Regression | 110728.211444 | 1.961227e+11 | 442857.411267 | 0.116443 | 0.109717 |
| 5 | Polynomial Regression | 1182.259271 | 9.462253e+08 | 30760.775980 | 0.995737 | 0.109717 |

## CONCLUSION

Several machine learning methods have been applied to agrarian data in this article in order to determine the best-performing procedure. Six distinct administered learning calculations were used. The proposed dataset includes a diversity of boundaries that are valuable for determining the harvest status and preparing gathered datasets. Additionally, this chapter illustrates the relationship between each of the six procedures. The consequences of these methods have been considered in light of various errors, and cross approval needs to be completed to ensure precision. When the target variable is "Yield" then the Gradient Boosting Regressor is more accurate with cross approval runs around 87.9%; however, when the target variable is "Creation" then the Random Forest Regressor is more precise with cross approval runs of 98.9%. This tool will assist ranchers in reducing the issues they face and will act as an agent, providing ranchers with the data they require acquiring high and increasing their profits.

## REFERENCES

Ankalaki, S., Chandra, N. & Majumdar, J. (2016). Applying Data Mining Approach and Regression Model to Forecast Annual Yield of Major Crops in Different District of Karnataka. *International Journal of Advanced Research in Computer and Communication Engineering, 5*(2), 25-29.

Chowdary, U. V. & Venkataramana, K. (2018). Tomato Crop Yield Prediction using ID3. *International Journal of Innovative Research in Technology, 4*(10), 663-62.

Everingham, Y., Sexton, J., Skocaj, D. & Bamber, G. I. (2016). Accurate prediction of sugarcane yield using a random forest algorithm. *Agronomy for Sustainable Development, 36*(2). doi:10.1007/s13593-016-0364-z

Fathima, G.N. & Geetha, R. (2014). Agriculture Crop Pattern Using Data Mining Techniques. *International Journal of Advanced Research in Computer Science and Engineering, 4*(5), 781-786.

Gandhi, N., Armstrong, L. J., Petkar, O., & Tripathy, A. K. (2016). Rice crop yield prediction in India using support vector machines. *13th International Joint Conference on Computer Science and Software Engineering (JCSSE)*, 1-5. 10.1109/JCSSE.2016.7748856

Katiyar, S., Khan, R., & Kumar, S. (2021). Artificial Bee Colony Algorithm for Fresh Food Distribution without Quality Loss by Delivery Route Optimization. *Journal of Food Quality*, *2021*, 1–9. Advance online publication. doi:10.1155/2021/4881289

Kaur, M., Gulati, H. & Kundra, H. (2014). Data Mining in Agriculture on Crop Price Prediction: Techniques and Applications. *International Journal of Computer Applications, 99*(12), 1-3.

Khan, R., Kumar, S., Dhingra, N., & Bhati, N. (2021). The Use of Different Image Recognition Techniques in Food Safety: A Study. *Journal of Food Quality*, *2021*, 1–10. Advance online publication. doi:10.1155/2021/7223164

Khan, R., Tyagi, N., & Chauhan, N. (2021). Safety of food and food warehouse using VIBHISHAN. *Journal of Food Quality*, *2021*, 1–12. Advance online publication. doi:10.1155/2021/1328332

KiranMai, C., Krishna, M., & Reddy, A. V. (2006). Data Mining o f Geospatial Database for Agriculture Related Application. *Proceedings of Map India*, 83-96.

Kumar, A. & Kumar, A. (2015). Prediction of production of crops using K-Means and Fuzzy Logic. *International Journal of Computer Science and Mobile Computing, 4*(8), 44-56.

Kumar, A., Kumar, N. & Vats, V. (2018). Efficient crop yield prediction using machine learning algorithms. *International Research Journal of Engineering and Technology, 5*(6), 3151-3159.

Kumar, N. (2018). *Regularized Active Set Least Squares Algorithm for Nonnegative Matrix Factorization in Near Infrared Spectral Data* [PhD thesis]. http://hdl.handle.net/10603/233051

Kushwaha, A. K. & Bhattacharya, S. (2015). Crop yield prediction using Agro Algorithm in Hadoop. *International Journal of Computer Science and Information Technology & Security, 5*(2), 271-274.

Paul, M., Vishwakarma, S. K., & Verma, A. (2015). Analysis of Soil Behaviour and Prediction of Crop Yield using Data Mining approach. *International Conference on Computational Intelligence and Communication Networks*. 10.1109/CICN.2015.156

Priya, P., Muthaiah, U. & Balamurugan, M. (2018). Predicting yield of the crop using machine learning Algorithm. *International Journal of Engineering Sciences and Research Technology, 7*(4), 2277-2284.

Ramesh, R., & Vardhan, B. (2015). Analysis of crop yield prediction using data mining techniques. *International Journal of Research in Engineering and Technology, 4*(1), 47-473.

Rub, G., Kruse, R., Schneider, M., & Wagner, P. (2008). Estimation of neural network parameters for wheat yield prediction. In M. Bramer (Ed.), *Artificial Intelligence in Theory and Practice II, IFIP International Federation for Information Processing. Springer.*

Sahni, V., Srivastava, S., & Khan, R. (2021). Modelling techniques to improve the quality of food using artificial intelligence, Journal of Food Quality. *Journal of Food Quality*, *2021*, 1–10. Advance online publication. doi:10.1155/2021/2140010

Shastry, A., Sanjay, H.A., & Bhanusree, E. (2016). Prediction of Crop Yield using Regression Analysis. *Indian Journal of Science and Technology, 9*(38), 1-5.

Singh, V., Sarwar, A. & Sharma, V. (2017). Analysis of soil and prediction of crop yield (Rice) using Machine Learning approach. *International Journal of Advanced Research in Computer Science, 8*(5), 1254-1259.

Sujatha, R., & Isakki, P. (2016). A study on crop yield forecasting using classification techniques. *International Conference on Computing Technologies and Intelligent Data Engineering (ICCTIDE'16),* 1-4.

Veenadhari, S., Misra, B., & Singh, S. (2014). Machine learning approach for forecasting crop yield based on climatic parameters. *International Conference on Computer Communication and Informatics*, 1-5. 10.1109/ICCCI.2014.6921718

Verheyen, K., & Adrianens, M. (2001). High resolution continuous soil classification using morphological soil profile descriptions. *Geoderma*, *101*(3-4), 31–48. doi:10.1016/S0016-7061(00)00088-4

Wu, J., Olesnikova, A., Song, C. H., & Lee, W. D. (2009). The Development and Application of Decision Tree for Agriculture Data. *IITSI*, 16-20. https://www.hilarispublisher.com/open-access/agriculture-role-on-indian-economy-2151-6219-1000176.pdf

# Chapter 8
# An Extensive Analysis of Flying Ad-Hoc Network Applications and Routing Protocols in Agriculture

**Ahona Ghosh**
https://orcid.org/0000-0003-0498-285X
*Maulana Abul Kalam Azad University of Technology, West Bengal, India*

**Caprio Mistry**
*Brainware University, India*

**Mousumi Biswas**
*Brainware University, India*

## ABSTRACT

*Agriculture and irrigation are sources of man's potential. For the sake of cost-effectiveness and the betterment of agricultural professionals, UAVs (unmanned aerial vehicles) can be deployed for surveillance, utilization of pesticides and insecticides, and detection of bioprocessing errors. With the proper collaboration and coordination of the clusters of UAVs forming a network, linked with a ground infrastructure or satellite will exceed the competencies of a single UAV system. However, one of the vital design issues FANETs deal with is in selecting the accurate routing protocol which is a prerequisite for the creation of FANET. In this chapter, the authors discuss the routing protocols of FANET in different platforms and different strategic manners. The open research challenges have been discussed and possible solutions have been attempted to be drawn from the conclusion. The main contribution lies in suggesting the most suitable routing protocol for each particular agriculture application based on the mobility model and requirement.*

DOI: 10.4018/978-1-6684-5141-0.ch008

## INTRODUCTION

With the speedy technological enhancement and decrement in manpower, technologies are being involved in every phase of our life, and their proper application leads to a healthy life indeed. Different sensors and electronics equipment are deployed to commercialize and maintain cost-effectiveness in several domains. Precision farming is a revolutionary approach used by farmers to maximize inputs like water and fertilizers, improving efficiency, quality, and yields. This definition also includes the minimization of pests and diseases by targeting spatially. Pesticide-level precise agriculture enables the farmer to deal correctly with each part of his field, enabling him to tackle crop diversification. Farmers being more suitable for planting, harvesting, and fertilization contributes to higher farm production and profitability while upholding environmental standards. Agriculture is an industry involving a vast volume of data and gets influenced by disease of plant and pest appearance. These data are also linked to quantitative environmental characteristics like ambient temperature, soil temperature, chemical composition, precipitation, solar radiation, and crop quality. A list like this one reporting, coupled with appropriate processing techniques, would bring new opportunities for rural advancement. The information and communication technology (I.C.T.) and its development will promote more data-oriented agriculture. Different advances such as the Internet of Things (IoT) and cloud computing are mixed in sequence (Mukherjee et al., 2020).

Intelligent data analysis and successful dissemination of knowledge to agricultural stakeholders and smart farming (Zainal et al., 2019) are required to introduce productive data collection. In turn, geospatial tools such as the Global Navigation and Global Positioning System in precision agriculture, even the satellite network (GNSS), will play an important role in indicating any component's exact position. Wireless sensor network (W.S.N.) (Yick et al., 2008) (Ghosh et al., 2020) and AdHoc Network (Shobana et al., 2013) (Godse and Mahalle, 2018), especially FANET (Flying AdHoc Network) (Bujari et al., 2017), a new form of Mobile Ad Hoc Network (MANET) (Khamayseh et al., 2019), is one of the efficient platforms which we can apply for betterment of agriculture. The increase in the production of UAVs is also a promising approach for collecting real-time details, as it is considered one of the most ambitious precision-farming technologies (Cauvery, 2020) (Alamsyah et al., 2019). UAVs can enforce a series of steps like data collection, data preparation, data interpretation, and data management. Single-mode and multi-mode UAV systems are both suitable for the application of FANET. Their willingness to travel makes them much more desirable since they can get useful knowledge and ground evaluation in a shorter period. The biggest downside is that the UAV is a complex model which requires the synchronization of multiple scientific logics with technologies for successful functionality.

The next section describes some related works already existing in the concerned area of research. Section 3 discusses different routing protocols that have been applied in the state-of-the-art agricultural applications of FANET. The application areas of agriculture have been demonstrated in Section 4. Finally, section 5 identifies areas suffering from lack of focus and finds out possible future works to address the existing challenges.

## RELATED WORKS

UAV implementation in agriculture results in six specific applications: crop scouting, crop survey, mapping, crop protection, crop planning and management, chemicals, geofencing, etc. Agriculture itself is one of the highly demanded domains due to the huge population density in our country. Among three different routing protocol categories of FANET (Sahingoz, 2014), Destination Sequenced Distance Vector (DSDV) (He, 2002), Optimized Link State Routing (OLSR) (Singh and Verma, 2014) come under the proactive category where all the network nodes share routing information with all the other nodes in a regular time interval which helps to create the path to the ultimate destination in turn. On the other hand, AdHoc On-Demand Distance Vector (AODV) (Singh and Verma., 2015), Associativity Based Routing (A.B.R.) (Toh, 2997), and Dynamic Source Routing (D.S.R.) (Khan, 2017) come under reactive protocol since they create path only when it is required and are the most popular and effective ones when applied to agriculture as they consume minimum bandwidth in the system. However, as the nodes are very frequently movable, resulting in frequently changing network topology, efficient routing is the most challenging part of FANET (Kashyap and Agarwal, 2018) and other challenges like Quality of Service (QoS), Security, Reliability, Network scalability, etc. A comprehensive study of the existing routing protocols has been presented in (Lakew et al., 2020), where it is clearly shown that the existing ones obtain a promising result. However, still, there is a scope for further research on more reliable and robust routing protocols which can harness the potential of UAV in different application areas. Self-organization of UAVs has been implemented in (Khare et al., 2018), being influenced by a swarm of birds to deal with the distributed and dynamic nature of FANET.

Packet forwarding technique, routing, mobility, and traffic density are the considered parameters to evaluate the performance of FANET, and simulation study on network delay and load and throughput has evaluated various routing protocols mentioned in (Khan et al., 2018). UAVs can be distinguished into two major divisions (i) fixed-wing and (ii) multirotor. A specified wing has better aerodynamics lower energy consumption leading to longer flights and faster speed but requires considerable space to take off and land. On the other hand, a multi-rotor can accommodate heavier payloads; piloting is smoother, as both the take-off and landing are carried out vertically. UAVs consist of six main sub-modules working together to create a beneficial platform.

i) The unmanned aerial vehicles airframe acts as the body of the UAV, which must be light enough to sustain low energy consumption and sufficiently strong to support the increased UAV payload for accepting fatal incidents and crashes when avionics have small space and no storage for the pilot.
ii) UAV's heart is the flight machine. It collects aerodynamic knowledge using a set of sensors (G.P.S., accelerometers, gyroscope, magnetometers, etc.) to direct the UAV ride according to the ride plan from the airframe-mounted control board (Kakamoukas et al., 2020).
iii) The payload consists of a collection of actuators and sensors responsible for processing the data after acquiring or transmitting it to the base station for further analysis.
iv) The payload/mission controller controls the sensors that are a part of the payload get controlled by the payload/mission controller.
v) The base station on the ground manages and controls the payload and UAV.
vi) The communication platform consisting of different channels, including microwave, radio, etc., helps establish the communication pathway between the base station and the UAVs in the network. Therefore, UAVs can be effective where the application of traditional aids is quite challenging, like sugarcane, pigeon pea, wet paddy field, etc. (Sinha et al., 2016).

Navigation-based UAVs weighing less than 14 Kg have been used to process high temporal and spatial resolution of agricultural field-based image data (Xiang and Tian, 2011). The experimental result showed only a 1.5% difference between the image assessed herbicide affected region and the herbicide affected region assessed by ground survey. A helicopter introduced in (Xiang 2006) could fly over previously set waypoints under the AFCS control, and the accuracy of the hovering reached $\pm 1$m. The controller could trigger the A.D.C. image sensor in terms of G.P.S. locations; the attitude and position of capturing images were recorded simultaneously for post-processing. FANET architectures should handle network fragmentation (Oubbati et al., 2019), and high mobility low density is the main challenges FANET in 3D spaces having different altitudes. Although existing works have focused on any of these issues, none has considered all of these combinedly as there is no physical guidance towards various flight schematics and control in a drone, different factors like the degree of autonomy, size and weight of the drone, difference in energy source (conventional aircraft or battery-driven or solar fuel-based energy) matter during designing of the same. Flight range, payload capacity, and cost are the other essential parameters since UAVs monitor the field protect plants, and apply pesticides according to the requirement (Marinello et al., 2016) (Salami et al., 2014).

## PROTOCOLS OF ROUTING USED IN AGRICULTURE

Implementation of routing protocols for FANET is a challenging task as it has to deal with three-dimensional movements due to the dynamic nature of FANET (Wen and Huang, 2018). Therefore, the routing techniques follow any combination of any of the following techniques, and the outline has been shown in Figure 1.

i. Store carry and forwarding method: Network with a high fragmentation consists of some nodes which sometimes fail to find out relay node for transmission and start carrying the packet until an appropriate relay node is found, which results in network delay in forwarding the required packet due to undesired movement of the node.
ii. Greedy method: The number of hops is the main criteria for selecting the destination path in this technique. High-dense networks apply this technique to use the closest node to relay the packets efficiently within a specified time.
iii. Single path method: In this technique, between two communicating nodes, there exists only one path where the maintenance of routing table for each node is easier than the other methods, but it results in packet loss when the only possible path exhibits some error.
iv. Multipath method: This method overcomes the loophole of the single path method by introducing multiple paths; in case of failure of one path, alternatives start functioning to avoid packet loss in the network, but routine table maintenance is a little bit complex in this as it has to manage the tables for all the possible paths existing.
v. Path exploration method: When the source network node does not contain any information about the path towards the destination node, the path needs to be explored using Route Request (RREQ), where either the source nodes get broadcasted to its neighbors, or it introduces flooding technique where each node upon receiving RREQ, copies it and transmits to its neighbors. Upon receiving the RREQ, the destination node unicasts the route reply to the source node. In this way, the route is discovered and applied in routing onwards.
vi. Forecasting-based method: The node's speed, direction, and geographical position are the three main criteria in this method to select the next relay node to a particular network node, reducing the network delay and packet delivery ratio.

*Figure 1. Mostly used routing architectures of FANET*

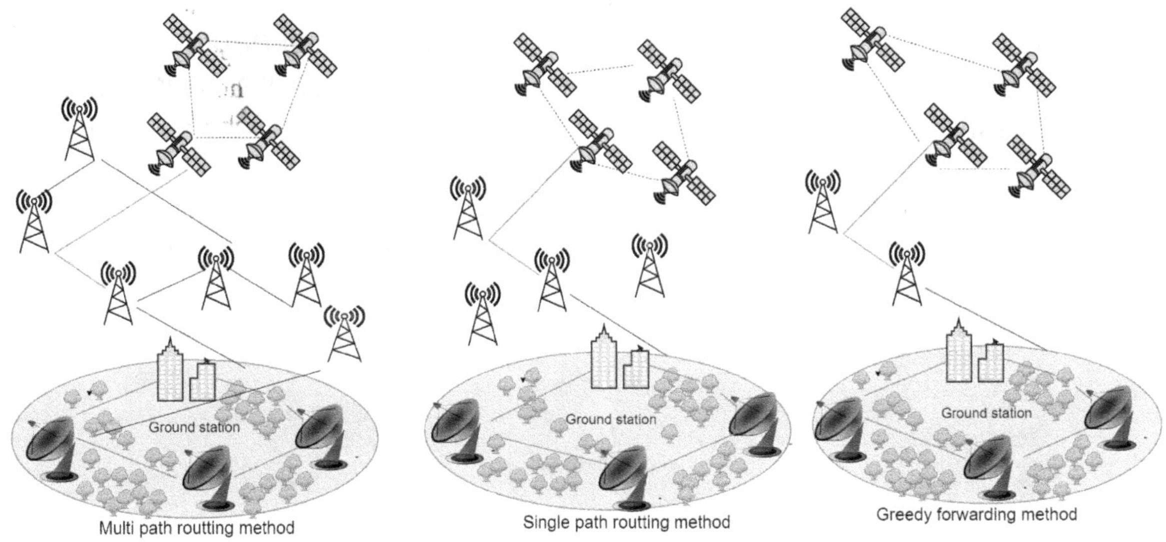


Two categories of network architecture have been used in agricultural applications of FANET discussed in the existing literature.

i) **Communication within air**

Infrastructure less architecture needs this type of protocol where packet forwarding is required by UAV to an unreachable node. However, in a scenario where ground setup is impossible, its wide application can be observed.

ii) **Communication between air and ground**

When every UAV in a network cannot communicate with an existing ground station due to their restriction in transmission range, relay nodes are applied to establish the connection within their transmission range.

The routing protocols are designed based on two parameters, the first one is its position where the location of each node is tracked by G.P.S., which does not show any additional advantage in highly mobile networks like FANET and the second one is the network topology which is extensively applied in the existing literature for remote sensing of an irrigation system using FANET. The two categories of routing are described as follows

## Routing Based on Position

AODV, D.S.R., TORA, M-OLSR, NC-OLSR, D-OLSR, CE-OLSR, DSDV, OLSR have been widely adopted in FANET's application in agriculture. The coordination mechanism for different nodes within FANET is shown in Figure 2.

*Figure 2. Coordination mechanism between FANET nodes*

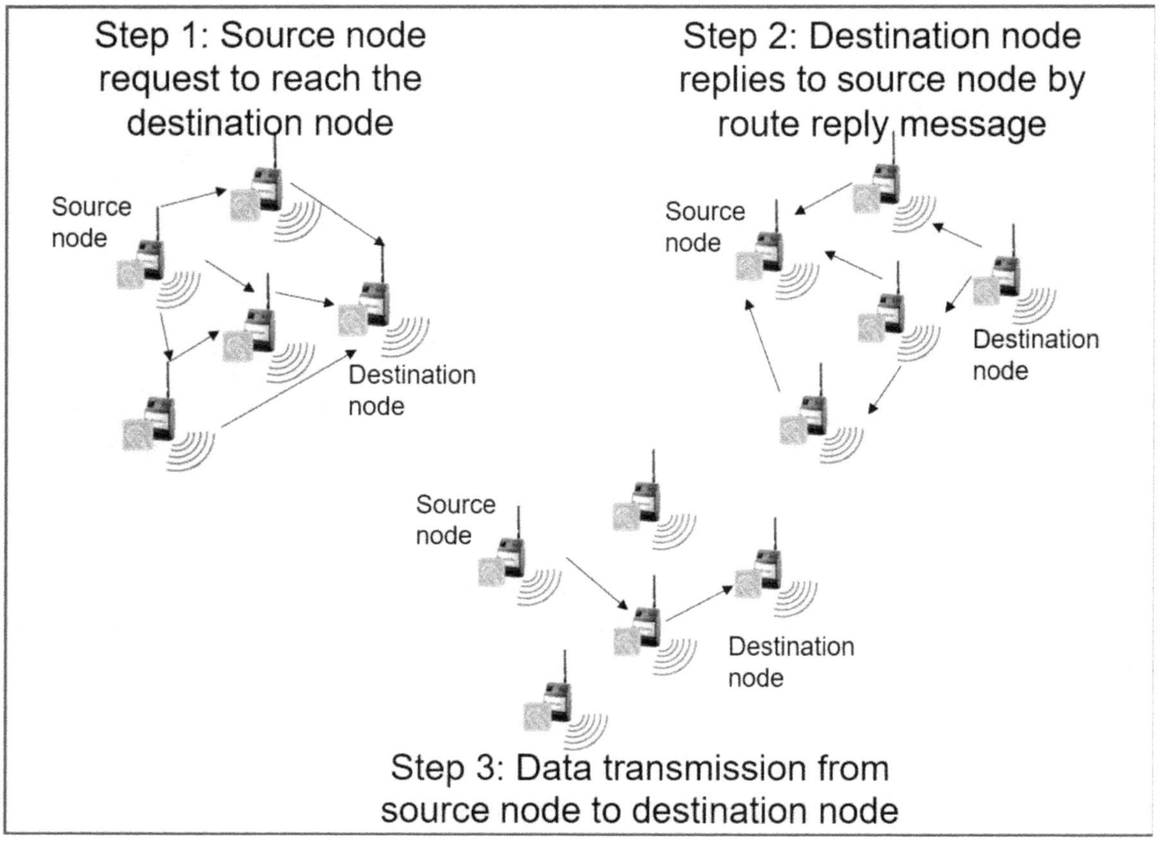


## Robust and Reliable Predictive Routing (RARP)

Growing compact and flexible communications requirements have led to rapid growth in networking between unmanned aerial vehicles (UAVs), also referred to as flying ad-hoc networks (FANETs). However, established handheld ad-hoc routing protocols are not appropriate for FANETs due to high-speed mobility, environmental conditions, and land structures—a new stable and efficient predictive routing system with lateral and dynamic angle change transmission for FANETs. In addition, several new aspects have been clarified in detail, such as estimation of the planned link time, route selection usefulness, directional transmission with a new upgrade method, dynamic angle change with the use of an adjustable antenna, alternate path configuration, and local path repair.

## Jamming Resilient Multipath Routing

Jamming attacks are extremely dangerous to reliability. Wireless correspondence, as can be easily interrupted Contact. Communication. Current jamming defenses rely mainly on restoring communication between neighboring nodes. As long as all routers do not collapse simultaneously, the availability of end-to-end paths is retained. Wireless networks interact through common media and are thus susceptible to interference with jamming and radio. Much research has been done to repair the jamming locally, i.e., restore communication between neighboring nodes. These measures include common physical-layer methods based on specialized transceivers (Proakis, 2000) (for example, frequency hopping) and MAC-layer mechanisms (Xu et al., 2017) (Ma et al., 2005) modifying error coding, channel adaption, or physical location. These measures are anti-jamming. Though we have to protect those strategies against jamming, different perspectives to emphasize defense Network level jamming, i.e., restoring end-to-end data transmission stability, can be considered.

## Ground Control System Based Routing

As per Ad Hoc network communication, the inherent shortcomings of MANET protocol that were not suitable for FANET were unavoidably inherited. Therefore, based on G.C.S. advantages and suggested G.C.S. routing specially designed for FANET to eliminate those limitations.

## Adaptive Beacon Scheme for Geographic Routing

The mobility prediction consists of a fuzzy controller for adjusting the beacon advertisement frequency. The error degree in prediction works as the input and the interval rate as the output. Its feats the time and location history of nearby FANETs applying a weighted linear regression model (Albu Salih, 2021). To reach the target, the nearest or the next to nearest UAV is chosen as the next hop destination in each step

## Routing Based on Topology

Topology-aware conventional routing protocols may be categorized into three types, proactive, reactive, and hybrid protocol. Continuously moving FANET systems require dynamic topology to avoid interruptions and delays in network communication (Sadiq and Salawudeen, 2020). Most MANET protocols are applied in FANET architecture by integrating G.P.S. positioning functionalities and dealing with complex situations. Sometimes, more than one routing protocol integration occurs as per the user requirement. The following discussion is based on recently applied routing protocols in FANET architecture (B.K. et al., 2020).

## Distributed Priority Tree-Based Routing Protocol

The issues faced by aero-ground ad-hoc networks for network segmentation have been considered in (Sharma et al., 2018) to introduce a protocol modifying the basic r-b structure suitable for communication and routing two self-organizing nodes within a coordinate system. At first, it identifies the ground nodes, then air nodes get detected by it. Finally, the interface between air and ground gets identified using a neural network algorithm that generates a routing table in a tree containing air and ground nodes.

## Continuous Hopfield Neural Network Optimization

Probabilistic models have efficiently handled and modeled different natural phenomena as close as possible to reality. For example, while most of the other routing protocols work based on the hops count (Aswini and Uma, 2019), the continuous Hopfield/Feedback neural network finds the shortest route visiting all the cities at least once based on the traveling salesman algorithm in the network. However, such a model's applicability has not been tested yet for more nodes. Furthermore, the utility function can also be modified based on characterizing the level of the link's satisfaction in FANET (Imran et al., 2021).

## Topology Construction Method

The topology construction method of routing has also proved its efficiency in reducing energy consumption for W.S.N. in the existing literature. Yu et al. projected a novel topology construction method based on cluster tree using particle swarm optimization method (Yu et al., 2019) abbreviated as PSO. An evaluation function reflecting network energy consumption is proposed to convert the network topology construction problem into an energy consumption optimization problem. The network topology is converted into an individual particle population for the PSO method. Two network topology rebuilding approaches with the PSO algorithm based on fixed and variable energy thresholds are also proposed to extend the network lifetime as much as possible.

## Topology Change Aware Routing Choosing Scheme

Mobility measuring metric, namely topology change degree, helps describe the topology changes in highly dynamic FANETs that optimize average jitter, mean end-to-end delay, network throughput, and packet delivery ratio. However, in the future, some research challenges are still to be addressed about the effective setting of the topology change perception interval, accurate calculation of the topology change threshold, and configuration of the weight factors among the influencing parameters, dealing with the issue of changing number of nodes.

## Routing Based on Metaheuristic Optimization

Different swarm intelligence-based routing protocols have been applied in existing precision agricultural frameworks (Maes and Steppe, 2019) (Primicero et al., 2012) (Sharma and Mishra, 2020). Honey bee, Particle swarm optimization, Ant colony optimization-based AODV are some of them which self-organize and efficiently divide labor to optimize the network life span, adaptability, fault tolerance, scalability, energy and latency, and the network. However, because of partial competences, the sensor nodes of existing W.S.N. architectures established for agricultural applications have suffered from energy limitations and complexity of the routing methods. The sensor-based agriculture fields, data transmission failure and delay are common. Because of these constraints, sensor nodes around the base station are always reliant on it, putting additional strain on the base station or rendering it worthless. Qureshi et al. presented a Gateway Clustering Energy-Efficient Centroid- (GCEEC-) based routing protocol to overcome these concerns, in which the cluster head is chosen from the centroid position and gateway nodes are chosen from each cluster. The data burden from cluster head nodes is reduced by the gateway node, which sends the data to the base station. The suggested protocol was compared to state-of-the-art protocols using simulation. The experimental results showed that the proposed approach performed better and made WSN-based temperature, humidity, and lighting monitoring in the agriculture sector more viable.

## Particle Swarm Optimization

In computational science, particle swarm optimization (PSO) is a computational method that optimizes a problem by iteratively improving a candidate solution concerning a given measure of quality. It solves a problem by having a population of candidate solutions, here dubbed particles, and moving them around in the search-space according to simple mathematical formula over the particle's position and velocity. Of course, each particle's movement is influenced by its local best-known position. Still, it is also guided toward the best-known positions in the search space, updated as other particles find better positions. This is expected to move the swarm toward the best solutions. Figure. 3 shows the working mechanism of PSO.

Agricultural robotics has become increasingly popular among agricultural researchers as an alternative to human workers in the future. However, the operational cost of agricultural mobile robots must be competitive with hiring human workers. Furthermore, it is difficult to determine an optimized sequential route with a minimal distance in agricultural mobile robot navigation. Mahmud et al. (Mahmud et al., 2018) employed binary particle swarm optimization (PSO) and a genetic algorithm (G.A.) to find the shortest routing path for spraying operations in a greenhouse. The agricultural robotics routing problem has been expressed in terms of the traveling salesman problem, commonly used in operational research. To solve the routing problem, a total path length objective was measured based on the path computed using a probabilistic roadmap path planner. The results indicated the performance of the G.A. was better for solution quality and computational time, while binary PSO performed better concerning convergence time.

*Figure 3. The working mechanism of Particle Swarm Optimization*

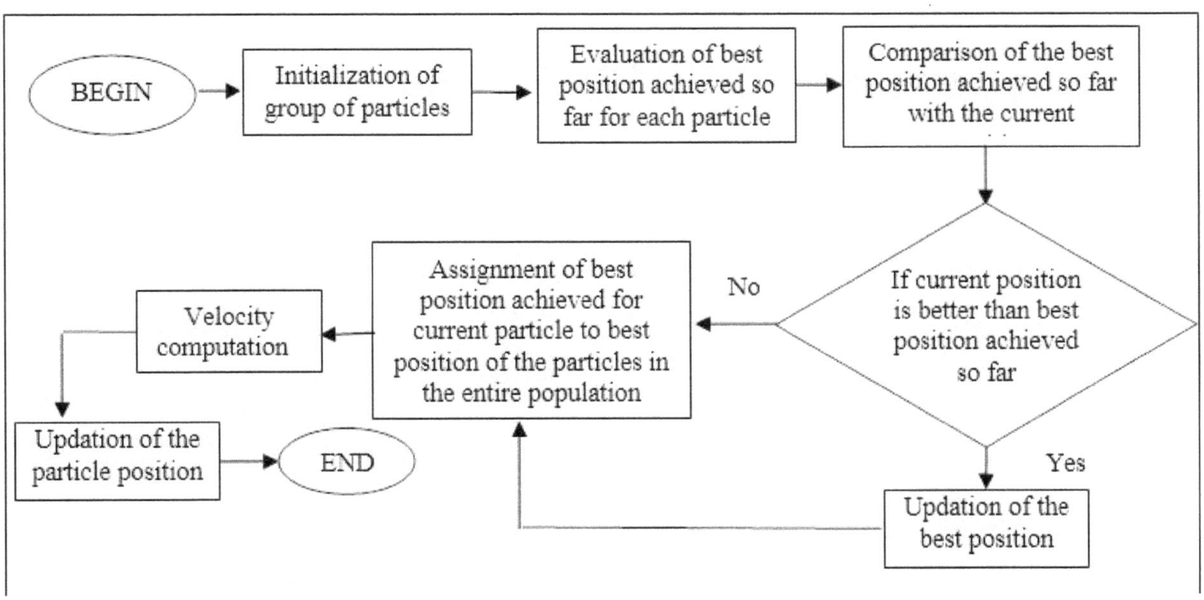


The purpose of the Farm as a Service system designed by Kin et al. (Kin et al., 2018) was to boost crop output and make merging technology uses more accessible to farm families who run smart farms. The

Internet of Things based network layer for wireless communication and the Farm as a Service middle layer for data collecting, processing, and analysis were used to create a strawberry disease prediction model.

## Ant Colony Optimization

A vital instance in the Swarm Intelligence area is the Ant Colony Optimization (A.C.O.) which is also a meta-heuristic approach and presents a common framework for approximating solutions to NP-hard optimization problems. A.C.O. has been successfully applied to balance the various routing related requirements in dynamic Mobile Adhoc Networks (Zhang et al., 2017). Agricultural wireless sensor networks (AWSNs) research is crucial for agricultural facility technology. Because the temperature and humidity nodes in AWSNs are so small, and are restricted in terms of computing, network management, data gathering, and storage. Under these circumstances, work allocation is critical for improving AWSN performance and lowering energy usage and computational restrictions. Task allocation optimization, on the other hand, is a nonlinearly restricted optimization problem where complexity rises due to the limited computer capabilities and power. The working mechanism of A.C.O. has been shown in Figure 4.

Li et al. present an elite immune ant colony optimization (EIACO) based job allocation strategy for maximising task execution efficiency (Li et al., 2020). The exact prototype of task assignment is given initially to analyse the efficiency of algorithms. The immune and elite selection mechanisms are then integrated into the A.C.O. process. The task allocation fitness function is intended to maximise task execution efficiency. The proposed immunity strategy and elite strategy increased EIACO's worldwide search ability, according to the findings. In the simulation, EIACO demonstrated a strong ability to identify ideal candidates. It's also capable of avoiding local optima. When the EIACO is incorporated into AWSNs, simulations show that it performs better.

*Figure 4. The working mechanism of Ant Colony Optimization*

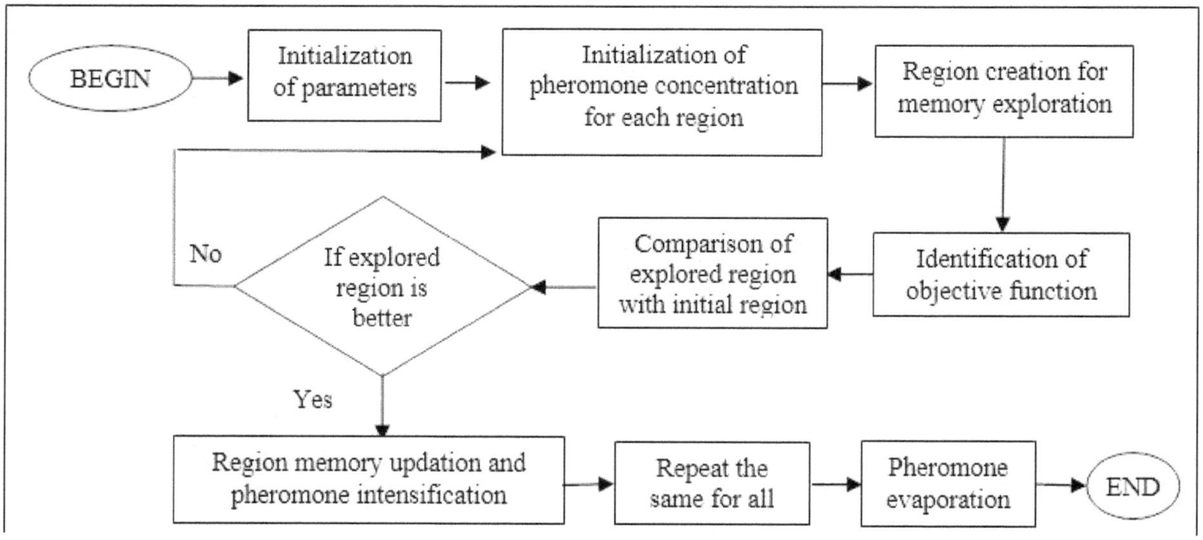

## Genetic Algorithm

Natural selection drives biological evolution, and the genetic algorithm provides a method for solving both confined and unconstrained optimization problems based on natural selection. A population of individual solutions is again and again modified using genetic algorithm. The genetic algorithm selects individuals from the present population to be parents at each phase and uses them to generate the following generation's children. The population "evolves" toward an optimal solution over generations. The genetic algorithm can be used to handle a variety of optimization problems which aren't well suited for traditional optimization techniques, such as issues with discontinuous, nondifferentiable, stochastic, or highly nonlinear objective functions. The evolutionary algorithm, for example, can be applied to have a solution of mixed integer programming issues in which certain components are just integer-valued. The general operating mechanism of the genetic algorithm is depicted in Figure 5. For creating the upcoming generation from present population, the genetic algorithm follows the following three rules.

- Selection rule to decide which individuals, known as parents, may be allowed for contributing to the population of the future generation. The selection is generally stochastic and can be influenced by the scores of the participants.
- Crossover rule to combine the children of two parents to create the following generation.
- To create children, mutation rules use random transformations to the parents separately.

After grain and other crops have been harvested, the bale collection problem (BCP) emerges. Its solution specifies the order in which bales should be collected across the area. Current location based schemes and auto-steering technologies for agricultural trucks and machines can provide precise data for reliable bale collection plans. Gracia et al. (Gracia et al., 2013) proposed a hybrid genetic strategy for addressing the BCP pursuing resource optimisation, like lowering un-productive time, fuel usage, or travel espace. The algorithmic path creation serves as the foundation for a loader and bale waggon navigation tool. The method is put to the test on scenarios that are similar to those encountered in real life. Comparative results, in instance, reveal an average improvement of 16 percent over earlier heuristics.

*Figure 5. The working mechanism of the genetic algorithm*

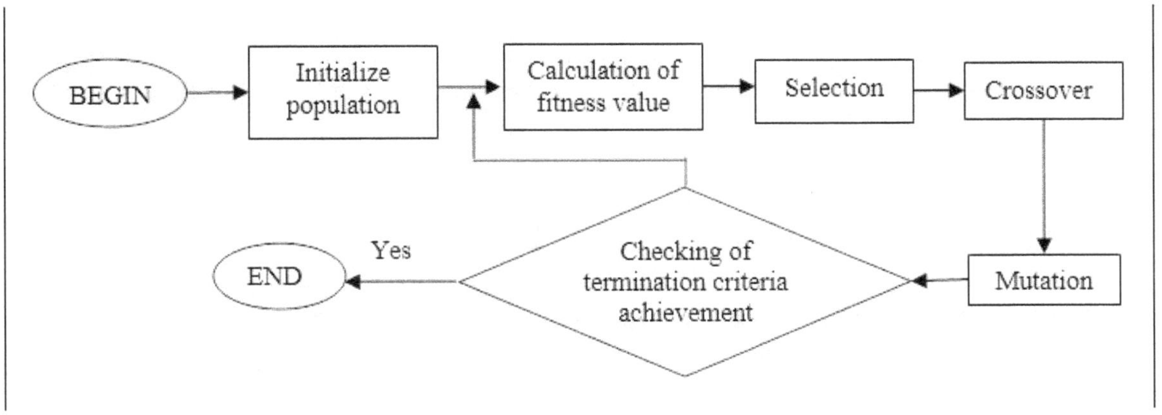

### Simulated Annealing

Simulated annealing (Aydemir et al., 1983) is a stochastic algorithm that is used to identify the best solution of an objective function as the global extreme of a function with local minimums (Mohiuddin et al., 2014). It is inspired by the physical annealing of metals (Grabusts et al., 2019). Physical constituents of the material gain enough energy to escape from their start positions (local minimums) when the process starts at a high temperature, and steady cooling leads to thermal equilibrium. Initial temperature, cooling rate, and termination policy are the three main characteristics of Simulated Annealing (Vahdanjoo et al., 2020). The Metropolis technique, which develops new solutions by modest probabilistic movements of the old solution, is the inner part of the algorithm; the optimal solution to be attained is decided via repetitive stages.

Modern approaches to agriculture machinery movement optimization with applications in sugarcane production were evaluated by Filip et al. (Filip et al., 2020). There are ways based on algorithms for spatial configuration, route planning, or path planning, as well as approaches based on cost criteria, such as energy, fuel, and time consumption. The cost/benefit relationship of savings and investments is evaluated using a combination of computational and economic approaches. Cerdeira-Pena et al. developed, tested, and implemented two heuristic algorithms based on tabu search and simulated annealing. They used their model to solve a real-world problem involving a farming cooperative that required to optimise the routes of numerous harvesters. The findings revealed that the cooperative was successful in implementing Simulated Annealing heuristics as part of an agricultural management tool (Cerdeira-Pena et al., 2017). For best route discovery between items utilising G.P.S. location, Grabusts et al. used Simulated Annealing heuristics. They created software that searches for and optimises the shortest path between two things. A simulated annealing approach was also utilised to optimise a forest stand's thinning timetable. It was improved by Moriguchi in terms of calculating speed and consistency (Moriguchi, 2020). Small unmanned aerial vehicles, which are becoming increasingly popular in precision agriculture, require complex path planning as well. The simulated annealing algorithm, as described in, can be used to do this (Behnck, 2015). Simulated annealing has also been used for agricultural monitoring with effectiveness. Leitold et al., for example, proposed using simulated annealing to assign extra sensors to dynamical monitoring systems (Leitold et al, 2018). Conesa-Muoz et almethod .'s resulted in a total travelled distance of 334.439 metres and a distance travelled on the headlands of 94.439 metres (Conesa-Muoz et al., 2016). These findings can be compared to those published in (Bochtis and Vougioukas, 2008). Bochtis and Vougioukas proposed a method for solving the herbicide application route planning problem based on simulated annealing. Experiments were carried out in four different fields. The overall distance covered was 335.767 metres, with 95.767 metres on the headlands. When comparing the two results, the variations between calculated distances are less than 1.4 percent.

## APPLICATION DOMAINS IN AGRICULTURE

Agricultural gear has evolved over time, much like human civilization itself. Sustainable growth requires desirable, optimized procedures. Metaheuristic algorithms, linear programming, equipment efficiency, cost and energy calculation, data collecting and analysis, and case studies have all been used in research on this topic. It's critical to look at each instance separately. Starting from crop monitoring to field and soil analysis, the various application domains of FANET in agriculture are described as follows.

## Crop Monitoring

Crop monitoring is required at every stage of a plant's growth, from sprouts to harvesting [30]. Farmers, especially commercial agriculture practitioners, face a challenge to monitor their crops efficiently and efficiently. After deploying autonomous quadcopter UAVs, the farmer can check them weekly or every month based on the need. It mainly functions to checkmate pest infestation on plants and helps the farmer identify plants that are not growing perfectly. In addition, the field inspection from the sky gives a better view and allows the farmer to judge the crop requiring uprooting or treatment in an improved way.

Even when everything is going as planned, crops must be surveyed and monitored to guarantee that the desired yield is available at harvest time. It's also crucial for long-term planning, whether it's identifying the best open market pricing or harvesting cyclical crops. Drones can provide precise data on every step of crop development and alert farmers to any changes before they become a disaster. Multispectral imaging can also reveal tiny distinctions between healthy and sick crops that the human eye would overlook. Stressed plants, for example, will reflect less near-infrared light than healthy plants. The human eye cannot always notice this difference. Drones, on the other hand, can supply this information at an early stage.

## Ranching Animal Husbandry

Cattle rearing can get easy by introducing ranching during inspecting animals by UAV on time. Due to their sensors' high-resolution infrared cameras, which can detect a sick animal and take action quickly, they can monitor and manage large cattle. As a result, drones' impact on precision dairy farming is quickly becoming the new normal.

## Avoiding Chemical Overuse

Drones are extremely useful for decreasing pesticide, insecticide, and other chemical abuse. These compounds do certainly aid in crop protection. However, if they are used excessively, they might be harmful. Drones can detect minute symptoms of pest attacks and provide precise data on the attack's severity and range. This can assist farmers in calculating the amount of chemicals that are required to protect rather than destroy crops.

## Getting Ready for Bad Weather

Weather conditions can be a farmer's best friend and worst opponent at the same time. Because these can't be foreseen with any accuracy, preparing for any shift in patterns becomes incredibly tough. Drones can be used to predict future weather patterns. Storm drones are already being utilised to improve forecasting accuracy. Farmers can utilise this information to better prepare themselves. Storms or a lack of rain can be predicted in advance, allowing you to select the best crop to plant for the season and how to care for your planted crops later.

## Geo-Fencing

Drones equipped with infrared cameras can readily detect animals or humans. As a result, drones can protect the crops from animal damage, especially at night.

### Spraying Crops

Agri-drones can spray chemicals because they have reservoirs that can be filled with fertilisers and insecticides for spraying on crops in a fraction of the time that traditional methods take. As a result, drone technology has the potential to bring in a new era of precision agriculture.

### Analysis of the Soil and the Field

Drones for agriculture can be used for soil and field studies, allowing for more efficient field planning. Sensors can be mounted on them to assess soil moisture content, topographical conditions, soil conditions, soil erosion, nutrients content, and soil fertility.

## CONCLUSION

This study has attempted to outline the improvement in overall product quality due to the significant applications of FANET in smart farming. The exploitation of land, energy, and other resources can be reduced to a great extent by deploying drone technology in the irrigation system. Developing new algorithms for peer-to-peer network configuration is suggested for collision control and congestion avoidance in such a platform. In the last decade, researchers have introduced almost a hundred routing protocols for FANET with their characteristics, benefits, and shortcomings. Since the mobility in the highly dynamic model plays a vital role, they have been compared using the taxonomy. Finally, some less explored areas have been identified as open research challenges for the future.

## REFERENCES

Alamsyah, A., Purnama, I. K. E., Setijadi, E., & Purnomo, M. H. (2019). M.P.R. selection to the OLSR quality of service in MANET using minmax algorithm. *Iranian Journal of Electrical and Computer Engineering*, *9*(1), 417.

Albu-Salih, A.T., & Khudhair, H.A. (2021). ASR-FANET: An adaptive SDN-based routing framework for FANET. *International Journal of Electrical & Computer Engineering, 11*(5).

Aswini, N., & Uma, S. V. (2019). Obstacle avoidance and distance measurement for unmanned aerial vehicles using monocular vision. *Iranian Journal of Electrical and Computer Engineering*, *9*(5), 3504.

Aydemir, E., & Karagul, K. (2020). Solving a periodic capacitated vehicle routing problem using simulated annealing algorithm for a manufacturing company. *Brazilian Journal of Operations & Production Management*, *17*(1), 1–13. doi:10.14488/BJOPM.2020.011

Behnck, L. P., Doering, D., Pereira, C. E., & Rettberg, A. (2015). A modified simulated annealing algorithm for SUAVs path planning. *IFAC-PapersOnLine*, *48*(10), 63–68. doi:10.1016/j.ifacol.2015.08.109

BK, C., Siddappa, M., & HS, J. (2020). Novel framework using dynamic passphrase towards secure and energy-efficient communication in Manet. *International Journal of Electrical & Computer Engineering, 10*(2).

Bujari, A., Calafate, C. T., Cano, J. C., Manzoni, P., Palazzi, C. E., & Ronzani, D. (2017). Flying ad-hoc network application scenarios and mobility models. *International Journal of Distributed Sensor Networks, 13*(10). doi:10.1177/1550147717738192

Cauvery, N.K. (2020). Trust-based secure routing against lethal behavior of nodes in wireless adhoc network. *International Journal of Electrical & Computer Engineering, 10*(2).

Cerdeira-Pena, A., Carpente, L., & Amiama, C. (2017). Optimised forage harvester routes as solutions to a traveling salesman problem with clusters and time windows. *Biosystems Engineering, 164*, 110–123. doi:10.1016/j.biosystemseng.2017.10.002

Chatterjee, S., Nandan, M., Ghosh, A., & Banik, S. (2022). DTNMA: Identifying Routing Attacks in Delay-Tolerant Network. In *Cyber Intelligence and Information Retrieval* (pp. 3–15). Springer. doi:10.1007/978-981-16-4284-5_1

Conesa-Muñoz, J., Pajares, G., & Ribeiro, A. (2016). Mix-opt: A new route operator for optimal coverage path planning for a fleet in an agricultural environment. *Expert Systems with Applications, 54*, 364–378. doi:10.1016/j.eswa.2015.12.047

Filip, M., Zoubek, T., Bumbalek, R., Cerny, P., Batista, C. E., Olsan, P., Bartos, P., Kriz, P., Xiao, M., Dolan, A., & Findura, P. (2020). Advanced computational methods for agriculture machinery movement optimization with applications in sugarcane production. *Agriculture, 10*(10), 434. doi:10.3390/agriculture10100434

Ghosh, A., Ho, C. C., & Bestak, R. (2020). *Secured Energy-Efficient Routing in Wireless Sensor Networks Using Machine Learning Algorithm: Fundamentals and Applications*. Deep Learning Strategies for Security Enhancement in Wireless Sensor Networks. doi:10.4018/978-1-7998-5068-7.ch002

Godse, S.P., & Mahalle, P.N. (2018). A Computational Analysis of E.C.C. Based Novel Authentication Scheme in VANET. *International Journal of Electrical & Computer Engineering, 8*(6).

Grabusts, P., Musatovs, J., & Golenkov, V. (2019). The application of simulated annealing method for optimal route detection between objects. *Procedia Computer Science, 149*, 95–101. doi:10.1016/j.procs.2019.01.112

He, G. (2002). *Destination-sequenced distance vector (DSDV) protocol*. Networking Laboratory, Helsinki University of Technology.

Imran, A.Z.M., Hakim, M.L., Ahmed, M., Islam, M.T., & Hossain, E. (2021). Design of microstrip patch antenna to deploy unmanned aerial vehicle as U.E. in 5G wireless network. *International Journal of Electrical & Computer Engineering, 11*(5).

Kakamoukas, G. A., Sarigiannidis, P. G., & Economides, A. A. (2020). FANETs in Agriculture-A routing protocol survey. Internet of Things, 100183.

Kashyap, K. K., & Agrawal, A. (2018). Fanet: Survey on Design Challenges, Application Scenario and Communication Protocols. *IJRAR-International Journal of Research and Analytical Reviews*, *5*(4), 1024–1032.

Khamayseh, Y., Yassein, M. B., & Abu-Jazoh, M. (2019). Intelligent black hole detection in mobile AdHoc networks. *Iranian Journal of Electrical and Computer Engineering*, *9*(3), 1968. doi:10.11591/ijece.v9i3.pp1968-1977

Khan, M. A., Khan, I. U., Safi, A., & Quershi, I. M. (2018). Dynamic routing in flying ad-hoc networks using topology-based routing protocols. *Drones (Basel)*, *2*(3), 27. doi:10.3390/drones2030027

Khan, M. A., Safi, A., Qureshi, I. M., & Khan, I. U. (2017). Flying ad-hoc networks (FANETs): A review of communication architectures, and routing protocols. *2017 First International Conference on Latest trends in Electrical Engineering and Computing Technologies (INTELLECT)*, 1-9. 10.1109/INTELLECT.2017.8277614

Khare, V. R., Wang, F. Z., Wu, S., Deng, Y., & Thompson, C. (2008). Ad-hoc network of unmanned aerial vehicle swarms for search & destroy tasks. *Proceedings of the 4th International IEEE Conference Intelligent Systems*. 10.1109/IS.2008.4670440

Kim, S., Lee, M., & Shin, C. (2018). IoT-based strawberry disease prediction system for smart farming. *Sensors (Basel)*, *18*(11), 4051. doi:10.339018114051 PMID:30463363

Lakew, D. S., Sa'ad, U., Dao, N. N., Na, W., & Cho, S. (2020). Routing in Flying Ad Hoc Networks: A Comprehensive Survey. *IEEE Communications Surveys and Tutorials*, *22*(2), 1071–1120. doi:10.1109/COMST.2020.2982452

Lee, W., Lee, J. Y., Lee, J., Kim, K., Yoo, S., Park, S., & Kim, H. (2018). Ground control system based routing for reliable and efficient multi-drone control system. *Applied Sciences (Basel, Switzerland)*, *8*(11), 2027. doi:10.3390/app8112027

Leitold, D., Vathy-Fogarassy, A., & Abonyi, J. (2018). Network distance-based simulated annealing and fuzzy clustering for sensor placement ensuring observability and minimal relative degree. *Sensors (Basel)*, *18*(9), 3096. doi:10.339018093096 PMID:30223464

Li, Y., Ghosh, A., Zhang, Z., Yu, Z., & Wang, J. (2021). Sensors and Applications in Agricultural and Environmental Monitoring. *Journal of Sensors*, *2021*. doi:10.1155/2021/9841470

Ma, K., Zhang, Y., & Trappe, W. (2005). Mobile network management and robust spatial retreats via network dynamics. *Proceedings of workshop on Resource Provisioning and Management in Sensor Networks (RPMSN05)*.

Maes, W. H., & Steppe, K. (2019). Perspectives for remote sensing with unmanned aerial vehicles in precision agriculture. *Trends in Plant Science*, *24*(2), 152–164. doi:10.1016/j.tplants.2018.11.007 PMID:30558964

Mahmud, M. S. A., Abidin, M. S. Z., & Mohamed, Z. (2018). Solving an agricultural robot routing problem with binary particle swarm optimization and a genetic algorithm. *International Journal of Mechanical Engineering and Robotics Research*, *7*(5), 521–527. doi:10.18178/ijmerr.7.5.521-527

Marinello, F., Pezzuolo, A., Chiumenti, A., & Sartori, L. (2016). Technical analysis of unmanned aerial vehicles (drones) for agricultural applications. *Engineering for Rural Development, 15*.

Mohiuddin, M. A., Khan, S. A., & Engelbrecht, A. P. (2014). Simulated evolution and simulated annealing algorithms for solving multi-objective open shortest path first weight setting problem. *Applied Intelligence*, *41*(2), 348–365. doi:10.100710489-014-0523-3

Moriguchi, K. (2020). Acceleration and enhancement of reliability of simulated annealing for optimizing thinning schedule of a forest stand. *Computers and Electronics in Agriculture*, *177*, 105691. doi:10.1016/j.compag.2020.105691

Mukherjee, D., Chakraborty, S., Sarkar, I., Ghosh, A., & Roy, S. (n.d.). A Detailed Study on Data Centre Energy Efficiency and Efficient Cooling Techniques. *International Journal of Advanced Trends in Computer Science and Engineering, 9*(5), 9221-9242.

Noubir & Lin. (n.d.). Low-power DoS attacks in data wireless lans and countermeasures. *SIGMOBILE Mob. Comput.*

Oubbati, O. S., Atiquzzaman, M., Lorenz, P., Tareque, M. H., & Hossain, M. S. (2019). Routing in flying ad hoc networks: Survey, constraints, and future challenge perspectives. *IEEE Access: Practical Innovations, Open Solutions*, *7*, 81057–81105. doi:10.1109/ACCESS.2019.2923840

Primicerio, J., Di Gennaro, S. F., Fiorillo, E., Genesio, L., Lugato, E., Matese, A., & Vaccari, F. P. (2012). A flexible unmanned aerial vehicle for precision agriculture. *Precision Agriculture*, *13*(4), 517–523. doi:10.100711119-012-9257-6

Proakis, J. (2000). *Digital Communications*. McGraw-Hill.

Psirofonia, P., Samaritakis, V., Eliopoulos, P., & Potamitis, I. (2017). Use of unmanned aerial vehicles for agricultural applications with emphasis on crop protection: Three novel case-studies. *International Journal of Agricultural Science and Technology*, *5*(1), 30–39. doi:10.12783/ijast.2017.0501.03

Qureshi, K. N., Bashir, M. U., Lloret, J., & Leon, A. (2020). Optimized cluster-based dynamic energy-aware routing protocol for wireless sensor networks in agriculture precision. *Journal of Sensors*, *2020*, 2020. doi:10.1155/2020/9040395

Sadiq, B. O., & Salawudeen, A. T. (2020). FANET optimization: A destination path flow model. *Iranian Journal of Electrical and Computer Engineering*, *10*(4), 4381–4389. doi:10.11591/ijece.v10i4.pp4381-4389

Sahingoz, O. (2014). Networking models in flying ad-hoc networks (FANETs): Concepts and challenges. *Journal of Intelligent & Robotic Systems*, *74*(1-2), 513–527. Advance online publication. doi:10.100710846-013-9959-7

Salamí, E., Barrado, C., & Pastor, E. (2014). UAV flight experiments applied to the remote sensing of vegetated areas. *Remote Sensing*, *6*(11), 11051–11081. doi:10.3390/rs61111051

Sharma, R., & Mishra, J. (2020). Review on Application of Drone Systems in Precision Agriculture. *Journal of Advanced Research in Electronics Engineering and Technology*, *7*(2), 5–7.

Sharma, V., Kumar, R., & Kumar, N. (2018). DPTR: Distributed priority tree-based routing protocol for FANETs. *Computer Communications, 122*, 129–151. doi:10.1016/j.comcom.2018.03.002

Shobana, M., & Karthik, S. (2013). A Performance Analysis and Comparison of various Routing Protocols in MANET. *Proceedings of the 2013 International Conference on Pattern Recognition, Informatics and Mobile Engineering (PRIME).* 10.1109/ICPRIME.2013.6496508

Singh, K., & Verma, A. K. (2015, March). Experimental analysis of AODV, DSDV, and OLSR routing protocol for flying adhoc networks (FANETs). In *2015 IEEE International Conference on Electrical, Computer and Communication Technologies (ICECCT)* (pp. 1-4). IEEE. 10.1109/ICECCT.2015.7226085

Singh, K., & Verma, A. K. Flying adhoc networks concept and challenges. In M. D. B. A. Khosrow-Pour (Ed.), *Advanced Methodologies and Technologies in Network Architecture.* Mobile Computing, and Data Analytics.

Singh, K., & Verma, A. K. (2014). Applying OLSR routing in FANETs. *2014 IEEE International Conference on Advanced Communications, Control, and Computing Technologies*, 1212-1215.

Sinha, J. P., Kushwaha, H. L., Kushwaha, D., Singh, N., & Purushottam, M. (2016). Prospect of Unmanned Aerial Vehicle (UAV) Technology for Agricultural Production Management. *International Journal of Emerging Trends in Engineering Research, 8*(10), 6738–6741.

Toh, C. K. (1997). Associativity-based routing for ad hoc mobile networks. *Wireless Personal Communications, 4*(2), 103–139. doi:10.1023/A:1008812928561

Vahdanjoo, M., Madsen, C. T., & Sørensen, C. G. (2020). Novel Route Planning System for Machinery Selection. Case: Slurry Application. *AgriEngineering, 2*(3), 408–429. doi:10.3390/agriengineering2030028

Wen, S., & Huang, C. (2018). Delay-constrained routing based on stochastic model for flying ad hoc networks. *Mobile Information Systems.*

Xiang, H., & Tian, L. (2006). Development of autonomous unmanned helicopter based agricultural remote sensing system. In *2006 ASAE Annual Meeting* (p. 1). American Society of Agricultural and Biological Engineers.

Xiang, H., & Tian, L. (2011). Development of a low-cost agricultural remote sensing system based on an autonomous unmanned aerial vehicle (UAV). *Biosystems Engineering, 108*(2), 174–190. doi:10.1016/j.biosystemseng.2010.11.010

Xu, W., Trappe, W., & Zhang, Y. (2007). Channel surfing: defending wireless sensor networks from interference. *Proceedings of conference on Information Processing in Sensor Networks (IPSN)*, 499–508. 10.1109/IPSN.2007.4379710

Yick, J., Mukherjee, B., & Ghosal, D. (2008). Wireless sensor network survey. *Computer Networks, 52*(12), 2292–2330. doi:10.1016/j.comnet.2008.04.002

Yu, Y., Xue, B., Chen, Z., & Qian, Z. (2019). Cluster tree topology construction method based on PSO algorithm to prolong the lifetime of ZigBee wireless sensor networks. *EURASIP Journal on Wireless Communications and Networking, 2019*(1), 1–13. doi:10.118613638-019-1523-0

Zainal, N., Mohamood, N., Norman, M.F., & Sanmutham, D. (2019). Design and implementation of smart farming system for fig using connected-argonomics. *International Journal of Electrical & Computer Engineering, 9*(6).

Zhang, H., Wang, X., Memarmoshrefi, P., & Hogrefe, D. (2017). A survey of ant colony optimization based routing protocols for mobile ad hoc networks. *IEEE Access: Practical Innovations, Open Solutions*, *5*, 24139–24161. doi:10.1109/ACCESS.2017.2762472

# Chapter 9
# Deep Learning-Based Plant Disease Detection Using Android Apps

**Mohd Shoaib**
*Department of Computer Engineering, Zakir Husain College of Engineering and Technology, Aligarh Muslim University, India*

**Firoz Ahmad**
*Department of Computer Engineering, Zakir Husain College of Engineering and Technology, Aligarh Muslim University, India*

**Mohd Ammad Rehman**
*Department of Computer Engineering, Zakir Husain College of Engineering and Technology, Aligarh Muslim University, India*

## ABSTRACT

*Disease detection in plants is crucial for preventing losses in yield and agricultural productivity. Historically, disease identification has been supported by agriculture extension organizations, which were difficult to access from villages. Farmers have to go to their field and manually monitor plant disease. The aim of this work is to develop an Android application that provides an easy-to-use platform for farmers to identify diseases in their crops. The mobile application will help to take responsive action according to the disease detected in their plants and can be easily used by anyone who is interested in analyzing the disease of the plants. This work reports on the classification of 26 diseases in 14 crop species using 54,306 images from PlantVillage dataset using a convolutional neural network approach. The models used are Inception-v3 and MobileNet. The correct prediction of the correct crop-diseases pair in 38 classes decides the criteria for performance measurement. The most accurate model achieves an overall accuracy of 96.32%.*

DOI: 10.4018/978-1-6684-5141-0.ch009

## INTRODUCTION

Today's innovations have enabled humans to produce enough food to meet the needs of over 7 billion people. Notwithstanding, food security stays undermined by various components including environmental change (Tai et al., 2014), the decrease in pollinators, plant diseases and others (Food and Agriculture Organization, 2019). Plant diseases can have catastrophic ramifications for small farmers who rely up harvest for livelihood and can be a danger to food security across the world (Farooqui & Ritika, 2019). In developing world, over 50% of the agricultural produced is lost due to various pests and diseases, where approximately 80% of the farming product is produced by small farmers (Harvey et al., 2014; Sanchez & Swaminathan, 2005). These small farmers are a weak group, and are extremely affected by disruption in agricultural produce and remain vulnerable to poverty.

Historically various effort and methodologies have always been deployed to prevent disease-related crop losses. Among the widely used approaches, application of incessant pesticides has been increasingly used, often complemented by integrated pest management (IPM) approaches (Zayan, 2019; Ehler, 2006). Essential measure of any technique is to efficiently ascertaining the disease at the very first appearance.

Community based, governments and many other institutes support the disease and pest identification through disseminating information, educating farmers about the possible control methods. With increase of internet penetration, information collection of disease and pests infection has become easy, thus providing support for disease diagnosis. Advent of tools based on mobile phones has rapidly transformed the traditional method and information of pest and disease control (International Telecommunication Union [ITU], 2015).

Many novel approaches had been developed by using computing power and high definition cameras and as well as displays of smartphones. As of year 2020, there are an estimated 5 to 6 billion unique users of smart phones across the globe. More than two third of the world population is using smart phones and is ever growing (Global Digital Overview, 2020). The high computing resources provides disease diagnosis based on automated image recognition. Along with the combination of high computing resources and smart phones having high definition cameras, high processing has made diagnosis a lot easier. Technological advances specifically in the field of computer vision and object recognition has made prediction of pest and disease diagnosis more efficient.

## MOTIVATION

The work is to provide a user-friendly, handy mobile application for analyzing the disease of the plants. As android-based smart phones are readily available in developing countries, therefore an android application is emphasized upon to make it farmer-friendly and the farmer do not need to wait for long distressing hours to know the disease that their plant is suffering. The detection of plant diseases is performed using the smartphone to check if the crops are healthy or have a disease, and predict the disease it suffers from. For a mobile application to classify and able to detect a disease, a smaller size model should be used as smart phones do not have a high computation GPUs. MobileNet-v2 is one such model, which has smaller size and has smaller complexity.

## LITERATURE REVIEW

This section briefly discusses the various machine learning techniques, and the various approaches used in plant disease detection. Recognizing the combination of smart phones and advances in computer vision, using convolutional neural network together to identify 14 crop species and 26 diseases (Mohanty et al. 2016). The trained model achieves an accuracy of 99.35%. Another approach of plant disease recognition model, developed by Berkley Vision and Learning Centre based on leaf image classification using deep learning framework called Caffe (Sladojevic, 2016; Nooraiyeen, 2020). ImageNet, a large-scale ontology of images database is used in three easy applications of object recognition, image classification and automatic object clustering (Deng, 2009).

Another method to regularly monitoring the agriculture fields and provide the automated disease detection using remote sensing images using Canny's edge detection algorithm (Badage, 2018). Clustering, multi-class SVM, and advanced neural network techniques were used to process data and develop a novel approach for detecting and identifying leaf diseases. After optimizing the data, the detection accuracy is improved and SVM classifier is used for disease classification (Farooqui & Ritika, 2019). In another method the dataset is created and healthy leaves are trained under Random Forest to classify the diseased and healthy images (Maniyath et al. 2018). Images features are extracted using Histogram of an Oriented Gradient. Aerial imagery is used to identifying the crop types and a hybrid neural network architecture is proposed with a combination of histograms and convolutional units (Rebetez, 2016). Another mobile application to detect lesions in coffee leaves was developed with accuracy more than 97% (Esgario et al., 2021). The various approaches of machine learning and deep learning for plant disease detection along with the various available datasets and their problems were discussed. Also, future progress to be made in this field was proposed (Khan et al., 2021).

An approach based on deep learning on dataset of banana leaves disease, using LeNet architecture as a convolutional neural network to automate the process of classifying (Amara et al., 2017). Performance analysis using different approaches of nine powerful architectures of deep neural networks was performed using the techniques of transfer learning and deep feature extraction methods (TÜRKOĞLU & Hanbay, 2019). 10,000 images of 100 ornamental plant species were collected using mobile phones, and a large-scale plant classification is done by proposing a deep learning model having residual building blocks (Sun et al.,2017). A method that uses convolutional neural network built from scratch is capable of recognizing plant species in colour images. Network is trained and tested on a total of 10,413 images containing 22 weed and crop species (Dyrmann et al., 2016). Paper uses the apple leaf disease dataset (ALDD) to detect diseases on apple leaf in real time. It proposed an improved convolutional neural network (CNNs) based deep learning approach. Also, a new disease detection model for apple leaves is proposed by the used of Deep-CNNs over GoogLeNet Inception structure and Rainbow concatenation (Jiang et al., 2019). Two CNN architectures, AlexNet and GoogleNet are evaluated over state-of-the-art CNN architectures using a public dataset of plant diseases and the result show improved accuracy (Brahimi et al., 2018). A smartphone application to detect disease in terrestrial plant. The distinction between the diseases was done using Deep Learning Neural network algorithms (Valdoria et al., 2019).

Many common datasets has been extensively used as a reference point for work in multiple areas of machine learning, image processing including object classification. Few such datasets include PASCAL VOC Challenge (Everingham et al., 2010), Large Scale Visual Recognition Challenge (ILSVRC) (Russakovsky et al., 2015) based on the ImageNet dataset (Deng et al., 2009) (Mohanty et al., 2016). The rapid classification of images from a trained model makes them suitable for easy-to-use smart phone applications but training large neural networks can take a great deal of time and needs high computing resources. Of late, different fields have developed an application of deep neural networks and is successfully utilised as models of end-to-end learning. Mapping a diseased plant image to a crop-disease pair as an output is performed using Neural Networks (Mohanty et al., 2016).

There are several techniques proposed for the plant disease detection using Deep Learning. One of the prominent work on Plant Disease Detection on which this work is based, used AlexNet and GoogLeNet architectures (Mohanty et al., 2016). The author have done the 60 experiment and comparative analysis by varying dataset and training mechanism. The deep learning architecture used was AlexNet (Deng et al., 2009), and GoogLeNet (Sladojevic, 2016). The training mechanism applied is transfer learning, and training from scratch is performed over the dataset types (Mohanty et al., 2016). The work attained accuracy of 85.53% in case of Alexnet architecture for training from scratch on Grayscale with train-test ratio 80-20 and 99.34% in GoogleNet architecture for transfer learning on Color images with train-test ratio 80-20 (Mohanty et al., 2016).

In the analysis training and validation of the dataset by Plant village (Hughes & Salathé, 2015) with 38 class labels and 14 crops analysis with 26 diseases with a total of 30 epochs. The paper focused on two popular deep learning architectures, AlexNet and GoogleNet, which were designed specifically for the "Large Scale Visual Recognition Challenge" (ILSVRC) (Russakovsky et al., 2015) for the ImageNet dataset (Deng et al., 2009). Five convolution layers, three fully connected layers, and a softmax layer at the end are the built-up of the AlexNet architecture. While with 22 Convolution Neural Network (CNN) layers in the GoogleNet architecture provide a depth to it.

In comparsion between the GoogleNet and AlexNet, the former performs consistently better than the later. Also, GoogleNet give better results according to the technique of training and transfer learning, as from the outcomes of the paper (Mohanty et al., 2016). In this work, GoogleNet (Inception-v3) and MobileNet-v2 Architecture is used for the comparative analysis of accuracy.

## Inception-v3 Architecture

Inception-v3 with 42 layers is a much deeper and wider architecture. On the ImageNet dataset, it attains greater than 78.1% accuracy and is widely used image recognition model. The Inception deep convolutional architecture was introduced as GoogleNet (Szegedy, 2016), named Inception. The Inception architecture has continuously evolved into better model, first by the addition of batch normalization which is known as Inception-v2. Later, Inception-v3 evolved with the introduction of additional factorization ideas in the third iteration. Figure 1 shows Inception V3 Architecture (Tsang, 2015).

*Figure 1. Inception-v3 architecture*

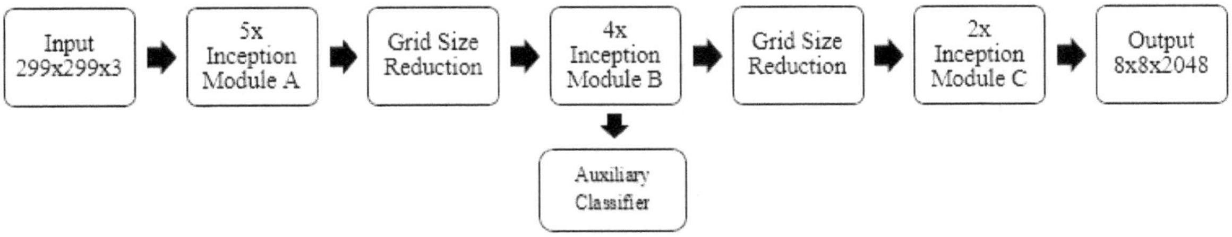

The model constitutes of both symmetric and asymmetric building blocks. These blocks comprise of convolutions, average pooling, max pooling, concats, dropouts, and fully connected layers. In the model, Batchnorm is extensively applied to activation inputs. Softmax is used to compute Loss. It became the 1st Runner Up for image classification in ILSVRC (ImageNet Large Scale Visual Recognition Competition) 2015 with 42 layers and obtains a lower error rate (Tsang, 2015).

## MobileNet-v2 Architecture

The MobileNet has a smaller model size, smaller complexity and is helpful for mobile and embedded vision applications. As an Android application is being developed to detect plant disease, Mobilenet model is particularly helpful. The MobileNet-v2 model has three building blocks: Depthwise Separable Convolutions, Linear Bottlenecks and Inverted Residuals (Sandler et al., 2018). Figure 2 (Sandler & Howard, 2020) shows MobileNet-v2 building block. Instead of a standard convolution, MobileNet-v1 (Howard et al., 2017) uses a form of factorized convolution called as depthwise separable convolution. The depthwise separable convolution block is faster, reduces computation, reduces model size and works approximately the same as traditional convolution. The two products of factorization are depthwise convolution and a pointwise convolution (1×1 convolution), also shown in Figure 3 (Hollemans, 2018). For MobileNet, each input channel is applied with a single filter in the depthwise convolution. The pointwise convolution layer creates new features by combining the filtered values, which are provided by depthwise convolution layer.

*Figure 2. MobileNetV2 building block*

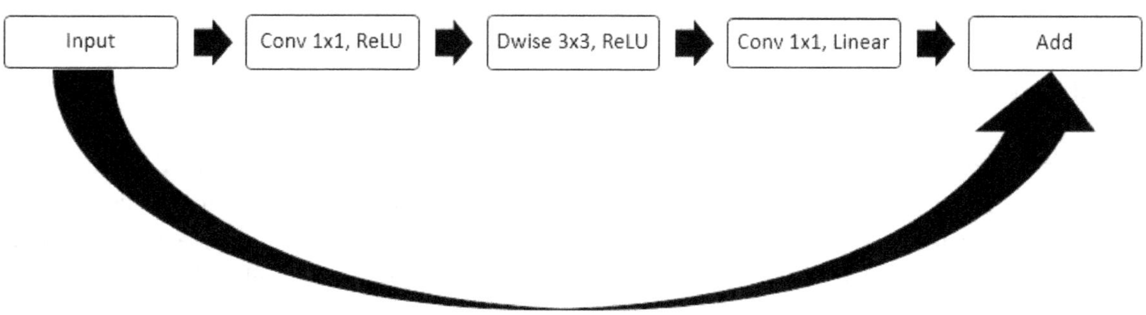


MobileNet-v2 has three convolutional layers in the block as shown in Figure 4 (Sandler et al, 2018; Hollemans, 2018):

- First layer is 1×1 convolution layer, also called as expansion layer
- Second layer filters the inputs called as a depthwise convolution
- Third layer is a 1×1 pointwise convolution layer, called as projection layer.

The purpose of expansion layer is to provide more output channels than input channels. The number of channels in the data are expanded to make the provision for more output channels. In v2, the pointwise convolution is known as projection layer as it reduces the number of channels. The shrinking of flow of data through the network, done by projection layer is called as bottleneck residual block and the output of each block is called bottleneck. Linear bottleneck layers are inserted into convolutional blocks. A shortcut connection is applied to connect the bottlenecks, called as inverted residuals.

In this work, each input channel (input depth) is applied with a single filter using depthwise convolutions. The output from depthwise layer is combined linearly using a simple 1x1 Pointwise convolution as shown in Figure 4. MobileNet uses both batchnorm and ReLU nonlinearities for both layers.

The complete MobileNet-v2 architecture consist of 17 building blocks in a row. The building blocks are followed by a global average-pooling layer, which is preceded by a regular 1×1 convolution, and later succeeded by a classification layer (Sandler et al, 2018). A building block of MobileNet-v2 is shown in Figure 4.

*Figure 3. Depthwise separable convolution block*

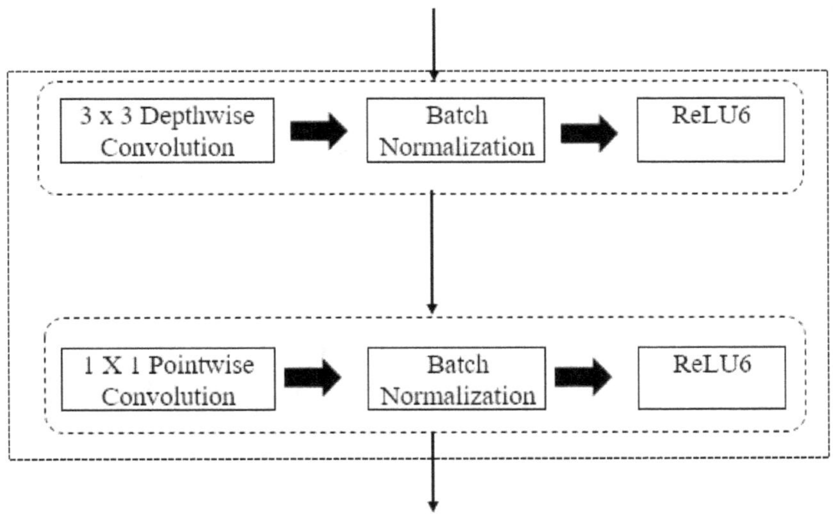


*Figure 4. MobileNet-v2 building block*

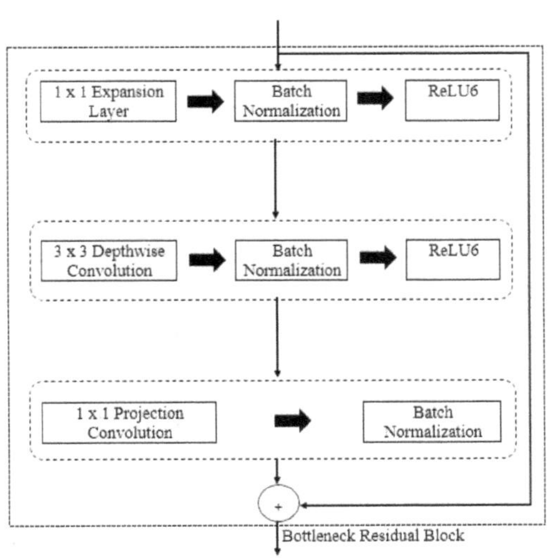

## PROPOSED WORK

This section discusses the dataset and its features, the approach that is followed to train and test the dataset.

### Dataset Used

A verified and large dataset is essential, consisting of diseased and healthy plants images to implement precise image classifiers for the identification of plant disease. In this work, Plant Village dataset (Hughes & Salathé, 2015) is work upon as it has thousands of healthy and diseased crop plants images, and is openly and freely available.

In Figure 5 is a report in form of chart, which shows the classification in 14 crop species for 26 diseases available in 54,306 colored images. The approximately correct prediction of the crop-disease pair in a given 38 possible classes which includes both diseased and healthy classes measures the performance of the models.

*Figure 5. Data set analysis report*

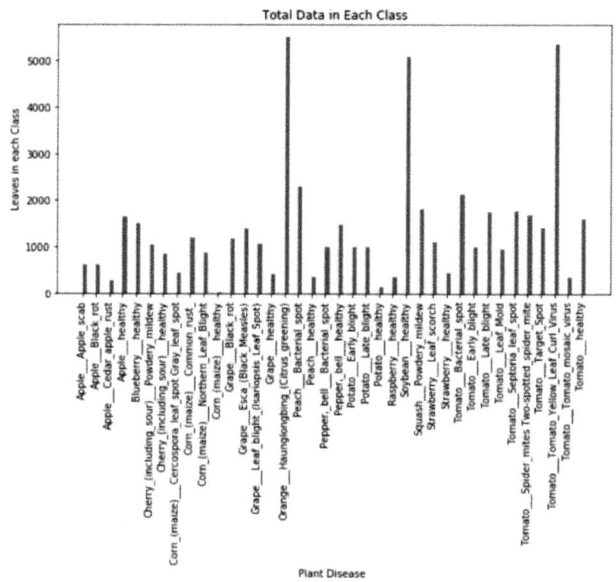


### Approach

The prediction of disease is done using two deep learning architecture, namely Inception-v3 and MobileNet-v2. On these two architecture, the dataset is divided into Train:Test =75: 25 for Inception-v3 architecture and Train:Test =78: 28 for MobileNet-v2 architecture. In this work model is trained with a total of 20 epochs for Inception-v3 architecture and a total of 20 epochs for MobileNet-v2 architecture.

Models already trained (trained on ImageNet dataset) are adapted on the dataset for analyzing the performance of both the architectures. This study also focuses on pursuing a comparative analysis of the two architectures over a dataset of 26 diseases in 14 crop species using 54,306 images. The performance measurement of the models in this work is based on the capability of the models to envisage the accurate crop-diseases pair, on 38 possible classes.

### Convert the Model using Converter

After the model is trained, it is converted to TensorFlow Lite version where this model is implemented in android application. TF lite model is a light-weight version which occupies less space, making it suitable to work on Mobile and Embedded Devices.

During the running of model, one of challenge of on mobile or embedded devices is the limited amount of resource such as memory, power and storage. Therefore, it is critical to the machine learning model that they are optimized and can run efficiently on such constraints. One of the most popular optimization techniques is called quantization.

Here, the model is optimized using Post-Training quantization. The conversion technique to reduce the model size is Post training quantization. It also improves CPU and hardware accelerator latency, with little degradation in model accuracy.

Suppose, the model consists of a deep neural network where all weights and activation outputs are represented by 32-bit floating-point numbers. This model can be quantize by converting all 32-bit floating-point numbers that represent weight and output to nearest 8-bit integers.

## DESIGN AND IMPLEMENTATION

This section discusses the implementation design for the disease detection, acquiring the dataset, splitting the dataset into training data and test data, and creating a model for transfer learning, training the model, lastly saving the model and converting it, so that it can be run over android. As the objective is develop a mobile application that can classify the plant and its disease by the image of leaves, Figure 6 shows the flow of the proposed system.

### Data Acquisition and Preprocessing

The first process shown in Figure 6 is acquiring data and preprocessing it. With the aim of implementing an accurate image classifier to diagnose plant disease, there is need for a large and substantiated dataset consisting of images of diseased as well as healthy plants. The Plant Village project has large and verified dataset and is freely available (Hughes & Salathé, 2015). To get an understanding of performance of the implemented approaches on the new hidden data, and checking the overfitting of any approaches, the model is run across a whole range as:

- The data is divided into two different datasets.
- One part of dataset is training set which is used to train the model and the second part of dataset is a validation set which is used to evaluate the performance of the model.
- For Inception Model, the ratio of distribution of the dataset into Train: Test = 75:25
- For MobileNet Model, the ratio of distribution of the dataset into Train: Test = 78:22

To get a maximum feature extraction, 32 batches of tensor image data are generated using ImageDataGenerator from TensorFlow. This is efficient in terms of memory.

## Creating Model

As two CNN architectures are used: Inception-v3 and MobileNet-v2. Here, have used Transfer learning from pre-trained models, which exploit a huge volume of visual knowledge already learned and extracted from ImageNet dataset. With these pre-trained neural networks, train the dataset and then classify the images from the 38 classes using a SoftMax layer. Using the created model feature extraction is performed.

*Inception-v3*: Inception-v3 consist of two parts, the first part is Feature Extraction which is performed with a convolutional neural network and the second part is Classification which performed with fully connected and SoftMax layers.

*Figure 6. Flow of the proposed system*

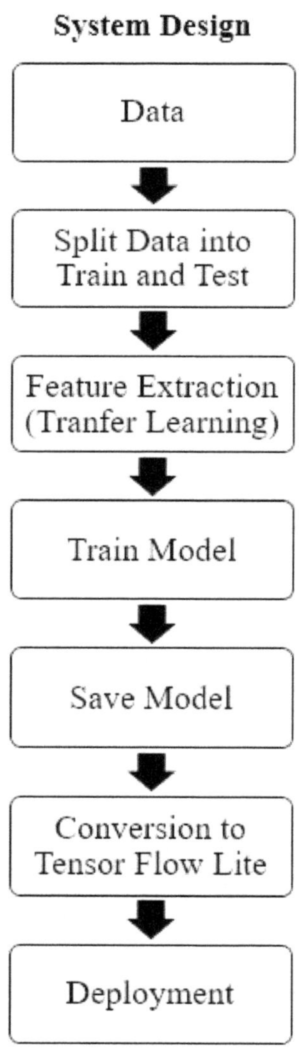

*Figure 7. Function to create a feature_extractor*

```
URL = "https://tfhub.dev/google/tf2-preview/inception_v3/feature_vector/4"
feature_extractor = hub.KerasLayer(URL,
input_shape=(IMG_SHAPE, IMG_SHAPE, 3))

feature_extractor.trainable=False
```

Figure 7 shows code for creating a feature extractor, to pass a correct input_shape. The final classifier layer is only modified by training, therefore the variables are freeze in the feature extractor layer. Now, creating a model with tf.keras.Sequential model, and adding the pretrained model and add the new classification layer with 38 number of output classes as shown in Figure 8.

*Figure 8. Summary of inception model with 38 output classes*

```
Model: "sequential"
_________________________________________________________________
Layer (type)                 Output Shape              Param #
=================================================================
keras_layer (KerasLayer)     (None, 2048)              21802784
_________________________________________________________________
dense (Dense)                (None, 38)                77862
=================================================================
Total params: 21,880,646
Trainable params: 77,862
Non-trainable params: 21,802,784
```

*MobileNet-v2:* Initialize the model with an identical input size as shown in Figure 9 to the preprocessed image data which is 224.

*Figure 9. Function to create a feature_extractor*

```
IMAGE_SHAPE = 224
URL = "https://tfhub.dev/google/tf2-preview/mobilenet_v2/feature_vector/4"
feature_extractor = hub.KerasLayer(URL,
  input_shape=(IMAGE_SHAPE, IMAGE_SHAPE, 3))
feature_extractor.trainable=False
```

Now, creating a model with tf.keras.Sequential model as shown in Figure 10, and adding the pretrained model and add the new classification layer with 38 number of output classes.

*Figure 10. Summary of inception of model with 38 output classes*

```
Model: "sequential"
_________________________________________________________________
Layer (type)                 Output Shape              Param #
=================================================================
keras_layer (KerasLayer)     (None, 1280)              2257984
_________________________________________________________________
dense (Dense)                (None, 38)                48678
=================================================================
Total params: 2,306,662
Trainable params: 48,678
Non-trainable params: 2,257,984
```

## Model Compilation: hyperparameter tuning

This work uses optimizer as Adam, loss function sparse_categorical_crossentropy and metrics is set as accuracy. The optimizer is implemented to optimized the maximum accuracy.

## Training and Evaluation

The fourth process in Figure 6 is train model. Therefore, the model is trained with 20 epochs each for Inception-v3 Architecture and MobileNet-v2 Architecture. Here, epochs mean number of times the model is needed to be evaluated during training.

## Save Model

Fifth process in Figure 6 is save model, to avoid training the model repeatedly, the model is saved using TensorFlow. It saves model's architecture, weights, and training configuration in a single file. During deployment or sharing with TFLite, execution of code of the original model is not required. The saved fully-functional model can be used for deploying the models easily using TensorFlow Lite.

## Converting Model to tflite Model

A TensorFlow model is converted using TensorFlow Lite converter as depicted in Figure 6. The converter generates a TensorFlow Lite FlatBuffer file (.tflite). This TensorFlow model will further deployed to the Android application. Using TensorFlow Lite performed quantization on model by converting 32-bit representation to 8-bit representation.

## Deployment into Android Application

The last process in the design as depicted in Figure 6 is deployment of android application and it is done in three stages.

1. **Adding TensorFlow Lite to Android Project**

Adding below code to build.gradle (Module:app)

```
aaptOptions {
                noCompress "tflite"
                noCompress "lite"
               }
Dependencies{
implementation 'org.tensorflow:tensorflow-lite:0.0.0-nightly'
               }
```

2. **Architecture of Android App:** The architecture of android app involves preprocessing of images and implementing classifier on TensorFlow lite as shown in Figure 11.
3. **Performing Inference on Plant Leaves Images**
    a. Initialize the Interpreter: Model is loaded in the interpreter at this stage.
    b. Preparing the Image Input: Getting the image and resize the images to specified size along with the pixel format known to the model.
    c. Perform Inference: Pass input to the interpreter and invoke the interpreter.
    d. Obtain and Map Results: Map the resulting confidence values to labels.

*Figure 11. TensorFlow Lite architecture*

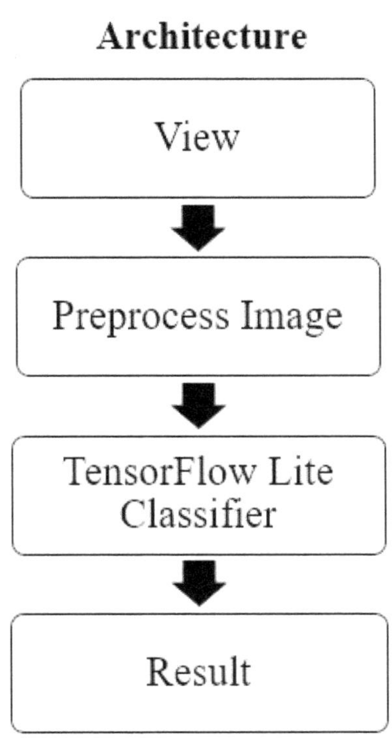

# RESULT

## Model Result

### INCEPTION-v3

The accuracy of inception model in Figure 12, is observed to be approximately 93.48% with epoch value 20. Figure 13 shows plotting the training and validation accuracy/loss graphs. The gap between the training loss and accuracy loss is also increasing with the increase in epochs as observed after epoch number 6. In addition, the validation accuracy is decreasing after epoch number 18. It is observe that the model achieved an accuracy of 93.48% in 20 epochs.

*Figure 12. Inception Model with accuracy 93.48%*

```
Epoch 1/20
991/991 [==============================] - 190s 157ms/step - loss: 0.7058 - accuracy: 0.8141 - val_loss: 0.3801 - val_accuracy: 0.8869
Epoch 2/20
991/991 [==============================] - 153s 154ms/step - loss: 0.3003 - accuracy: 0.9117 - val_loss: 0.2905 - val_accuracy: 0.9110
Epoch 3/20
991/991 [==============================] - 153s 154ms/step - loss: 0.2244 - accuracy: 0.9327 - val_loss: 0.2484 - val_accuracy: 0.9209
Epoch 4/20
991/991 [==============================] - 153s 154ms/step - loss: 0.1785 - accuracy: 0.9456 - val_loss: 0.2490 - val_accuracy: 0.9203
Epoch 5/20
991/991 [==============================] - 153s 155ms/step - loss: 0.1504 - accuracy: 0.9544 - val_loss: 0.2327 - val_accuracy: 0.9247
Epoch 6/20
991/991 [==============================] - 153s 154ms/step - loss: 0.1307 - accuracy: 0.9606 - val_loss: 0.2080 - val_accuracy: 0.9321
Epoch 7/20
991/991 [==============================] - 153s 154ms/step - loss: 0.1150 - accuracy: 0.9651 - val_loss: 0.2171 - val_accuracy: 0.9287
Epoch 8/20
991/991 [==============================] - 152s 153ms/step - loss: 0.1034 - accuracy: 0.9693 - val_loss: 0.2105 - val_accuracy: 0.9301
Epoch 9/20
991/991 [==============================] - 152s 154ms/step - loss: 0.0913 - accuracy: 0.9730 - val_loss: 0.2135 - val_accuracy: 0.9309
Epoch 10/20
991/991 [==============================] - 152s 154ms/step - loss: 0.0825 - accuracy: 0.9757 - val_loss: 0.1965 - val_accuracy: 0.9359
Epoch 11/20
991/991 [==============================] - 153s 154ms/step - loss: 0.0749 - accuracy: 0.9785 - val_loss: 0.1987 - val_accuracy: 0.9364
Epoch 12/20
991/991 [==============================] - 152s 154ms/step - loss: 0.0709 - accuracy: 0.9790 - val_loss: 0.1991 - val_accuracy: 0.9375
Epoch 13/20
991/991 [==============================] - 152s 154ms/step - loss: 0.0632 - accuracy: 0.9816 - val_loss: 0.2138 - val_accuracy: 0.9319
Epoch 14/20
991/991 [==============================] - 152s 154ms/step - loss: 0.0580 - accuracy: 0.9834 - val_loss: 0.1962 - val_accuracy: 0.9374
Epoch 15/20
991/991 [==============================] - 152s 154ms/step - loss: 0.0517 - accuracy: 0.9862 - val_loss: 0.1864 - val_accuracy: 0.9399
Epoch 16/20
991/991 [==============================] - 152s 154ms/step - loss: 0.0497 - accuracy: 0.9865 - val_loss: 0.1992 - val_accuracy: 0.9384
Epoch 17/20
991/991 [==============================] - 153s 154ms/step - loss: 0.0445 - accuracy: 0.9882 - val_loss: 0.1931 - val_accuracy: 0.9395
Epoch 18/20
991/991 [==============================] - 152s 154ms/step - loss: 0.0427 - accuracy: 0.9883 - val_loss: 0.2008 - val_accuracy: 0.9374
Epoch 19/20
991/991 [==============================] - 153s 154ms/step - loss: 0.0462 - accuracy: 0.9869 - val_loss: 0.2117 - val_accuracy: 0.9367
Epoch 20/20
991/991 [==============================] - 153s 154ms/step - loss: 0.0368 - accuracy: 0.9901 - val_loss: 0.2183 - val_accuracy: 0.9348
```

*Figure 13. Plot of Accuracy and loss on train and validation dataset for inception model*

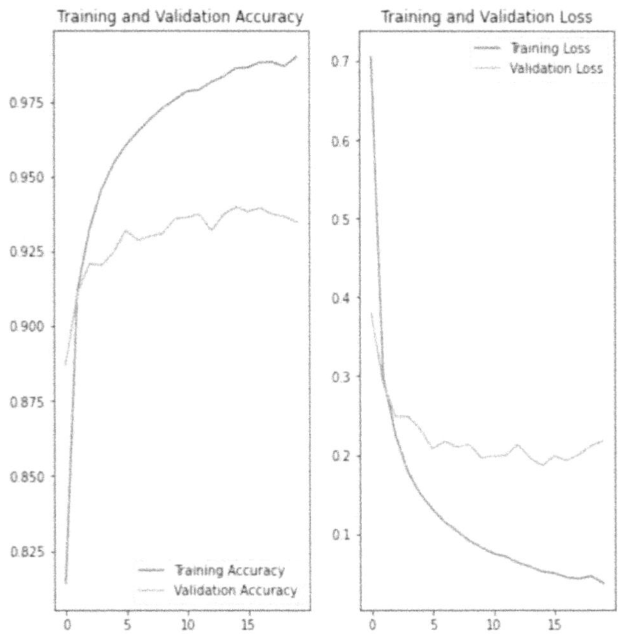


## MOBILENET-v2

The accuracy of MobileNet-v2 model in Figure 14, is observed to be approximately 96.32% with epoch value 20. Plotting of the training and validation accuracy/loss graphs is shown in Figure 15. The gap between the training loss and accuracy loss is also increasing with the increase in epochs. It is observe that the model achieved an accuracy of 96.32% in 20 epochs. Therefore, it can be concluded that MobileNet-v2 has achieved higher accuracy than Inception-v3.

*Figure 14. MobileNet model with accuracy 96.32%*

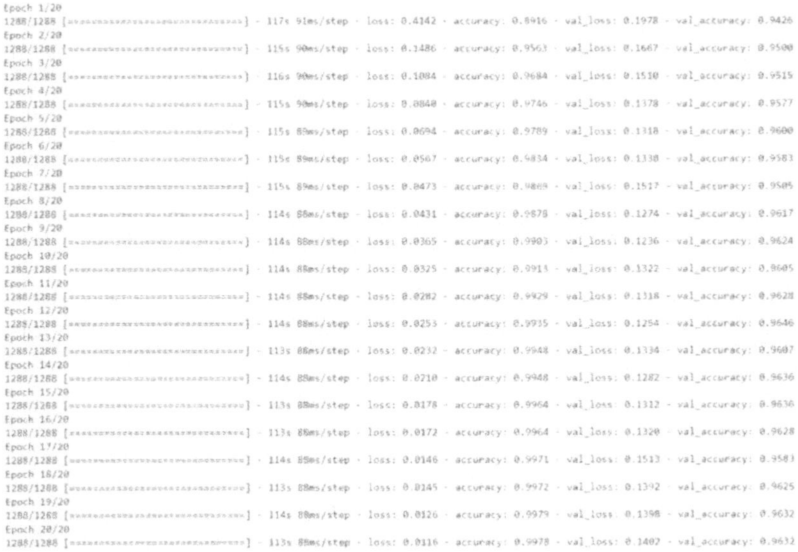

```
Epoch 1/20
1288/1288 [==============================] - 117s 91ms/step - loss: 0.4142 - accuracy: 0.8916 - val_loss: 0.1978 - val_accuracy: 0.9426
Epoch 2/20
1288/1288 [==============================] - 115s 90ms/step - loss: 0.1486 - accuracy: 0.9563 - val_loss: 0.1667 - val_accuracy: 0.9500
Epoch 3/20
1288/1288 [==============================] - 116s 90ms/step - loss: 0.1084 - accuracy: 0.9684 - val_loss: 0.1510 - val_accuracy: 0.9515
Epoch 4/20
1288/1288 [==============================] - 115s 90ms/step - loss: 0.0840 - accuracy: 0.9746 - val_loss: 0.1378 - val_accuracy: 0.9577
Epoch 5/20
1288/1288 [==============================] - 115s 89ms/step - loss: 0.0694 - accuracy: 0.9789 - val_loss: 0.1318 - val_accuracy: 0.9600
Epoch 6/20
1288/1288 [==============================] - 115s 89ms/step - loss: 0.0567 - accuracy: 0.9834 - val_loss: 0.1338 - val_accuracy: 0.9583
Epoch 7/20
1288/1288 [==============================] - 115s 89ms/step - loss: 0.0473 - accuracy: 0.9889 - val_loss: 0.1517 - val_accuracy: 0.9505
Epoch 8/20
1288/1288 [==============================] - 114s 88ms/step - loss: 0.0431 - accuracy: 0.9878 - val_loss: 0.1274 - val_accuracy: 0.9617
Epoch 9/20
1288/1288 [==============================] - 114s 88ms/step - loss: 0.0365 - accuracy: 0.9903 - val_loss: 0.1236 - val_accuracy: 0.9624
Epoch 10/20
1288/1288 [==============================] - 114s 88ms/step - loss: 0.0325 - accuracy: 0.9913 - val_loss: 0.1322 - val_accuracy: 0.9605
Epoch 11/20
1288/1288 [==============================] - 114s 88ms/step - loss: 0.0282 - accuracy: 0.9929 - val_loss: 0.1318 - val_accuracy: 0.9628
Epoch 12/20
1288/1288 [==============================] - 114s 88ms/step - loss: 0.0253 - accuracy: 0.9935 - val_loss: 0.1254 - val_accuracy: 0.9646
Epoch 13/20
1288/1288 [==============================] - 113s 88ms/step - loss: 0.0232 - accuracy: 0.9948 - val_loss: 0.1334 - val_accuracy: 0.9607
Epoch 14/20
1288/1288 [==============================] - 114s 88ms/step - loss: 0.0210 - accuracy: 0.9948 - val_loss: 0.1282 - val_accuracy: 0.9636
Epoch 15/20
1288/1288 [==============================] - 113s 88ms/step - loss: 0.0178 - accuracy: 0.9964 - val_loss: 0.1312 - val_accuracy: 0.9636
Epoch 16/20
1288/1288 [==============================] - 113s 88ms/step - loss: 0.0172 - accuracy: 0.9964 - val_loss: 0.1320 - val_accuracy: 0.9628
Epoch 17/20
1288/1288 [==============================] - 114s 88ms/step - loss: 0.0146 - accuracy: 0.9971 - val_loss: 0.1513 - val_accuracy: 0.9583
Epoch 18/20
1288/1288 [==============================] - 113s 88ms/step - loss: 0.0145 - accuracy: 0.9972 - val_loss: 0.1392 - val_accuracy: 0.9625
Epoch 19/20
1288/1288 [==============================] - 114s 88ms/step - loss: 0.0126 - accuracy: 0.9979 - val_loss: 0.1398 - val_accuracy: 0.9632
Epoch 20/20
1288/1288 [==============================] - 113s 88ms/step - loss: 0.0116 - accuracy: 0.9978 - val_loss: 0.1402 - val_accuracy: 0.9632
```

*Figure 15. Plot of accuracy and loss on train and validation dataset for MobileNet model*

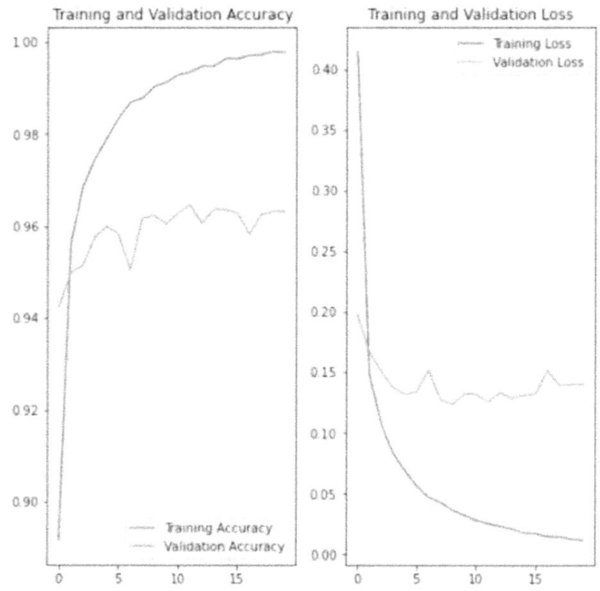


## Android App

An android application has been created using Java programming language. The trained model is loaded into the app. The home screen shows the options for either uploading a photo of a leave or capture a leave photo as shown in Figure 16.

- **Upload Photo**- Clicking on UPLOAD PHOTO takes the user to upload plant leaves from the user's internal storage.
- **Capture Photo** - Clicking on CAPTURE PHOTO takes the user to capture plant leaves from the user's camera and after this, it will give the option to crop images and it will save the image to the user's internal memory.

After capturing or uploading the photo, the user classifies image by clicking on classify button shown on screen. This screen contains buttons for classifying and for the back.

- **CLASSIFY!**- After clicking this event, it will show the user top three labels with the confidence of each predicted disease.
- **BACK**- On click, the user goes to the home screen.

An image of corn plant is given as input to the application. The application classifies the image and the disease in shown in terms of confidence measure along with disease name under the label header. The results window after classify is shown in Figure 17.

*Figure 16. Home screen*

*Figure 17. Classify image*

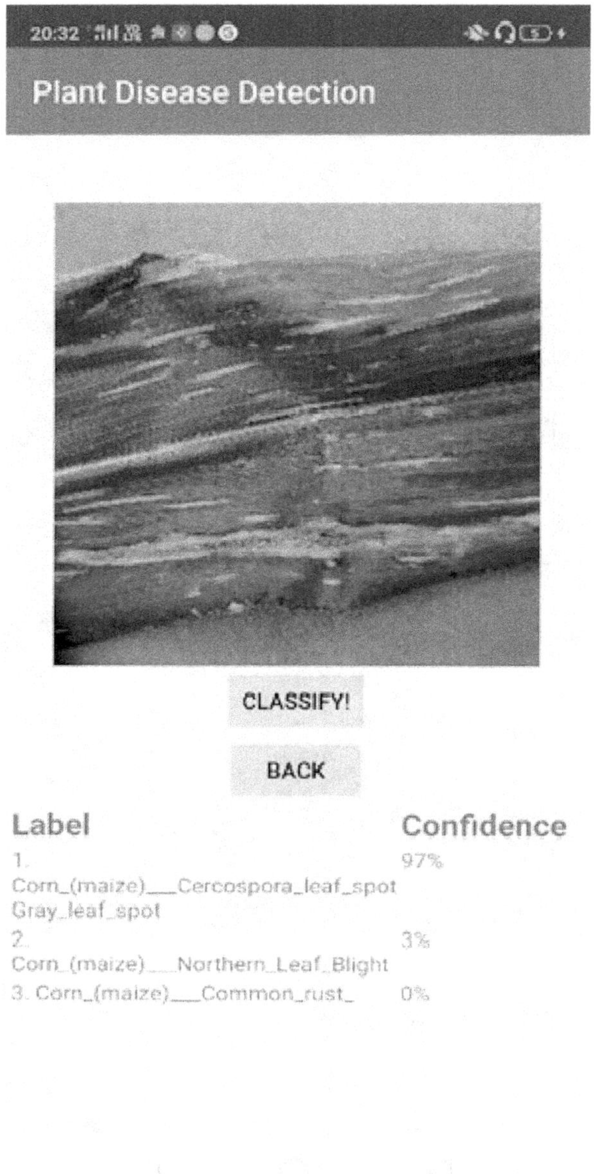


## CONCLUSION AND FUTURE WORK

A number of factors threaten food security. Few of the threatening factors include climate change, decline in pollinators, plant diseases and others. According to the reports, crop yield loss is more than 50% due to pest and disease and is a major concern. The existing methodology of plant disease detection is simply naked eye observation by experts. To continue using existing methodology a large team of experts and monitoring of plants is required. In developing countries like India, farmers do not have facilities and even the information about ways that they can use to contact to experts.

In such conditions, a mobile application proves to be advantageous in monitoring large fields of plants. Mobile application makes it easier to see and detect the symptoms on the plant leaves and as well as it is cheaper. Nowadays, using smartphones offers novel approaches to help identify diseases with their computing power, high-resolution display and advanced HD cameras. Also, the computer vision and object recognition have tremendous achievements in past few years, along with the Deep neural network.

The features were extracted from the dataset using two CNN architectures: InceptionV3 and MobileNetV2. Through transfer learning, the exploitation of pre-trained visual knowledge over Inception and MobileNet models is extracted from ImageNet. The MobileNet V2 has given more accuracy on dataset (96.32%). Also, MobileNet is a lightweight Neural Network for vision application on low computing power devices like mobile phones. As smartphones are having low computing power, model needs to saved and converted to .tflite format. Later, the interpreter is initialized in the android application.

This application can detect plant-disease, by uploading the disease of the leaf from local media or by capturing photo and it will classify the plant and disease with its confidence. The application is for 14 species and 26 diseases, and can accommodate more plant species and their diseases. Also, can accommodate suggestions related to disease treatment and details of curing the disease. Researchers are invited to contribute to this project on Git Hub with new ideas:

https://github.com/rehman0211/android_plant_disease_detection

## REFERENCES


Amara, J., Bouaziz, B., & Algergawy, A. (2017). A Deep Learning-based Approach for Banana Leaf Diseases Classification. In Datenbanksysteme für Business, Technologie und Web (BTW 2017) - Workshopband. Bonn: Gesellschaft für Informatik e.V.

Badage, A. (2018). Crop Disease Detection using Machine Learning: Indian Agriculture. *International Research Journal of Engineering and Technology*, *5*(9).

Brahimi, M., Arsenovic, M., Laraba, S., Sladojevic, S., Boukhalfa, K., & Moussaoui, A. (2018). Deep Learning for Plant Diseases: Detection and Saliency Map Visualisation. In J. Zhou & F. Chen (Eds.), *Human and Machine Learning. Human–Computer Interaction Series*. Springer. doi:10.1007/978-3-319-90403-0_6

Deng, J., Dong, W., Socher, R., Li, L.-J., Li, K., & Li, F.-F. (2009). ImageNet: A large-scale hierarchical image database. *2009 IEEE Conference on Computer Vision and Pattern Recognition*, 248–255. 10.1109/CVPR.2009.5206848

Dyrmann, M., Karstoft, H., & Midtiby, H. S. (2016). Plant species classification using deep convolutional neural network. *Biosystems Engineering*, *151*, 72–80. doi:10.1016/j.biosystemseng.2016.08.024

Ehler, L. E. (2006). Integrated pest management (IPM): Definition, historical development and implementation, and the other IPM. *Pest Management Science*, *62*(9), 787–789. doi:10.1002/ps.1247 PMID:16786545

Esgario, J. G., de Castro, P. B., Tassis, L. M., & Krohling, R. A. (2021). *An app to assist farmers in the identification of diseases and pests of coffee leaves using deep learning. Information Processing in Agriculture*. doi:10.1016/j.inpa.2021.01.004

Everingham, M., Van Gool, L., Williams, C. K. I., Winn, J., & Zisserman, A. (2009). The Pascal Visual Object Classes (VOC) Challenge. *International Journal of Computer Vision*, *88*(2), 303–338. doi:10.100711263-009-0275-4

Farooqui, N. A., & Ritika, R. (2019). An Identification and Detection Process for Leaves Disease of Wheat Using Advance Machine Learning Techniques. *Bioscience Biotechnology Research Communications*, *12*(4), 1081–1091. doi:10.21786/bbrc/12.4/31

Food and Agriculture Organization of the United Nations. (2019). *Agriculture and climate change Challenges and opportunities at the global and local level Collaboration on Climate-Smart Agriculture*. FAO. https://www.fao.org/3/CA3204EN/ca3204en.pdf

Global Digital Overview. (2020). *DataReportal – Global Digital Insights*. https://datareportal.com/global-digital-overview

Hanson, A., Joy, A., & Francis, J. (2017). *Plant Leaf Disease Detection using Deep Learning and Convolutional Neural Network*. Academic Press.

Harvey, C. A., Rakotobe, Z. L., Rao, N. S., Dave, R., Razafimahatratra, H., Rabarijohn, R. H., Rajaofara, H., & MacKinnon, J. L. (2014). Extreme vulnerability of smallholder farmers to agricultural risks and climate change in Madagascar. *Philosophical Transactions of the Royal Society of London. Series B, Biological Sciences*, *369*(1639), 20130089. doi:10.1098/rstb.2013.0089 PMID:24535397

Hollemans, M. (2018, April 22). *MobileNet version 2*. Blog of Matthijs Hollemans about Machine Learning and AI. https://machinethink.net/blog/mobilenet-v2/

Howard, A. G., Zhu, M., Chen, B., Kalenichenko, D., Wang, W., Weyand, T., Andreetto, M., & Adam, H. (2017). *Mobilenets: Efficient convolutional neural networks for mobile vision applications*. arXiv preprint arXiv:1704.04861.

Hughes, D., & Salathé, M. (2015). *An open access repository of images on plant health to enable the development of mobile disease diagnostics through machine learning and crowdsourcing*. ArXiv, abs/1511.08060.

International Telecommunication Union (ITU). (2015). *ICT Facts and Figures – the World in 2015*. https://www.itu.int/en/ITU-D/Statistics/Documents/facts/ICTFactsFigures2015.pdf

Jiang, P., Chen, Y., Liu, B., He, D., & Liang, C. (2019). Real-Time Detection of Apple Leaf Diseases Using Deep Learning Approach Based on Improved Convolutional Neural Networks. *IEEE Access: Practical Innovations, Open Solutions*, *7*, 59069–59080. doi:10.1109/ACCESS.2019.2914929

Khan, R. U., Khan, K., Albattah, W., & Qamar, A. M. (2021). Image-Based Detection of Plant Diseases: From Classical Machine Learning to Deep Learning Journey. *Wireless Communications and Mobile Computing*, *2021*, 1–13. doi:10.1155/2021/5541859

Mohanty, S. P., Hughes, D. P., & Salathé, M. (2016). Using Deep Learning for Image-Based Plant Disease Detection. *Frontiers in Plant Science*, *7*, 1419. doi:10.3389/fpls.2016.01419 PMID:27713752

Nooraiyeen, A. (2020, July). Robotic Vehicle for Automated Detection of Leaf Diseases. *2020 IEEE International Conference on Electronics, Computing and Communication Technologies (CONECCT)*. 10.1109/CONECCT50063.2020.9198326

Ramesh, S., & Hebbar, R. (2018, April). Plant Disease Detection Using Machine Learning. *International Conference on Design Innovations for 3Cs Compute Communicate Control (ICDI3C)*. 10.1109/ICDI3C.2018.00017

Rebetez, J., Satizábal, H., Mota, M., Noll, D., Büchi, L., Wendling, M., Cannelle, B., Pérez-Uribe, A., & Burgos, S. (2016). *Augmenting a convolutional neural network with local histograms - A case study in crop classification from high-resolution UAV imagery*. ESANN.

Russakovsky, O., Deng, J., Su, H., Krause, J., Satheesh, S., Ma, S., Huang, Z., Karpathy, A., Khosla, A., Bernstein, M., Berg, A. C., & Fei-Fei, L. (2015). ImageNet Large Scale Visual Recognition Challenge. *International Journal of Computer Vision*, *115*(3), 211–252. doi:10.100711263-015-0816-y

Sanchez, P. A., & Swaminathan, M. S. (2005). Public Health: Enhanced: Cutting World Hunger in Half. *Science*, *307*(5708), 357–359. doi:10.1126cience.1109057 PMID:15661994

Sandler, M., & Howard, A. (2018, April 3). *MobileNetV2: The Next Generation of On-Device Computer Vision Networks.* Google AI Blog. https://ai.googleblog.com/2018/04/mobilenetv2-next-generation-of-on.html

Sandler, M., Howard, A., Zhu, M., Zhmoginov, A., & Chen, L.-C. (2018). MobileNetV2: Inverted Residuals and Linear Bottlenecks. *2018 IEEE/CVF Conference on Computer Vision and Pattern Recognition*. 10.1109/CVPR.2018.00474

Sladojevic, S., Arsenovic, M., Anderla, A., Culibrk, D., & Stefanovic, D. (2016). Deep Neural Networks Based Recognition of Plant Diseases by Leaf Image Classification. *Computational Intelligence and Neuroscience*, *2016*, 1–11. doi:10.1155/2016/3289801 PMID:27418923

Sun, Y., Liu, Y., Wang, G., & Zhang, H. (2017). Deep Learning for Plant Identification in Natural Environment. *Computational Intelligence and Neuroscience*, *2017*, 1–6. doi:10.1155/2017/7361042 PMID:28611840

Szegedy, C., Vanhoucke, V., Ioffe, S., Shlens, J., & Wojna, Z. (2016, June). Rethinking the Inception Architecture for Computer Vision. *2016 IEEE Conference on Computer Vision and Pattern Recognition (CVPR)*, 2818–2826. 10.1109/CVPR.2016.308

Tai, A. P. K., Martin, M. V., & Heald, C. L. (2014). Threat to future global food security from climate change and ozone air pollution. *Nature Climate Change*, *4*(9), 817–821. doi:10.1038/nclimate2317

Tsang, S. (2019, March 23). *Review: Inception-v3 — 1st Runner Up (Image Classification) in ILSVRC 2015*. Medium. https://sh-tsang.medium.com/review-inception-v3-1st-runner-up-image-classification-in-ilsvrc-2015-17915421f77c

Türkoğlu, M., & Hanbay, D. (2019). Plant disease and pest detection using deep learning-based features. *Turkish Journal of Electrical Engineering and Computer Sciences*, *27*(3), 1636–1651. doi:10.3906/elk-1809-181

Valdoria, J. C., Caballeo, A. R., Fernandez, B. I. D., & Condino, J. M. M. (2019, October). iDahon: An Android Based Terrestrial Plant Disease Detection Mobile Application Through Digital Image Processing Using Deep Learning Neural Network Algorithm. *2019 4th International Conference on Information Technology (InCIT)*. 10.1109/INCIT.2019.8912053

Zayan, S. A. (2019). *Impact of climate change on plant diseases and IPM strategies*. IntechOpen. doi:10.5772/intechopen.87055

# Chapter 10
# A Sustainable Approach of Artificial Neural Network for Prediction of Irrigation, Pesticides, Fertilizers, and Crop Yield

**Pandurangan Prakash**
*Sathyabama Institute of Science and Technology, India*

**G. Elanthendral**
*Sathyabama Institute of science and Technology, India*

**V. Vedanarayanan**
*Sathyabama Institute of Science and Technology, India*

**J. Aravindh Kumar**
*Saveetha School of Engineering SIMATS (Deemed), India*

**Antony V. Samrot**
*Mahsa University College, Malaysia*

**R. Jaiganesh**
*SRM Institute of Science and Technology, India*

## ABSTRACT

*Food demand increases as the world's population expands; the world populace will reach 9.9 billion by 2050. India will be the world's largest population by 2024. Agriculture, therefore, should be fruitful and affordable for subsistence. Organic methods of agriculture are still effective for healthier crops. However, production shrinks as it depends wholly on manual labour. Since traditional farming, agriculture has seen various revolutions and developments. Currently, in the era of Agriculture 5.0, precision agriculture principles using artificial intelligence, machine learning, and IoT are being used. India still relies heavily on manual work. Educational level, inadequate training, and indigent farmers put India at a disadvantage. Technology may lead to sustainable agriculture, which means integration of plant and animal production that leaves unshakable benefits on the environment, farmers, and society that essential for the climate change- and disaster-prone world. Machine learning techniques that can possibly cater to various agricultural challenges faced by famers in India are reviewed.*

DOI: 10.4018/978-1-6684-5141-0.ch010

## INTRODUCTION

Indian economy is greatly dependent on agriculture which forms the backbone of major business activities taking place in the country. Agriculture can be affected by various factors, but environmental factors play an explicit role in all kinds of agricultural activities (Mishra et al., 2016). Agriculture directly influences the national income and plays pivotal role in the employment sector. In 1951, around 69.7 per cent of working population was involved in the agriculture in which the percentage of cultivator is higher than the agricultural labour (Salgotra et al., 2018). Traditional methods that are prominent in India face many challenges in the agricultural industry, such as overuse of chemical fertilisers and pesticides, change in soil quality, weed control, irrigation, and climate (Kumar & Prakash, 2020). There are many factors that determines the success of agriculture in India that is depicted in Figure 1.

*Figure 1. Factors concerned with agriculture*

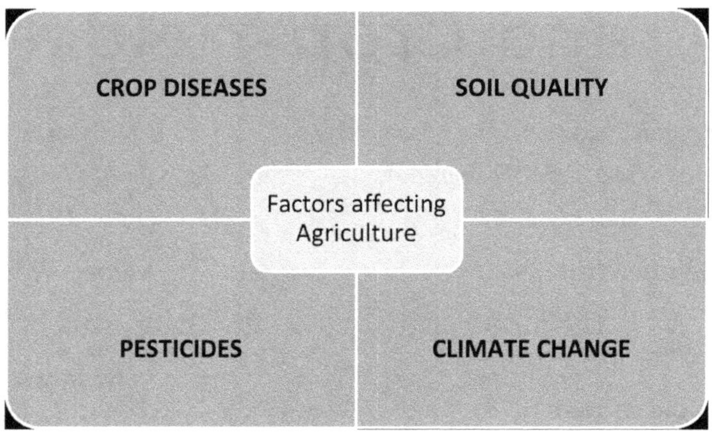


Soil has the principal responsibility in agriculture. Maintenance of soil should be cautious and can be strenuous. Soil can get easily degraded which is of risk. Innumerable natural disasters and man-made aspects lead to this situation. Inappropriate agricultural practices, such as, uncontrolled tilling, using heavy machines, abuse of pesticides and inorganic fertilizers, poor planning of crop cycle and deprived irrigation results in the detrimental effect (Bhattacharyya et al., 2015).

Climate change enigma is another factor of concern. Indian farmers are highly dependent on monsoon. As climate changes worldwide, a farmer is unable to predict the right time to cultivate. Unforeseen natural disasters add to the pain. Through various studies the temperature of the world is expected to rise around 3.16 degree Celsius by 2050. In India it will be a maximum of 4.8-degree Celsius increase by 2080 (Palanisami & Kakumanu, 2019). Precipitation is expected to decrease by 2050. Weather patterns also affects soil quality, pest and disease spread in crops thus, taking a toll on productivity (Gupta & Yadav, 2020). Climate change directly affects water. Flood and drought results from flood or shortage (Palanisami & Kakumanu, 2019). Because India's prime resource for cultivation is water, farmers mainly culture during monsoon. But there is excessive water flow that brings about wastage and loss. Fresh water is also used in agriculture. However, it has a scarce amount left to use. Less than half of the farming lad in India are properly irrigated. Fresh water is vulnerable to various pollutions which degrades the quality of crops (Dharminder & Ram, 2019). Moreover, using fresh water questions sustainability because population growth means more demand for freshwater (Jain, 2019). Furthermore, the main reason for growth or death of crops is the improper supply of water. Absence of irrigation systems contribute to this (Dela Cruz, 2017).

Pesticide overuse is a problem in Indian agriculture. These pesticides which are non-biodegradable, are used to control pests. Overuse of pesticides affects the crop, consumers, and farming land. The chemicals used in pesticides penetrate the crops. Intake of even trace amounts of these chemicals is dangerous. Farmers manually spray these pesticides and are at high risk of diseases (Boudh, S & Singh, 2018).

Crop diseases is one of the significant challenges in Indian agriculture. Nematodes contribute to this largely. They are free living and parasitic and hence, they affect the crops directly and indirectly by various complicated diseases combined with other pathogens (Kumar, 2020). Fusarium species cause some of the largest numbers of crop diseases in India. This fungus gives rise to diseases in cardamom, tea, ginger, and maize. There has been no solution to fully eradicate this pathogen till date (Gopi, 2019). This takes a toll on the productivity. Together with these lack of knowledge with existing technology, financial instability to invest in infrastructure and technology lead to fall in productivity and economy eventually (Upadhyay, 2020).

The aim of sustainable agriculture is environmental health, economic profitability and ethical soundness. It should be able to maintain its production through time (Arulbalachandran et al., 2017). Sustainable agriculture protects and conserves natural resources. It protects crops and minimises losses while increasing productivity. This boosts quality of life for the farmers as well. Water and soil are the foundation (McNabb, 2019). Technology plays an important role in influencing agriculture in ways such as, equipment, pest control, management of crops (Arulbalachandran et al., 2017). The resources required decreases with the use of technology, which decreases the money and labour. The quality of the crop is increased, and farmers can sell their product at a higher price (Walter et al., 2017).

Precision agriculture aims to improve production and quality of crops (Chlingaryan, 2018). Precision agriculture improves productivity by identifying problems and catering to it. (Dela Cruz, 2017). It should decrease environmental problems and be cost efficient (Chlingaryan, 2018). The use of precision agriculture requires adequate knowledge and financial ability to invest in technologies (Shannon, 2018). Precision agriculture increases the profit. It helps in sustainable agriculture (Saiz-Rubio, 2020). It is the basic principle in agriculture 5.0. The new agriculture 5.0 uses artificial intelligence (AI), machine learning (ML) and IoT (Jha et al., 2019). It uses technology to produce more food while maintaining sustainability (Fraser, 2019). As the world grows more urbanised, farming labour decreases. Technology can match up to this shortage at a higher pace. Automation is faster than human and laborious work such as seeding and harvesting is easier (Saiz-Rubio, 2020). Precision farming is also able to reduce greenhouse gas emissions (Achim Walter, 2017). AI systems are computational systems that are designed to take inputs and present output to the user after processing the data (Smith et al., 2020). Artificial intelligence aids in developing a computer system capable of working in place of human intelligence. This system is able to identify patterns and make decisions (Misra et al., 2020). AI has remote monitoring capabilities and is also able to forecast weather. Therefore, it is beneficial in predicting the situation of the farm after appropriate recognition and analysis (Smith et al., 2020) AI has been used in agriculture to determine crop production, weed management and also in the marketing of products (Dutta et al., 2020). A prominent advantage of using AI in agriculture is reduction in stress on the available resources as things are carried out in a well-planned sequential manner. An eminent application of AI includes better crop growth and yield through strict monitoring of the crops based on their nutritional level, and also strategically differentiating weed from crops (Jha et al., 2019). Using AI in Agriculture makes it a more profitable and sustainable field (Real et al., 2020).

Machine learning is a division of AI that can be categorised into three groups, supervised, unsupervised and reinforcement learning. The three most commonly used paradigms are supervised learning, unsupervised learning and reinforcement learning and can conceivably be used by any given type of artificial neural network architecture. Each learning model has various training algorithms. Supervised learning is a machine learning technique with a known output called the labelled data. This algorithm maps the inputs with the desired output. The training algorithms used in supervised learning are linear regression, logistic regression, and design tree. The unsupervised learning algorithm trains ANN unlabelled data. K-means clustering, principal component analysis and hierarchical clustering are examples of this type of learning. Reinforcement learning, sets parameters of an artificial neural network, where data is not given, but generated through interchange with the environment (Zakaria et al., 2014). Components of artificial intelligence and the different deep learning approaches are described in Figure 2 and Figure 3 respectively.

Artificial Neural Network (ANN) is currently being used in many sectors like education, security, art and agriculture (Abiodun et al., 2018). The foremost model of ANN was introduced in 1943 by McCulloch and Pitts (Zakaria et al., 2014). Artificial Neural Network imitates the human brain in structure and behaviour. They are able to perform tasks like pattern recognition and classification. Just like the human brain, there are neurons in ANN which are its processing units. Basically, the connection between these processing units only makes up the ANN. The structure of ANN comprises of three layers. The input, hidden and output layers. Neurons can receive numerous inputs and give a single output, or they can receive multiple inputs and provide with multiple outputs. The output layer usually has a single neuron with values ranging between 0 to 1. The output of a neuron from the given set of inputs can be termed activation function of that neuron (Kukreja et al., 2016). The inputs are in numerical form (Dharwal & Kaur, 2016). ANN takes in the input, adapts to it, and predicts the output of a similar set of inputs. Sequentially, first the inputs are received by the input layer and these inputs are then transferred to the

*Figure 2. Definition of artificial intelligence (Inspired by Dutta et al., 2016)*

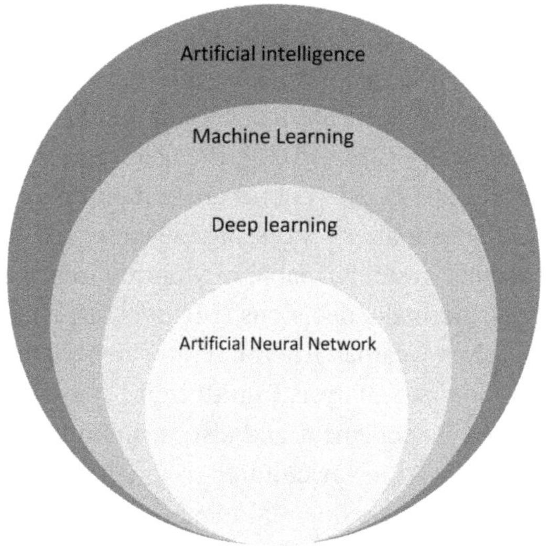

*Figure 3. different deep learning approaches (Inspired by Alom et al., 2019)*

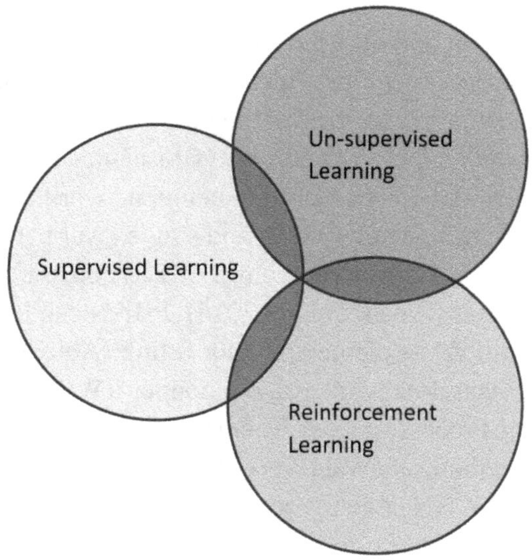


hidden layer that processes them and finally transfers the result to the output layer (Zakaria et al., 2014). Hidden layer processes all the information. Hidden layers in a neural network varies depending upon the problem. Multiple hidden layers are more complex (Abiodun et al., 2018). The outer layer provides the outcome of the problem to the user. The sum of the weighted input and the activation function gives the output. Figure 4 represents the structure of an ANN network. For neurons present in the same layer the same activation function is used (Dharwal & Kaur, 2016). Various functions can be used for activation

*Figure 4. Architecture of artificial neural network (Inspired by Brea et al., 2018)*

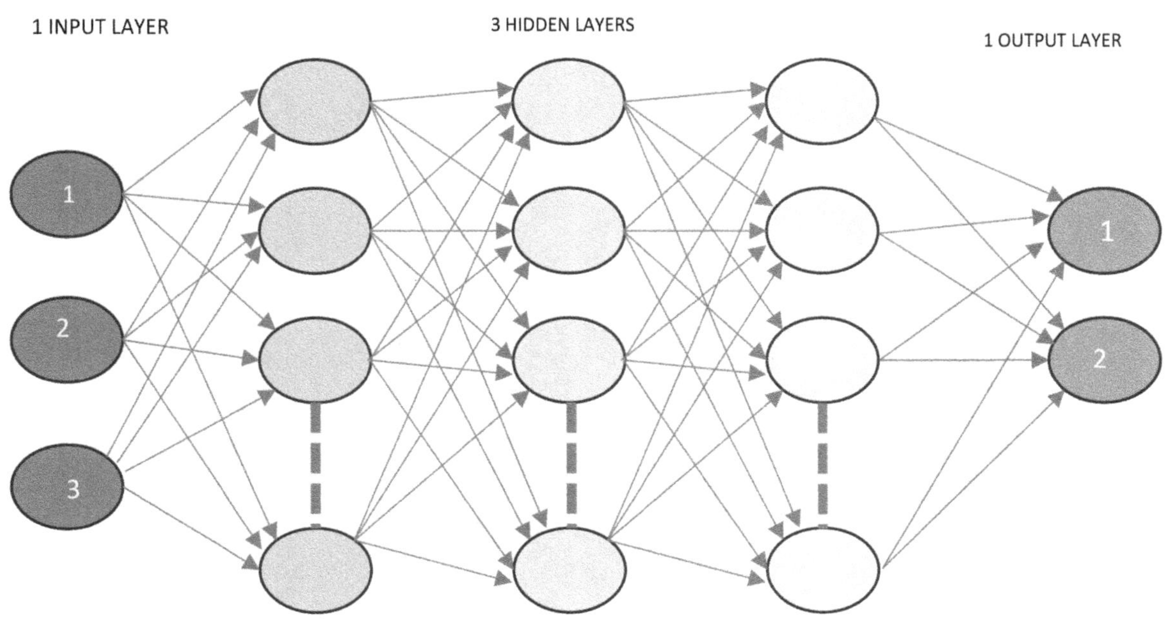

like the Step function, Linear function, Ramp function and the most commonly used Sigmoid function (Kukreja et al., 2016).

Once the output is released, it becomes the input for the next layer and the process continues (Haglin et al., 2019). Similar to the biological neurons which learn and adapt based on input from the environment, the ANN processing units too can learn with the help of learning paradigm. ANN is categorized into feed forward neural networks (FFNN), feed backward neural network (FBNN) (Abiodun et al., 2018), convolutional neural network (CNN), recurrent neural network, etc. In FFNN all the neurons are interrelated but are unequal in weight. Multi (MLP) and single layer perceptron (SLP) are examples of Feed Forward Neural Network. Multi-Layer Perceptron (MLP) consists of three consecutive layers: an input, a hidden, and an output layer (Zakaria et al., 2014). FBNN can be used for storing information, usually used for pattern recognition, medicine and data fitting (Abiodun et al., 2018). Among different ANN types, convolutional neural network is most popular (Waheed et al., 2020). CNN is able to recognise patterns without much processing time. There are different types of CNN architectures such as VGGNet, EfficientNet, and XceptionNet (Waheed et al., 2020). It is the leading unsupervised learning model (Golhani et al., 2018). The CNN contains convolutional layers, MaxPooling layers, ReLu activation function layer and fully connected layers (Hari et al., 2019). Single layer perceptron, multi-layer perceptron, recurrent network and feed forward network is illustrated in Figures 5, 6, 7, and 8 sequentially. Convolutional neural network architecture is described in Figure 9.

*Figure 5. Single layer perceptron (Inspired by Camuñas-Mesa et al., 2019)*

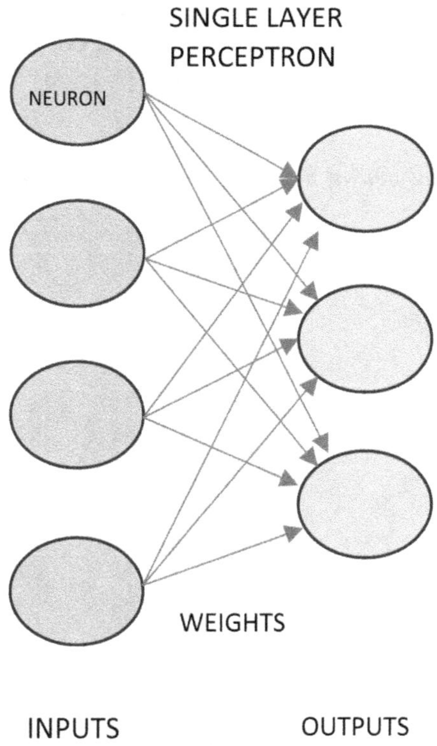

*Figure 6. Multi-layer perceptron (Inspired by Camuñas-Mesa et al., 2019)*

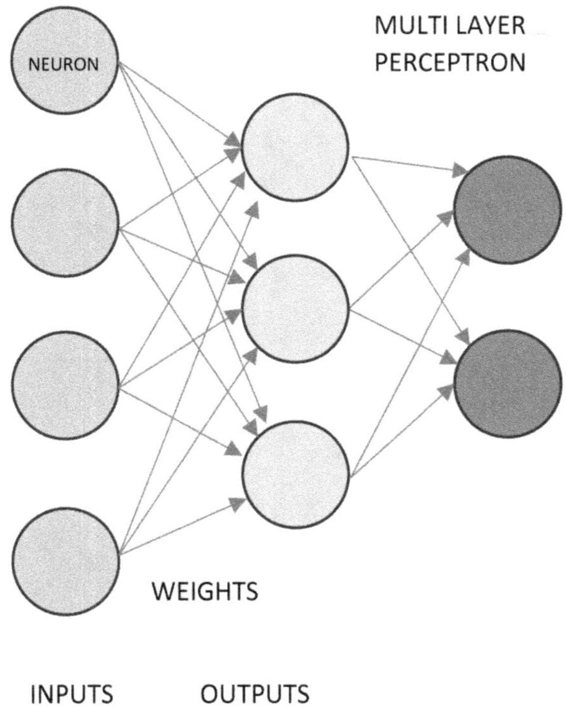


*Figure 7. Recurrent neural network (Inspired by Eliasy et al., 2020)*

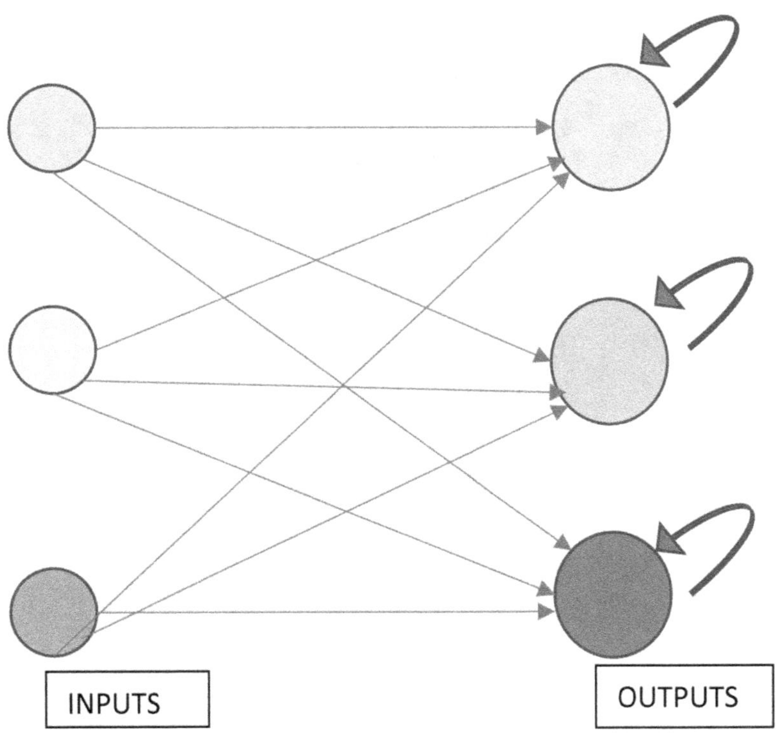

*Figure 8. Feed forward neural network (Inspired by Eliasy et al., 2020)*

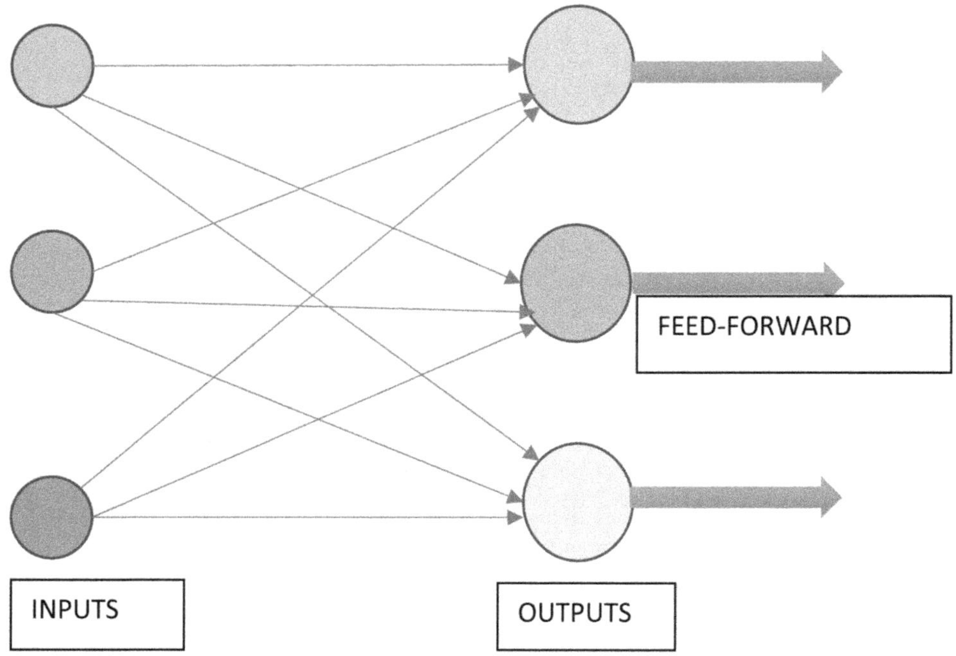


*Figure 9. Convolution neural network (Inspired by Zaniolo, 2020)*

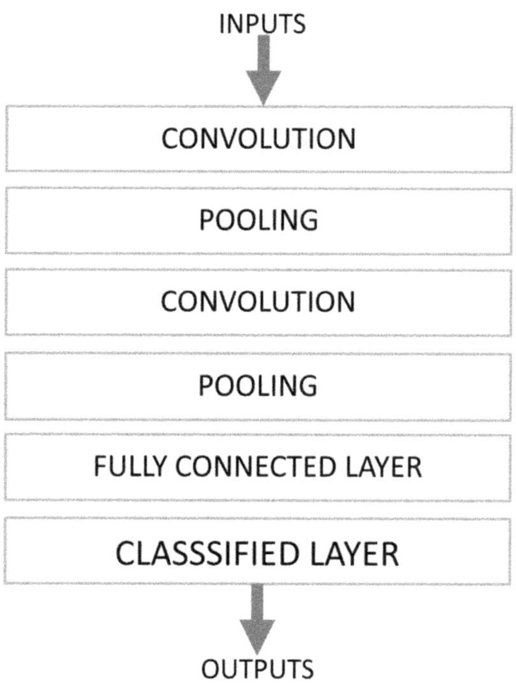

Generally, across all the types of ANN there are a few basic steps that are followed

1. Image or Data acquisition
2. Training
3. Prediction

The ANN model is trained for purposes such as identification of diseases, pests and so on after the image samples are obtained from the source. The samples are separated into learning and test data sets. After training with the training sets, the testing set images are fed and the accuracy of detection and classification is measured.

The advantages of ANN

Prediction accuracy, adaptability, and dependability.
Able to take huge data as inputs
Able to work with labelled and unknown datasets.

ANN helps in image processing, language translation, speech control devices and predicts agriculture. In agriculture, ANN can be employed to precision farming which helps in replacing laborious labour-intensive work with more conventional and convenient ones (Pathan et al., 2017). Factors crucial to the agricultural production include soil characteristics, crop production, plant related diseases and weeds. Analysis and monitoring of these factors through the application of ANN can help in easy and speedy detection of symptoms that can cause a negative impact on overall production (Eli-Chukwu et al., 2019).

## Prediction of Crop Yield

Crop yield prediction is essential for food security. Crop production can be affected due to climate change. It also depends on factors such as soil nutrition, sunlight, rainfall, fertilizers, and irrigation (Chlingaryan, 2018). Use of manual or conventional methods for data collection and yield prediction can be time consuming and expensive. Also, the information gathered through these methods is very less. Human error becomes unavoidable. Through crop yield prediction, many methods can be deduced for the improvement of crops (Manjula & Narsimha, 2016). The yield of a crop depends on soil factors, weather, cultivating and harvesting technology, plant protection and crop rotation. Prediction of yield becomes the foundation to estimate the production levels (Niedbała et al., 2019). Artificial neural network can help in predicting rainfall and soil nutrition levels. In this study a data driven algorithm is created, that is able to predict rice crop yield using soil and rainwater data. Here the neural network model recognises the optimal concentration of the soil parameters and rainfall pattern to predict the crop yield. There are many types of algorithms in the supervised learning and the algorithm that was PEST used is Time series prediction. Prediction is done with the help of Recurrent Neural Network. The prediction obtained from the studies has been found successful and the work also confirms that crop yield depends on soil and the rainfall (Kulkarni et al., 2018). In another study, crop yield prediction was done for Cotton, Sugarcane, Groundnut, Soyabean and Jawar. The various inputs used are pH, temperature, rainfall, calcium, etc. Here, the Feed Forward Back Propagation type is used. Feed Forward Back Propagation is an algorithm that comes under the supervised learning. In this type both the input and anticipated output should be given to the system. The prediction of crop yield is also done using the soil and the rainfall parameters (Dahikar & Rode, 2014). The process of crop yield prediction is portrayed in Figure 10.

*Figure 10. Crop yield prediction process (Inspired by Zaniolo, 2020)*

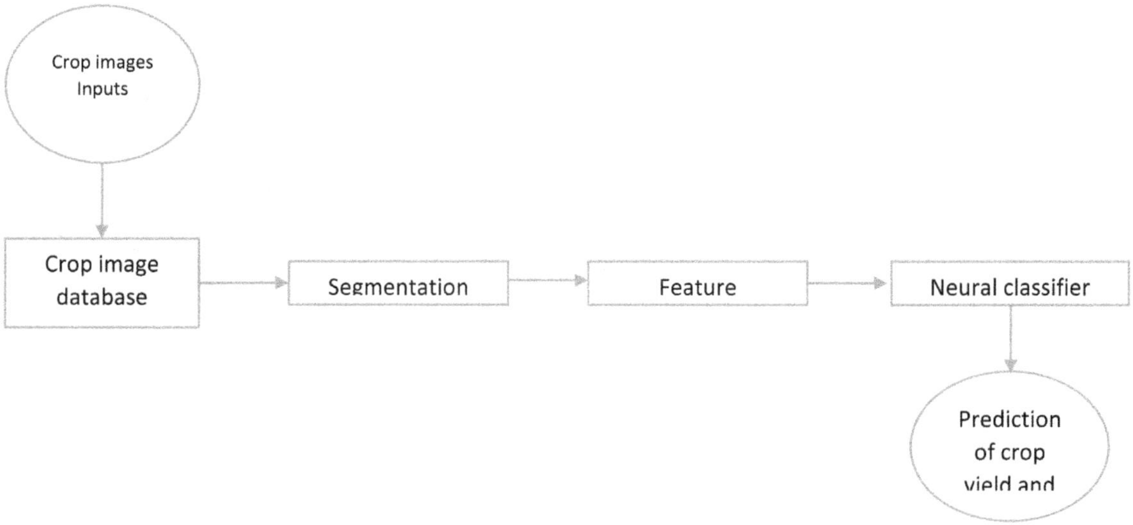


## Water Irrigation

Water irrigation systems efficiently sprays crops with the appropriate amount of water. But allocation of water is still a challenge. Irrigation mostly uses fresh water hence sustainable usage Is important. Precision irrigation schedules irrigation, accurately tending to water needs of the crop. This improves efficiency, enhances health and productivity while reducing energy. In artificial neural network feed forward neural networks (FFNN) is a highly capable model. however, they have inadequate capability in preserving information. Recurrent neural network (RNN) can also be used in such cases. It is similar to FFNN. It has self-feedback nodes that gives this model its memory and dynamic usage. Long short-term memory (LSTM) is a very powerful subset of RNN which comes under RNN which has memory. It requires lesser date pre-processing too. The study aimed to develop LSTM for soil moisture prediction. The performance was then evaluated based on the comparison between FFNN and LSTM. Data was gathered depending on windspeed, rainfall, air temperature, relative humidity etc. The data was then pre-processed for training both LSTM and FFNN. For LSTM soil moisture data and rainfall needed to be standardised and trained. For proper irrigation scheduling the soil moisture content has to be predicted a day earlier. The trained model predicts soil moisture. In another study this LSTM model was used to predict irrigation schedule by drip method. Multilayer neural network is a highly capable model which recognizes and regulates sprinkler systems (Kamyshova, 2020). In the study the authors proposed a system model that combines recurrent neural network and IoT for soil moisture prediction and irrigation. The irrigation aspect of the model contains 4 components. The inputs (temperature, rainfall, air humidity, etc.,) that are measured using various sensors. The next component is the anchor and server nodes. Server nodes save the data. RNN LSTM model comes next. It predicts the soil moisture from the previous day's data. The irrigation scheduling system determines the water needed for the crop. The water sprinklers are automated to sprinkle the desired amount of water (Kashyap, 2021). Water sprinklers can be trained to with multi-layer neural network models to increase its irrigation efficiency. Irrigation systems consumes high energy. Decrease in pressure lead to decrease in energy consumption. However,

*Figure 11. Irrigation using neural network block diagram model (Singh et al., 2018)*

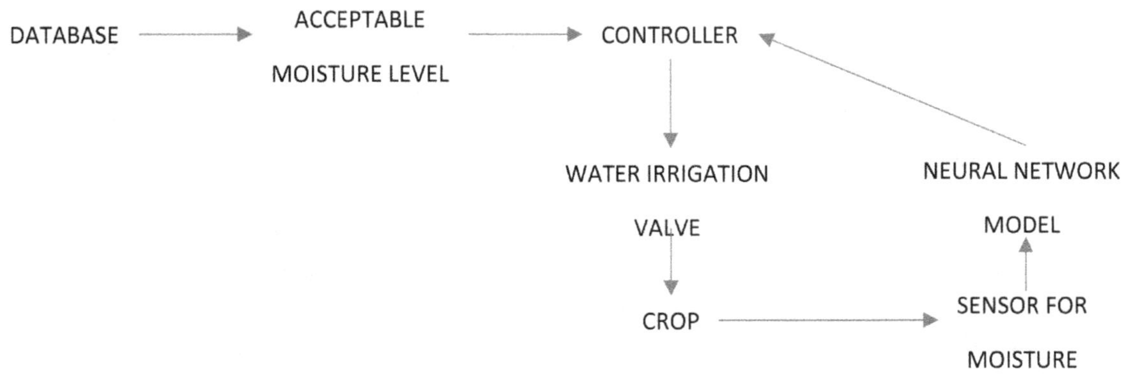


*Figure 12. Irrigation for crop using multi-layer perceptron neural network (Inspired by Neto et al., 2015)*

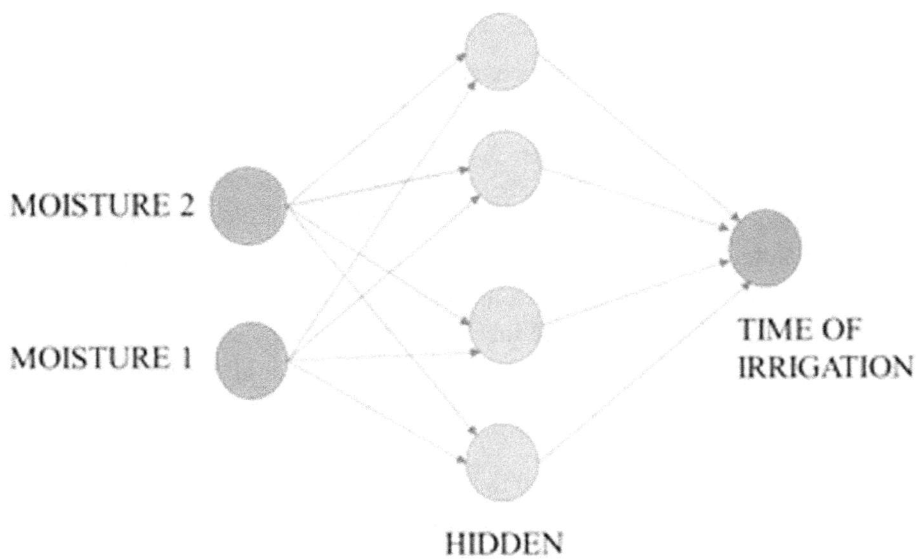


as pressure decreases the efficiency of the system fails. A multilayer neural network model can regulate the pressure and speed of the irrigation system which conserves water and energy (Kamyshova, 2020). Soil moisture was estimated through CNN model using thermal infrared images (TIR). Temperature can be reflected in TIR images. CNN is used to connect these two factors to evaluate the moisture content. After training the model was applied to real fields. A drone attached with thermal camera was set to picture different areas of the fields. The CNN was compared with a Deep neural network model for comparison of the output. CNN proved better results with better precision (Sobayo, 2018). The process of irrigation through neural network is pictured in Figure 11 and 12.

## Disease Detection and Identification

Identification and detection of diseases is a key feature in practising sustainable agriculture. Diseases are encountered by plants during the growing and the production stage. Presence of moisture at the surface in contact with air is the key factor that causes development of a disease. Learning capabilities of artificial neural network can be useful in the detection and identification of diseases (Singh et al., 2016). Early detection is necessary to avoid much human interference. Diseases can be symptomatic or asymptomatic. The artificial neural network systems identify diseases based on combining features such as the colour and texture (Golhani et al., 2018). Corn is a highly valued crop in India. The overall production of corn is more than wheat and rice. However, there are many diseases that affect corn which can significantly affect the yield and the nutritional level. Manual crop monitoring is very costly, less reliable and time consuming. In the proposed work an optimised version of the DenseNet has been used. This architecture is cost-friendly and provides accurate results as compared to other models. The dense net connects with the other layers in a feed forward manner, the connection allows optimal flow of information that enhances ability of network. The feature map of one-layer acts as the input of the following layer. DenseNet reduces the number of variables and encourages reuse of the features which contributes to less computational time. The connectivity pattern of the architecture is dense and hence called Dense CNN. DenseNet can be made by connecting many blocks that are all attached to each other. The DenseNet model was first trained for leaf disease recognition and classification. The first layer of the CNN extracts the features and give out a feature map and passes it to the first dense block. The dense blocks are made up of 5 convolutional layers. Each convolutional layer consists of many other layers like the activation layer. The first convolutional layer produces 4 feature maps. In a dense block, the convolutional layers must produce feature maps in exact dimension. In between 2 dense blocks there exists a transition layer that is able to change the size of the feature map according to average pooling. The dataset used contains 12,332 images. The architecture showed identification accuracy of 98.06% for 3 disease types. The approach manages to learn by itself and reach an effective recognition rate (Waheed et al., 2020).

Digital images that can come in handy and the proposed study classifies the diseases through CNN with the help of mobile phone cameras that are affordable to the farmers (Shrivastava et al., 2019). The CNN usually requires a large, labelled dataset like the ImageNet because it does not display efficient performance with small scale datasets (Shrivastava & Pradhan, 2020). Rather than training the network from scratch, the use of transfer learning can have effective results with small datasets. Transfer learning reuses models applied for other tasks as the starting point for another model of a new interest. It borrows labelled data to attain the highest performance in the field of interest (Abade et al., 2020). One of the models is AlexNet. It does not use a classification layer, instead, the features are fed into Support Vector Machine (SVM) (Shrivastava & Pradhan, 2020). The architecture used in this study was AlexNet that is a deep CNN model pre-trained on large ImageNet dataset. The first layer of the model defines the dimension of inputs. Whatever images that are given as inputs have to be in that particular dimension. The final layer of the CNN model is called the classification layer. In this studypaper the author does not use the classification layer and therefore, removes it. Instead, the extracted features are fed into a classifier SVM for training (Shrivastava et al., 2019).

Different types of disease detection tools are described with an example of the model is provided in Table 1. Convolutional neural network can be considered as the successful classification tool. The plant disease identification and general process is represented by Figures 13 and 14.

*Table 1. Different plant diseases identified using artificial neural network*

| S. No | Plant | Disease Tested | Classification Tool | Accuracy Achieved | Author(s) |
|---|---|---|---|---|---|
| 1. | Tomato Leaf | Bacterial leaf spot, septorial leaf spot, yellow leaf curl | Convolutional neural network | 94-95% | Prajwala et al. |
| 2. | Groundnut Leaf | Cercospora | Back propagation neural network | 97.41% | Ramakrishnan et al. |
| 3. | Tomato | Fungal late blight, septorial leaf spot, bacterial canker, leaf curl | Multi-layer feed forward back propagation network | 87.2% | Sabrol et al. |
| 4. | Tomato Leaf | Target spot, mosaic virus, bacterial spot, late blight, leaf mold, yellow leaf curl, spider mites | Convolutional neural network | 91.2% | Agarwal et al. |
| 5. | Oil palm | Ganoderma basal stem rot disease | Multi-layer back propagation neural network | 83.3-100% | Ahmadi et al. |
| 6. | Mango Leaf | Anthracnose, alternaria leaf spots, leaf gall, leaf webber, leaf burn | Convolutional neural network | 96.67% | Arivazhagan et al. |
| 7. | Soyabean Leaf | Carijo leaf, charcoal rot, southern blight, mildio, mela, rust, murcha sclerocio | Convolutional neural network (SoyNet) | 98.4% | Karlekar et al. |

*Figure 13. Plant disease identification block diagram model (Inspired by Ramakrishnan et al., 2015)*

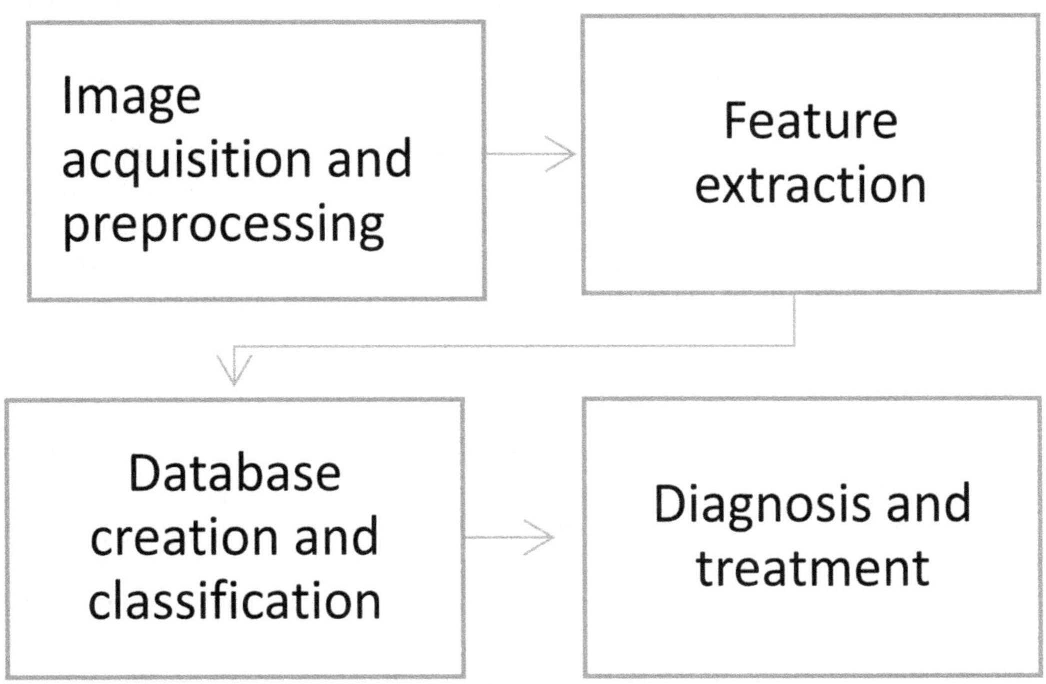

*Figure 14. General process of disease detection (Inspired by Pawar et al., 2016)*

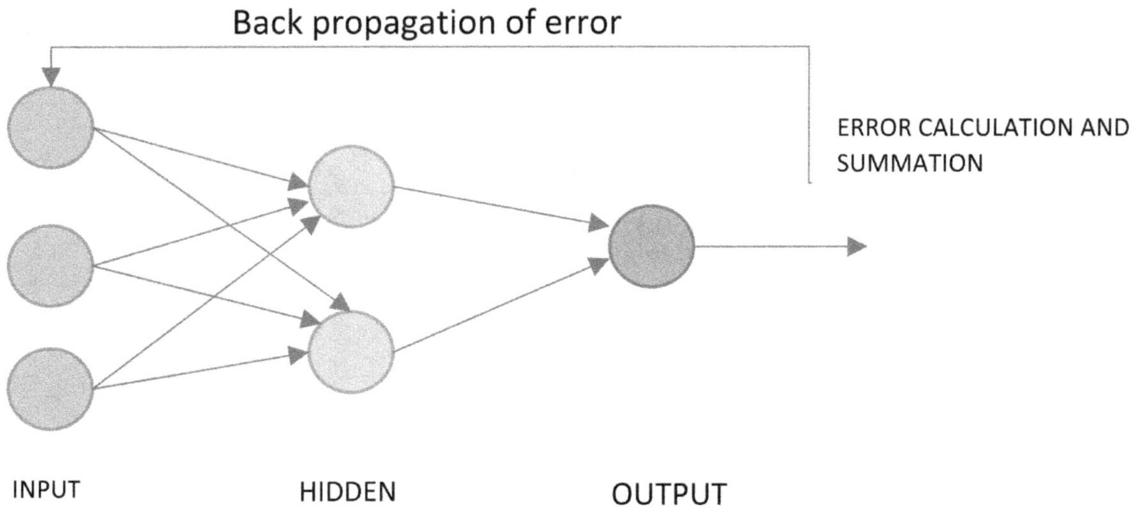


## Pest Manifestation

ANN can be used in the detection of pest manifestation. The study was based on the pest in coconut trees. The pests were identified using the drone, the image of the pest manifestation was directly sent to the farmers' mobile devices using wifi network. Coconut trees are mainly infested with Rhinocerous Beetle, Red palm weevil, termites, mites, and black headed caterpillar. Some of the methods used for plant disease detection are Back propagation neural network, feed-forward neural network, image processing and probabilistic neural network (Chandy, 2019). To properly train the ANN and receive accurate results, high quality images are required. Detection of pests through ANN is done by means of image acquisition and processing. The images of the sample are obtained and then processed. The object is first detected and then segmented. Segmentation is the division of an image into parts or into sets of pixels. The image has to be made simple for easy analysis. Features are then to be extracted. The features can be morphological or colour properties. Here multilayer perception feed forward ANN is used. The accuracy for identification of pest has been recorded to be 80% (Nair et al., 2017).

## Classification Tool

Cashew kernels are acquired from raw cashew nuts which is a widely celebrated ingredient in an Indian household. To date these kernels are processed manually as quality is a major concern. Grading is an important aspect in the local and international market. Usually, the sorting and grading process involves use of equipment that may harm the cashew kernels. However, mechanisation at some stages may help in the process which may in turn result in human error in the process or fatigue in labourers. The mechanism used in the process are image acquisition, pre-processing, feature extraction and classification. The images were captured by 2 types of cameras. MATLAB was used for the image processing after which they were segmented. Multi- layer perceptron model was used in the process with 21 input and 4 output neurons. This system helped sort and grade 4 types of cashew kernels: WW-180, WW-210, WW-240 and

*Figure 15. Pest manifestation identification block diagram model (inspired by Sneha et al., 2019)*

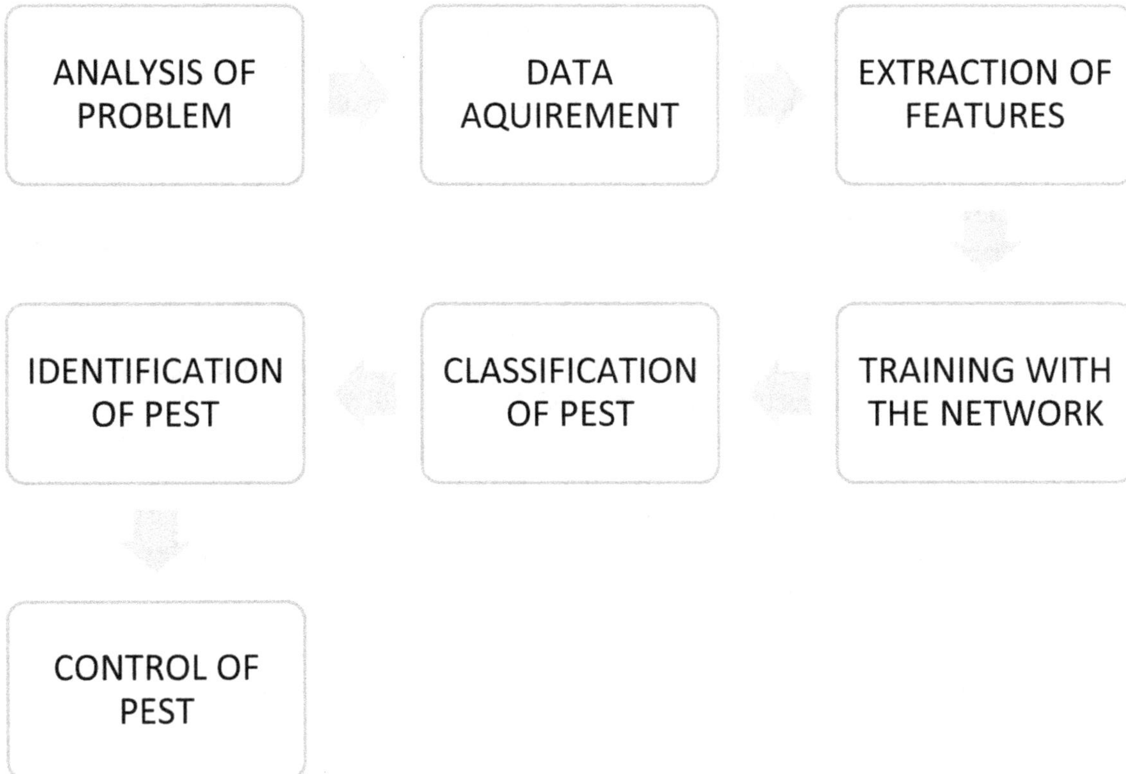


WW-320. The features extracted were the length, shape and colour. Segmentation separates the cashew from its background into non-overlapping region that is useful for further analysis (Ganganagowdar et al., 2016). Artificial neural network greatly aids farmers from manual sorting which may increase the efficiency of production and hence fulfilling an aspect of sustainable agriculture.

## Weed Detection and Herbicide Application

Traditionally, weed is eliminated by herbicides or by manual means. However, these actions have their own drawbacks. Using these methods develops soil infertility and pests (Patidar et al., 2020). Weeds reduce crop productivity by decreasing the quality and quantity of the crop leading to economic losses (Ekwealor et al., 2019). Weed causes weight drop, poor nutrient uptake, and decrease in grain yield in food crops (Tiwari et al. 2019). They compete for essentials such as sunlight and nutrients (Yu et al., 2019). Automation is very important for this. In weed detecting seedlings are usually used to avoid confusion. CNN model was used to automatically identify and remove weed. Images of weed was captured through raspberry pi-based module (Mounashree et al., 2021). CNN was used to train and identify with sample images first, followed by newly captured images. If weed is detected it is cut off using a cutter attached to the module. Convolutional neural network model is used, called the Mask R CNN. This algorithm correctly identifies the distinct objects. The input is an image, and the output is a bounding box with mask formed on the object specifically. It highlights specific areas where the target might lie. Later, class label is determined for each region of interest and mask is predicted for these regions using pretrained data set. The dataset used is the V2 plants seedlings dataset that contain more than 5,500 weed seedling images. They can be grouped into different classes such as Cleavers, Common wheat, Black-grass, and Fat-Hen. The mask Regional CNN is the newest version of Region Based convolutional network (R-CNN). After the completion of the training process, a random image is selected and mask R CNN is applied to it, which detects the weed. The accuracy was recorded to be 98%. As the manual removal and the excessive and non-specific use of herbicides have degrading effects on the soil and also requires high labour and time (Patidar et al., 2020). Specific spraying of herbicides is also possible with artificial neural network. Weeds of perennial ryegrass were identified using Deep convolutional neural network. Deep CNN was noted to be reliable, portrays high precision in feature extraction and is generally fast with an almost perfect accuracy. Different architectures under DCNN were compared. GoogleNet, VGG-16 Net, AlexNet and DetectNet were compared. Hence, use of artificial neural network increases sustainability in the field of agriculture. Images were acquired and trained on the basis of single weed and multiple weed images. Which means that in one picture there can be only single species or multiple species of weeds pictured. The architectures showed different potential levels. The F1 scores were the best for VGGNet and Alexnet in terms of precision measurements but exhibited a dip in the recalling ability. The values were approximately 0.99 and 0.60 respectively (Yu et al., 2019).

## CONCLUSION

Technology is expensive, which is the most basic challenge for a farmer in shifting to precision agriculture. Farmers in India are still refusing to break their comfort zone to adopt to technology in farming. The government plays a significant role in influencing its citizen. It is a responsibility of the government to educate and promote technology.

## REFERENCES

Abade, A., Ferreira, P., & Vidal, F. (2020). Plant Diseases recognition on images using Convolutional Neural Networks: A Systematic Review. *Computers and Electronics in Agriculture*, *185*, 106125. Advance online publication. doi:10.1016/j.compag.2021.106125

Abiodun, O. I., Jantan, A., Omolara, A. E., Dada, K. V., Mohamed, N. A., & Arshad, H. (2018). State-of-the-art in Artificial Neural Network Applications: A Survey. *Heliyon*, *4*(11), e00938. doi:10.1016/j.heliyon.2018.e00938 PMID:30519653

Agarwal, M., Singh, A., Arjaria, S., Sinhad, A., & Gupta, S. (2019). ToLeD: Tomato Leaf Disease Detection using Convolution Neural Network. *Procedia Computer Science*, *167*, 293–301. doi:10.1016/j.procs.2020.03.225

Ahmadi, P., Muharam, F. M., Ahmad, K., Mansor, S., & Seman, I. A. (2017). Early detection of Ganoderma basal stem rot of oil palm using artificial neural network spectral analysis. *Plant Disease*, *101*(6), 1009–1016. doi:10.1094/PDIS-12-16-1699-RE PMID:30682927

Alom, M. Z., Taha, T. M., Yakopcic, C., Westberg, S., Sidike, P., Nasrin, M. S., Hasan, M., Van Essen, B. C., Awwal, A. A. S., & Asari, V. K. (2019). A State-of-the-Art Survey on Deep Learning Theory and Architectures. *Electronics (Basel)*, *8*(3), 292. doi:10.3390/electronics8030292

Arivazhagan, S., & Ligi, S. V. (2018). Mango Leaf Diseases Identification Using Convolutional Neural Network. *International Journal of Pure and Applied Mathematics*, *120*, 11067–11079.

Arulbalachandran, D., Mullainathan, L., & Latha, S. (2017). Food Security and Sustainable Agriculture. *Sustainable Agriculture Towards Food Security*, 3–13. doi:10.1007/978-981-10-6647-4_1

Bhattacharyya, R., Ghosh, B. N., Mishra, P. K., Mandal, B., Rao, C. S., Sarkar, D., Das, K., Anil, K. S., Lalitha, M., Hati, K. M., & Franzluebbers, A. J. (2015). Soil degradation in India: Challenges and potential solutions. *Sustainability*, *7*(4), 3528–3570. doi:10.3390u7043528

Boudh, S., & Singh, J. S. (2018). Pesticide Contamination: Environmental Problems and Remediation Strategies. *Emerging and Eco-Friendly Approaches for Waste Management*, 245–269. doi:10.1007/978-981-10-8669-4_12

Brea, F., Gimeneza, J. M., & Fachinottia, V. D. (2018). Prediction of wind pressure coefficients on building surfaces using Artiðcial Neural Networks. *Energy and Building*, *158*, 1429–1441. doi:10.1016/j.enbuild.2017.11.045

Camuñas-Mesa, L. A., Barranco, B. L., & Gotarredona, T. S. (2019). Neuromorphic Spiking Neural Networks and Their Memristor-CMOS Hardware Implementations. *Materials (Basel)*, *12*(17), 2745. doi:10.3390/ma12172745 PMID:31461877

Chandy, A. (2019). Pest Infestation Identification in Coconut Trees Using Deep Learning. Journal of Artificial Intelligence, *1*(01), 10–18. doi:10.36548/jaicn.2019.1.002

Chlingaryan, A., Sukkarieh, S., & Whelan, B. (2018). Machine learning approaches for crop yield prediction and nitrogen status estimation in precision agriculture: A review. *Computers and Electronics in Agriculture*, *151*, 61–69. doi:10.1016/j.compag.2018.05.012

Dahikar, S. S., & Rode, S. V. (2014). Agricultural Crop Yield Prediction Using Artificial Neural Network Approach. International Journal of Innovative Research in Electrical, Electronics, Instrumentation and Control Engineering, 2(1), 683–686.

Dela Cruz, J. R., Baldovino, R. G., Bandala, A. A., & Dadios, E. P. (2017). Water usage optimization of Smart Farm Automated Irrigation System using artificial neural network. *2017 5th International Conference on Information and Communication Technology (ICoIC7)*. 10.1109/ICoICT.2017.8074668

Dharminder, R. K. S., Kumar, V., Devedee, A. K., Mruthyunjaya, M., & Bhardwaj, R. (2019). The clean water: The basic need of human and agriculture. *International Journal of Chemical Studies*, *7*, 1994–1998.

Dharwal, R., & Kaur, L. (2016). Applications of Artificial Neural Networks: A Review. *Indian Journal of Science and Technology*, *9*(1), 1–8. doi:10.17485/ijst/2016/v9i47/106807

Dutta, S., Rakshit, S., & Chatterjee, D. (2020). Use of Artificial Intelligence in Indian Agriculture. *Food and Sci. Rep.*, *1*, 65–72. doi:10.13140/RG.2.2.24973.97762

Ekwealor, K. U., Echereme, C. B., Ofobeze, T. N., & Okereke, C. N. (2019). Economic Importance of Weeds: A Review. *Asian Plant Research Journal*, *3*, 1–11. doi:10.9734/aprj/2019/v3i230063

Eli-Chukwu, N. C. (2019). Applications of Artificial Intelligence in Agriculture: A Review. *Eng. Technol. Appl. Sci. Res.*, *9*(4), 4377–4383. doi:10.48084/etasr.2756

Eliasy, A., & Przychodzen, J. (2020). The role of AI in capital structure to enhance corporate funding strategies. *Array.*, *6*, 100017. Advance online publication. doi:10.1016/j.array.2020.100017

Fraser, E. D. G., & Campbell, M. (2019). Agriculture 5.0: Reconciling Production with Planetary Health. *One Earth*, *1*(3), 278–280. doi:10.1016/j.oneear.2019.10.022

Ganganagowdar, N. V., & Siddaramappa, H. K. (2016). Recognition and classification of White Wholes (WW) grade cashew kernel using artificial neural networks. *Acta Scientiarum. Agronomy*, *38*(2), 145–155. doi:10.4025/actasciagron.v38i2.27861

Golhani, K., Balasundram, S. K., Vadamalai, G., & Pradhan, B. (2018). A review of neural networks in plant disease detection using hyperspectral data. *Information Processing in Agriculture*, *5*(3), 354–371. doi:10.1016/j.inpa.2018.05.002

Gopi, R., Singh, S., & Raj, C. (2019). Status of Fusarium diseases of crop plants in North East India. *Indian Phytopathology*, *72*(4), 637–646. doi:10.100742360-019-00148-3

Haglin, J. M., Jimenez, G., & Eltorai, A. E. M. (2019). Artificial neural networks in medicine. *Health Technology (Hong Kong)*, *9*(1), 1–6. doi:10.100712553-018-0244-4

Hari, S. S., Sivakumar, M., Renuga, P., & Suriya, S. (2019). Detection of plant disease by leaf image using convolutional neural network. *2019 International Conference on Vision Towards Emerging Trends in Communication and Networking (ViTECoN)*, 1-5. 10.1109/ViTECoN.2019.8899748

Jain, S. K. (2019). Water resources management in India-challenges and the way forward. *Current Science*, *117*(4), 569–576. doi:10.18520/cs/v117/i4/569-576

Jha, K., Doshi, A., Patel, P., & Shahd, M. (2019). A comprehensive review on automation in agriculture using artificial intelligence. *Artif. Intell. Agric.*, *2*, 1–12. doi:10.1016/j.aiia.2019.05.004

Kamyshova, G., Solovyev, D., Zatinatsky, S., Kolganov, D., & Terekhova, N. (2020). Improving the efficiency of circular irrigation machines based on models of neural network irrigation control. *E3S Web of Conferences*, 175. doi:10.1051/e3sconf/202017505007

Karlekar, A., & Seal, A. (2020). SoyNet: Soybean leaf diseases classification. *Computers and Electronics in Agriculture*, *172*, 105342. Advance online publication. doi:10.1016/j.compag.2020.105342

Kashyap, P. K., Kumar, S., Jaiswal, A., Prasad, M., & Gandomi, A. H. (2021). Towards Precision Agriculture: IoT-Enabled Intelligent Irrigation Systems Using Deep Learning Neural Network. *IEEE Sensors Journal*, *21*(16), 17479–17491. doi:10.1109/JSEN.2021.3069266

Kukreja, H., Bharath, N., Siddesh, C.S., & Kuldeep, S. (2016). An Introduction to Artificial Neural Network. *International Journal of Advance Research and Innovative Ideas in Education, 1*(5), 27-30.

Kulkarni, S., Mandal, S. N., Sharma, G. S., & Mundada, M. R. (2018). Predictive analysis to improve crop yield using a neural network model. In *2018 International Conference on Advances in Computing, Communications and Informatics (ICACCI)* (pp. 74-79). IEEE. 10.1109/ICACCI.2018.8554851

Kumar, T., & Prakash, N. (2020). Adoption of AI in agriculture: The game changer for Indian farmers. *International Conferences ICT, Society, and Human Beings*, 204-208.

Kumar, V., Khan, M. R., & Walia, R. K. (2020). Crop Loss Estimations due to Plant-Parasitic Nematodes in Major Crops in India. *National Academy Science Letters*, *43*(5), 409–412. doi:10.100740009-020-00895-2

Manjula, A., & Narsimha, G. (2016). Crop Yield Prediction with Aid of Optimal Neural Network in Spatial Data Mining: New Approaches. International Journal of Information and Computation Technology, 6(1), 25–33.

McNabb, D. E. (2019). Pathways to Sustainable Agriculture. Global Pathways to Water Sustainability, 185-199. doi:10.1007/978-3-030-04085-7_12

Mishra, S., Mishra, D., & Santra, G. H. (2016). Applications of Machine Learning Techniques in Agricultural Crop Production: A Review. *Indian Journal of Science and Technology*, *9*(38). Advance online publication. doi:10.17485/ijst/2016/v9i38/95032

Misra, N. N., Dixit, Y., Al-Mallahi, A., Bhullar, M. S., Upadhyay, R., & Martynenko, A. (2020). IoT, big data and artificial intelligence in agriculture and food industry. *IEEE Internet of Things Journal*. Advance online publication. doi:10.1109/JIOT.2020.2998584

Mounashree, J. P., Sanjay, N., Sushmitha, B. S., Usha, B. G., & Shivamurthy, M. A. (2021). Automated weed removal system using convolutional neural network. *International Journal of Engineering Applied Sciences and Technology.*, *6*(2), 231–235.

Nair, D., Parte, A., Pokharkar, Y., & Pande, N. (2017). Recognition of Silverleaf Whitefly and Western Flower Thrips Via Image Processing and Artificial Neural Network. International Research Journal of Engineering and Technology, *4*(3), 1773–1776.

Neto, O. C., Teixeira, A. S., Braga, A. P., Santos, C. C., & Leao, R. A. (2015). Application of artificial neural networks as an alternative to volumetric water balance in drip irrigation management in watermelon crop. Journal of the Brazilian Association of Agricultural Engineering, *35*(2), 266–279. doi:10.1590/1809-4430-Eng.Agric.v35n2p266-279/2015

Niedbała, G. (2019). Application of Artificial Neural Networks for Multi-Criteria Yield Prediction of Winter Rapeseed. *Sustainability*, *11*(2), 533. doi:10.3390u11020533

Palanisami, K., Kakumanu, K.R., Nagothu, U.S., & Ranganathan, C.R. (2019). Climate Change and Agriculture in India. *Climate Change and Future Rice Production in India*, 1–6. doi:10.1007/978-981-13-8363-2_1

Pathan, M., Patel, N., Yagnik, H., & Shah, M. (2020). Artificial Cognition for Applications in Smart Agriculture: A Comprehensive Review. *Artif. Intell. Agric.*, *4*, 81–95. doi:10.1016/j.aiia.2020.06.001

Patidar, S., Singh, U., & Sharma, S. K. (2020). Weed Seedling Detection Using Mask Regional Convolutional Neural Network. *International Conference on Electronics and Sustainable Communication Systems (ICESC)*, 311-316. 10.1109/ICESC48915.2020.9155701

Pawar, P., Turkar, V., & Patil, P. (2016). Cucumber disease detection using artificial neural network. *International Conference on Inventive Computation Technologies (ICICT)*, 1-5. 10.1109/INVENTIVE.2016.7830151

Prajwala, T. M., Pranathi, A., Ashritha, K. S., Nagaratna, B. C., & Shasidhar, G. K. (2018). Tomato leaf disease detection using convolutional neural networks. *2018 eleventh international conference on contemporary computing (IC3),* 1-5. 10.1109/IC3.2018.8530532

Ramakrishnan, M., & Nisha, S. A. A. (2015). Groundnut leaf disease detection and classification by using back propagation algorithm. *2015 international conference on communications and signal processing (ICCSP),* 964-968. 10.1109/ICCSP.2015.7322641

Real, J. L., Toril, J. U., Arriaza, J. A., & Valenciano, J. P. (2020). A Look at the Past, Present and Future Research Trends of Artificial Intelligence in Agriculture. *Agronomy (Basel)*, *10*(11), 1839. doi:10.3390/agronomy10111839

Sabrol, H., & Kumar, S. (2016). Fuzzy and neural network based tomato plant disease classification using natural outdoor images. *Indian Journal of Science and Technology*, *9*(44), 1–8. doi:10.17485/ijst/2016/v9i44/92825

Saiz-Rubio, V., & Rovira-Más, F. (2020). From Smart Farming towards Agriculture 5.0: A Review on Crop Data Management. *Agronomy (Basel)*, *10*(2), 207. doi:10.3390/agronomy10020207

Salgotra, A. K., Manha, A. S., & Singh, P. D. (2018). Agricultural growth and productivity in India. *Internat. J. Appl. Soc. Sci.*, *5*, 192–199.

Shannon, S., Brennan, D., Hanna, D., Younger, Z., Hassan, J., & Breslin, G. (2018). The effect of a school-based intervention on physical activity and well-being: A non-randomised controlled trial with children of low socio-economic status. *Sports Medicine-Open, 4*(1), 1-12.

Shrivastava, V. K., & Pradhan, M. K. (2021). Rice plant disease classification using color features: A machine learning paradigm. *Journal of Plant Pathology*, *103*(1), 17–26. doi:10.100742161-020-00683-3

Shrivastava, V. K., Pradhan, M. K., Minz, S., & Thakur, M. P. (2019). Rice Plant Disease Classification Using Transfer Learning of Deep Convolution Neural Network. *International Archives of Photogrammetry and Remote Sensing*, 42(3), 631–635. doi:10.5194/isprs-archives-XLII-3-W6-631-2019

Singh, K., Kumar, S., & Kaur, P. (2016). Detection of Powdery Mildew Disease of Beans in India: A Review. Oriental Journal of Computer Science and Technology, *9*(3), 226–234. doi:10.13005/ojcst/09.03.08

Smith, M. J. (2018). Getting value from artificial intelligence in agriculture. *Animal Production Science*, *60*(1), 46–54. doi:10.1071/AN18522

Sneha, A. J., & Chakravarthi, R. (2019). Neural network-based pest identification and control in cauliflower crop using sounds of pest. *International Journal of Innovative Technology and Exploring Engineering*, *8*(9), 3127–3134. doi:10.35940/ijitee.I8546.078919

Sobayo, R., Wu, H.-H., Ray, R., & Qian, L. 2018. Integration of Convolutional Neural Network and Thermal Images into Soil Moisture Estimation. *1st International Conference on Data Intelligence and Security (ICDIS)*, 207-210. 10.1109/ICDIS.2018.00041

Tiwari, O., Goyal, V., Kumar, P., & Vij, S. (2019). An experimental set up for utilizing convolutional neural network in automated weed detection. *4th International Conference on Internet of Things: Smart Innovation and Usages (IoT-SIU)*, 1-6. 10.1109/IoT-SIU.2019.8777646

Upadhyay, M. (2020). Problems and Solutions to Low Agricultural Productivity in India. *Int. J. Adv. Multidiscip. Res.*, *3*, 40–49.

Waheed, A., Goyal, M., Gupta, D., Khanna, A., Hassanien, A. E., & Pandey, H. M. (2020). An Optimized Dense Convolutional Neural Network Model for Disease Recognition and Classification in Corn Leaf. *Computers and Electronics in Agriculture*, *175*, 105456. Advance online publication. doi:10.1016/j.compag.2020.105456

Walter, A., Finger, R., Huber, R., & Buchmann, N. (2017). Opinion: Smart farming is key to developing sustainable agriculture. *Proceedings of the National Academy of Sciences of the United States of America*, *114*(24), 6148–6150. doi:10.1073/pnas.1707462114 PMID:28611194

Yu, J., Schumann, A. W., Cao, Z., Sharpe, S. M., & Boyd, N. S. (2019). Weed detection in perennial ryegrass with deep learning convolutional neural network. *Frontiers in Plant Science*, *10*, 1422.

Zakaria, M., Al-Shebany, M., & Sarhan, S. (2014). Artificial Neural Network: A Brief Overview. *International Journal of Engineering Research and Applications*, *4*, 7–12.

Zaniolo, L., & Marques, O. (2020). On the use of variable stride in convolutional neural networks. *Multimedia Tools and Applications*, *79*(19), 13581–13598. doi:10.100711042-019-08385-4

# Chapter 11
# Impact of Artificial Intelligence and Machine Learning in the Food Industry:
## A Survey

**Archana Sharma**
https://orcid.org/0000-0002-7386-1931
*ABES Institute of Technology, India*

**Kajol Mittal**
*ABES Institute of Technology, India*

**Sunil Kumar**
*Institute of Management Studies (University Courses Campus), India*

**Utkarsh Sharma**
*Indraprastha Institute of Information Technology, Delhi, India*

**Prashant Upadhyay**
*ABES Institute of Technology, India*

## ABSTRACT

*In recent years, the food sector or industry has escalated to prominence as the most important industry to receive widespread attention. It encompasses various industrial activities related to food production, distribution, processing, preparation, preservation, transportation, and packaging. Machine learning (ML) is a subpart of artificial intelligence (AI), and it is widely used in the food sector for industrial automation and predictive modeling with the world's growing demand and population. AI assists in improving package shelf life, menu selection, food cleanliness, and safety. Because of AI and machine learning, smart agriculture, drones, and robotics in the area of the food sector are becoming the need of the modern era. This chapter discusses how AI and machine learning have the potential to be used in the food business to save money while simultaneously increasing resource efficiency. It highlights the food industry's achievements and challenges with specific attention to the role of machine learning and artificial intelligence.*

DOI: 10.4018/978-1-6684-5141-0.ch011

## INTRODUCTION

Technology is an essential component of food manufacturing and distribution in today's food industry. Packaging might be considerably improved with technology, extending shelf life and boosting food safety. Other technologies, such as machines, drones, robotics, and 3D printing, improve food quality and also cut down manufacturing costs of a food product. AI and ML technologies deal with a variety of ways to expedite and process automation, save revenue, remove human mistakes, reduce waste of plentiful items, happier consumers, streamlined and automated operations, and more individualized orders in a variety of food businesses such as restaurants, bars, cafes, and food manufacturers. In production lines, AI systems exceed human efforts in terms of accuracy, speed, and consistency. Artificial Intelligence solutions have much potential for enhancing hygiene and cleaning activities, which are the most critical variables in food safety. Intelligent algorithms can help businesses enhance the quality of their food and services, resulting in healthier meals for customers. AI makes more efficient use of enormous quantities of comprehensive agricultural data for our food crops to record levels of productivity. While AI has a lot of benefits in the food sector, it also has many disadvantages. AI has failed to become widespread due to financial constraints and a scarcity of experienced specialists. The industry's market growth is being stifled by the increased cost of large-scale implementation. It's never easy to integrate new technologies like AI and ML in its early stage of deployment into the food industry. Humans will always be needed to supervise operations, repair, and maintain old equipment in the food industry. Therefore AI will never be able to replace them. Technology can effectively collaborate with humans to boost operational efficiency. Plant pests and diseases, which vary from place to region and season to season, are a major concern in the food sector (FAO, April 2017). AI algorithms have a significant impact on predicting plant diseases. Machine Learning encompasses deep learning as a subpart (Figure 1). It simulates how humans make judgments using machine learning and neural networks (Zhang et al., 2021). It is pretty expensive and demands a considerable number of data points.

*Figure 1. AI, ML, and DL relationships*

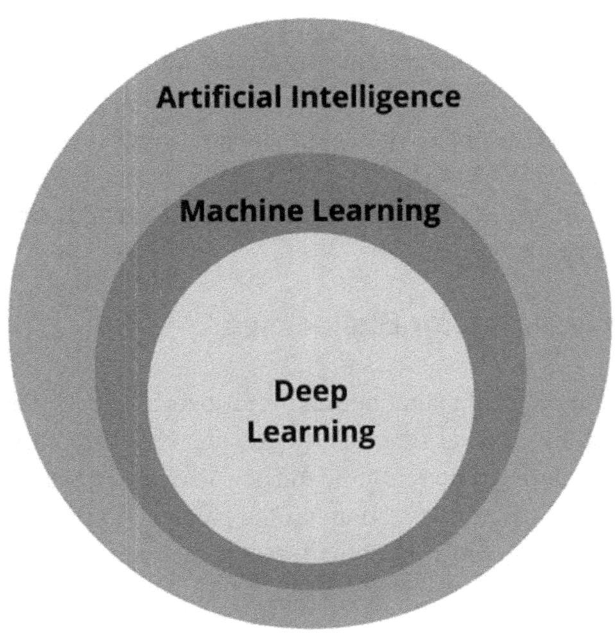

## The Machine Learning Procedure

Data is collected by a computer, which allows systems to learn new things from it. For the system, ML creates new self-learning algorithms and makes predictions as shown in figure 2. Machine learning employs statistics to detect patterns in large amounts of data. Essentially, anything that can be converted to digital data, such as numbers, images, and clicks, can be put into a sophisticated Machine Learning algorithm. Machine Learning is one of the essential technologies on the planet right now. Netflix, Spotify, and YouTube may now be able to provide suggestions. There are numerous other ML instances, such as feeds from multiple social media sites like Twitter and Facebook or voice assistant programs such as Siri and Alexa.

*Figure 2. An ML model*

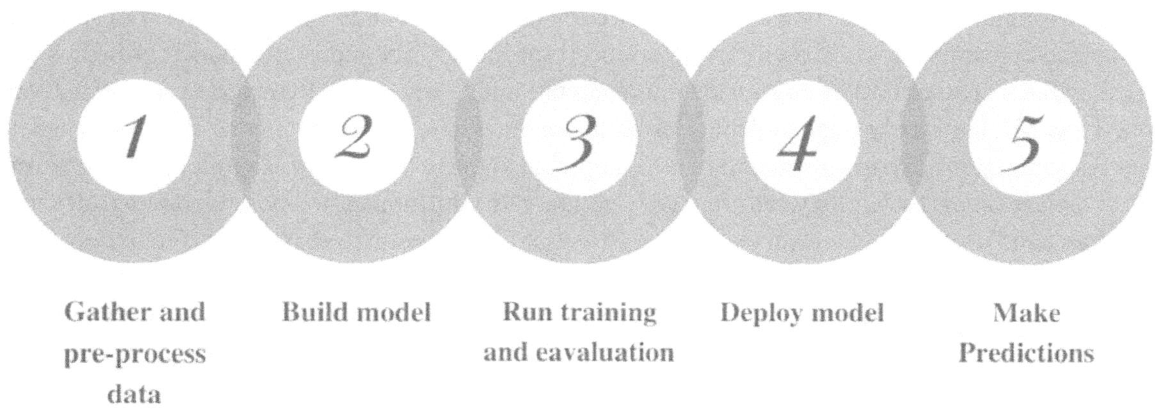


Machine learning techniques are classified into two types: supervised learning (Parvin et al., 2013) and unsupervised learning (Minaei-Bidgoli et al., 2014).

Supervised Learning - To train computers, this method employs examples of labeled data. In supervised learning, classification is a common technique Rokach, 2010). Neural networks, support vector machines, and decision trees are examples of classification techniques (Parvin et al., 2015).

Unsupervised Learning - This method allows algorithms to recognize patterns in data and group them into groups based on similarities. The K-means clustering method (Ahmad et al., 2007) is the most widely used clustering algorithm.

## Benefits of Machine Learning for Businesses

Machine Learning algorithms can help automate and prioritize corporate decisions, allowing for faster decision-making and adaptability by real-time data processing. For example, while driving a car, Google Maps adapts the optimum path based on real-time traffic. Advanced machine learning models provide a higher level of automation. This transformation has the potential to the user in entirely new business models, services, and products. Because machine learning can work with large amounts of complex streaming data, it may help discover insights beyond human competency and initiate appropriate action patterns. Business Procedures enabled by Machine Learning Algorithms could greatly boost efficiency. Business procedures enabled by machine learning algorithms could significantly improve the company's productivity. Exact forecasts, automated tasks, cost savings, and the elimination of human involvement are all advantages.

*Figure 3. ML's advantages*

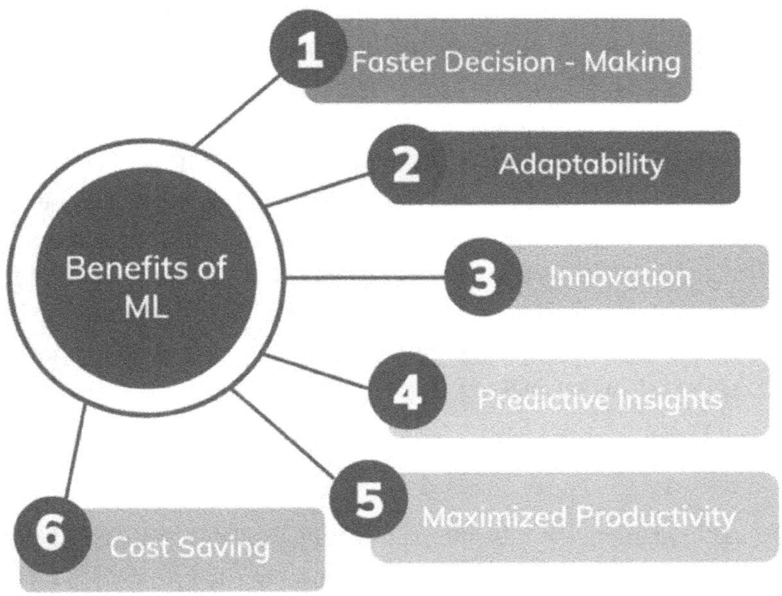


## Machine Learning applications in the Restaurants

Machine learning helps restaurant businesses in a variety of ways like customer experience, online sell and purchases, recommendations, revenue prediction, and food demand and delivery.

Machine Learning can **help improve client experience and provide analytic solutions.** There are currently a variety of programs that may assist in forecasting the number of orders required for a certain time or date by estimating customer visits, the number of orders placed, and related inventory demands. These programs and solutions gather historical data in order to better engage consumers by analyzing their behavior and preferences, leading to increased repeat visits and purchases.

Machine learning can also help **sell food on Websites and Mobile apps.** In this digital age, it is convenient to find the menu online through E-commerce. Consumer segmentation and automated customer service can improve the accuracy and efficiency of administrative operations like preparing reports, placing orders, deploying teams, and forming new jobs.

**Restaurant search on the internet** becomes easy with the help of Google Maps and Search engines. With the integration of AI, it provides the user suggestion for cafes or restaurants that may appeal to their tastes and are near their location. Restaurants may use tools like Amazon Alexa to allow customers to place orders instantly without clicking, allowing them to place orders swiftly. A self-serving systema (point-of-sale system) allows customers to control the ordering process, carefully examine their options, and sometimes even check the number of flavorings and spices in a meal. As human worker wages rise exponentially, these devices may become more widespread. As a result, robotics may save restaurants more money in the long term. The use of drones, robots, and street bots in delivery service is making task fast and easy.

## Artificial Intelligence (AI)

The goal of AI is to learn and apply information. In other words, the goal is to increase the likelihood of success rather than to achieve precision. As a result, a system that replicates human response and behavior is created. Artificial Intelligence is gaining the interest of businesses in a variety of fields and industries, including food processing and handling (FP&H). According to McKinsey. The FP&H sector is now valued at $100 billion and is predicted to increase at a CAGR of 5% by 2021. Artificial intelligence has a direct and indirect impact on the FP&H industry. Indirectly, it is beneficial to farmers in terms of weather forecasting, which helps them produce raw resources of superior grade in the course of food processing firms. AI also helps transportation businesses reduce shipping expenses, resulting in decreased transportation costs for food processors. However, when it comes to the benefits, AI aids the FP&H industry in five key areas: Sorting packages and goods, food safety compliance, hygiene, product development, and consumer decision-making.

## Artificial Intelligence in Food Processing and Handling

Facilities for processing food employ AI to automate sorting food or raw materials from the farm, maintaining machines and equipment, and ensuring quality until the final product is ready to transport. The following are AI applications that have a direct impact on food processing companies and help them save money and improve customer experience.

1. **Sorting Packages and Products using TOMRA**

A major issue for food processing enterprises is the need to separate raw materials before processing them into finished products. Because every kind of fruit and vegetable is unique, meticulous sorting is necessary. Every food processing firm has to maintain a particular level of quality in order to compete. In the absence of AI and other cutting-edge technologies like the Internet of Things (IoT), this strategy will need a large number of human employees. Unlike existing food sorting robots that only sort bad grade fruits and veggies from the good, TOMRA uses X-ray, NIR (Near Infrared) spectroscopy, LASER, cameras, and a novel machine-learning algorithm to assess the Sorting different characteristics of a fruit or veggie To recognise foodstuff, a TensorFlow machine created by Kewpie Corporation, a Japanese food processing firm, is utilised (Carey, 2017).

2. **Cameras with artificial intelligence**

Cameras equipped with artificial intelligence may assist restaurant management in monitoring their personnel to determine if they are wearing the proper food safety equipment. It is also critical to keep an eye on the culinary staff to ensure that the restaurant is clean and well-organized. Using Surveillance systems like KanKan AI, organizations can ensure that their employees keep their hands and other items clean. (Misra et al., 2020), which is an important factor affecting food safety.

3. **Food Hygiene - Keeping Things Clean**

Even if the operations are automated and unaffected by human participation, cleanliness is a big issue in food processing factories. The fact that a method is automated does not imply that the end result is safe to consume. It's possible that the equipment will become polluted. On a regular basis, both producers and major eating places require exclusive and complex machines to process a vast number of things. As a result, a large number of impurities flow through the cleaning equipment. As a result, there needs to be a better method as disassembling it every time is prohibitively expensive (Göbel et al., 2015).

4. **Development of Customized Products**

A single company can create a diverse array of unique items. Coca-Cola, for example, has acquired over 500 trademarks and provides its customers with over 3500 different drinks. So how does the company decide which flavor to produce next? Prior to AI, the corporation relied on surveys and advertising to learn about the interests of its customers. Many Coca-Cola stores now include self-serve soda stores where customers may mix their own beverages using Coca-Cola liquids. These stores were put in tens of thousands of locations around the United States. At each of these fountains, hundreds of people made their own drinks. They examined the data using artificial intelligence and determined that the buyers used to mix cherry-flavored Coke and sprite in the majority of cases. This information was used by Coca-Cola to introduce a new product Cherry Sprite.

5. **Assisting customers in making decisions**

Customers can use AI to make better purchasing decisions, Kellogg's created Bear Naked Bespoke, allowing consumers to customise Granola from over 50 ingredients. This technology not only expedites the production of batches of granola, but it also benefited the firm to figure out what their next product line should be, like Coca-Cola.

6. **Food Market Analysis**

Increasing revenue requires knowing which meals to include on the restaurant's menu. Because customer and market expectations are continuously changing, it has never been more important, to keep a step ahead of its competitors. AI/ML collects and categorises information to investigate user perceptions of flavour and preferences, dividing users into demographic subgroups and modeling their interests or predicting their desires.

7. **Production Optimization**

AI can optimise output and find the best-operating zones in industrial facilities to meet or surpass key performance parameters. Faster production changeovers, a reduction in the time it takes to transition from one product to another, and the early detection of production bottlenecks are just a few of its possible applications.

8. **Waste Reduction**

AI/ML based ways to measure and monitor can substantially improve waste reduction. Rather than evaluating the output quality at the end of a batch or cycle, Real-time AI monitoring may be able to detect abnormalities as they occur.

9. **Supply Chain Management**

Artificial Neural Network-based algorithms can monitor and confirm AI food delivery and product monitoring process at each level. It generates forecasts for price and inventory, which saves money. When it comes to food safety, food producers can know more about the travel of their products through the supply chain. In this scenario, AI in food manufacturing assists in monitoring every stage of the process, producing price and inventory management predictions, and tracking product flow from where it is grown to where it is received by consumers, ensuring transparency.

10. **Culinary Applications in the Real World**

Chef Watson, a product of IBM, is an AI-enabled digital culinary research helper. It has access to a database of flavour profiles and recipe ratios that helps to construct new dish combinations. Select the products to use in the software, and the cooking method wants to employ, and then review the suggested combinations by the algorithm. Watson constructs a model of the chef's tastes and delivers instructions based on it. In this case, machine learning allows real chefs to break free from monotonous cooking routines and come up with new ideas that will develop something unique.

## Statistics on AI in the Food and Beverage Industry

Due to rapidly changing customer patterns, technological breakthroughs, and severe regulations, the food, and beverage business has changed dramatically in the last decade. Such considerations have created a slew of challenges for the food and beverage business. The implementation of AI in food-and-drinks market is being propelled forward by factors such as dynamic changes in customer buying patterns. Artificial Intelligence (AI) leverages data from previous records, allowing sales outcomes to be anticipated for a given period. AI benefits food makers and merchants by assisting them in better understanding their clients. Companies will be able to discover the tastes and preferences of their customers, which will aid them in forecasting possible sales patterns for their items. The Food and Beverage industry is predicted to grow at a CAGR of approximately 65.3 percent from 2019 to 2024. Industry leaders are already reinventing their firms by incorporating cutting-edge technologies into their operations (Spanaki et al., 2021).

## AI's Advantages in the Food Industry

Because this is a low-margin, high-volume industry, increasing the use of AI in the food sector could help in many ways. Even a slight boost in efficiency can have a major impact on a company's success.

1. In addition to improving supply chain management, AI is increasingly being utilised to provide transparency and improve logistics.
2. Digitalization of the supply chain increases revenue and provides a clearer picture of the situation. Artificial intelligence (AI) is capable of analysing vast amounts of data in a manner that humans just cannot.
3. AI helps organisations reduce time to market and manage unpredictability.
4. Automated sorting reduces labor costs, speeds up production, and improves product quality.

*Figure 4. Application of ML and AI*

The following are some examples of applications of AI and ML in the food industry:

- Artificial intelligence aids in price forecasting, optimization of manufacturing processes, inventory management, and logistics management. It aids in determining where a specific crop has thrived.
- Combining machine learning algorithms with relevant cameras and sensors assists in the sorting of fresh produce as manual sorting is inefficient in terms of time, money, and accuracy.
- Machine learning helps forecast large, complex systems used in the food production process. It reduces operating expenses, reduces the workforce, and increases output.

- It aids in the forecasting of new products. Customers may create hundreds of distinct cocktails using these self-service gadgets by mixing additional flavors to their basic drinks. Artificial intelligence is used to evaluate hundreds of different beverages per day.
- Machine learning facilitates food delivery by using smart logistics. It assists in the tracking and identification of agricultural products and vegetables. In hotels, robots are also used to serve food (Göbel et al., 2015).

### Sustainable Food System (SFS)

Frameworks and approaches that can help build and implement AI-powered solutions in SFS in a sustainable and ethical manner are desperately required. Transdisciplinary AI strategies are required to create scenarios that balance the social, technical, and environmental realms. One research approach will be to build visioning techniques, in which stakeholders construct AI-powered solutions and discuss their potential ramifications.

Research of ethics in real-world AI solutions for food systems might be done. The use of AI in the design of systems transitions, commodities, and services has raised ethical problems, as has the employment of algorithms. A design-led case study keeps the conversation focused on practice-based SFS treatments rather than philosophical discussions. These findings might be used to improve future SFS systems by analysing the impacts of AI (benefits, drawbacks, and trade-offs). AI-powered system connections take on an unpredictable structure. They defy our notions about what makes a community or an interest group. They can disrupt existing production and consumption patterns.

### Problems/Challenges in Adapting AI and ML in the Food Industry

Some of the challenges faced by the Food Industry in Adapting AI and ML are as follows (Chidinma-Mary-Agbai, 2020):

Cost: The cost of deploying AI in the food industry is quite high, making it unaffordable for small-scale business owners; only large corporations can afford it.

Dread (cultural shift): As technology advances, there is an increase in machine power, which leads to a fear of unemployment. The trend to more advanced features such as using robots or drones to serve or distribute food results in the labour is being delegated to machines. They can be influenced negatively and are potentially detrimental to consumers.

Recipe Secrets: Companies do not want to divulge their recipes, and AI devices rely on transparency and user input to function properly.

## RELATED WORK

Machine learning is concerned with the use of intelligent software to aid machines in doing tasks more efficiently. The goal of this research is to provide a high-level overview of machine learning and to demonstrate the implementation of various unique ML models. Machine learning are applied in a wide number of applications, including data mining, predictive analytics, and image processing. Statistical learning approaches are at the heart of intelligent software used to develop machine intelligence. Machine learning datasets are in great demand these days due to the expanding amount of accessible datasets.

Computational techniques are used to automate knowledge acquisition through experience enhancement enabled by machine learning. Artificial intelligence (AI) is a subfield of computer science that is concerned with simulating human cognitive processes, learning capacities, and knowledge storage (Kumar et al., 2020). Artificial intelligence is classified as either strong AI or weak AI. The weak AI concept requires that the computer operate intelligently and reproduce human judgments, but the vital AI principle requires that the machine be capable of reflecting the human mind. On the other hand, strong artificial intelligence does not yet exist and is currently under investigation (Mavani et al., 2021).

"Machine learning may be used to address delivery route difficulties by optimising the delivery agent's position in respect to existing or upcoming traffic conditions and then advising them on the ideal route in a timely way." It becomes simpler to provide consistent orders to address difficulties in late deliveries. Additionally, using machine learning increases the quantity of data gathered over time, which can then be analysed using additional AI-based algorithms to produce a more intelligent system. This research might be conducted using more sophisticated AI-based methodologies, such as deep learning (DL), which would give an organization a competitive edge (Kumar et al., 2021). The advancements in the area of machine learning over the past several years have been nothing short of remarkable. Machine-learning models may be used to augment or replace human capabilities. They have a huge influence on our life in each of these instances. Math Destruction, a book by Cathy O'Neil, details various instances of machine learning algorithms gone astray. At the conclusion, she compared her efforts to Progressive Era muckrakers Upton Sinclair and Ida Tarbell. Upton Sinclair's famous 1906 novel The Jungle was about the processed food business. For computer scientists and engineers, learning about the origins and progress of processed food from inside the industry may be more informative (Camaréna, 2020). Food sales forecasting is the process of estimating future sales of food-related businesses. It may limit stocked and expired items within shops while also avoiding revenue loss by using accurate short-term sales forecasting (Ceballos-Magaña et al., 2012). The ambiguity surrounding records, citing a variety of factors, including growing high incomes in underdeveloped countries and rising measures of financial injustice, among others, might result in unbiased worldwide estimations. Finally, food supply remains a difficult challenge in the demand-supply chain, and the approach of picking an acceptable path from among sustainable contemporary technologies may provide better outcomes in terms of maintaining productivity while meeting demand. In comparison to other commercial industries, the food business is conceptually mature and slow-growing, with very little investment in research and development (Kumar et al., 2021).

Food demand increases in lockstep with the global population. Food processing is also thriving. Food waste should be minimised, supply chain operations streamlined, and food logistics, delivery, and safety all enhanced. ML enables a computer algorithm's accuracy at predicting outcomes to be improved without explicitly altering it. There are two types of machine learning techniques: supervised and unsupervised learning. This article discusses the use of machine learning and artificial intelligence in the food business (Göbel et al., 2015).

The purpose of this article is to discuss issues faced by customers in the food sector. Sorting fresh fruit, managing the supply chain, monitoring food safety compliance, implementing efficient wash-in-place systems, estimating customer demand, and developing new products have all been successfully implemented with increased efficiency and cost savings. Cost, cultural upheavals, expert skill needs, worries about transparency, and single-track minds all act as impediments to AI adoption. Despite these obstacles, research into the use of AI in manufacturing processes is continuing. It's worth mentioning, however, that the benefits of AI in the food industry much exceed the disadvantages (Deng et al., 2021).

AI plays a critical role in completing the functioning of the complete processing unit. While it has vital applications in the food processing and handling industries, its value is not limited to these sectors. It may be beneficial for food processing, storage, and transportation. Additionally, intelligent equipment like robots and intelligent drones may assist reduce packaging costs. Additionally, it will assist in the transportation of food, work completion in dangerous places, and the availability of high-quality items. AI's primary responsibilities in the food industry are grouped into two categories: food security management and food quality control (Kumar et al., 2021).

Artificial intelligence (AI) is already changing how the food and beverage sector thinks about food production, manufacturing, processing, quality, distribution, and consumption. This article discusses the potential applications of artificial intelligence in the food and beverage industry. It is worth mentioning, however, that the benefits of AI in the food business significantly exceed the disadvantages (Chidinma-Mary-Agbai, 2020). It can extract operational human knowledge, establish interpretable rules for categorizing samples despite non-linear human behavior or method, and assess the impact of each objective aspect of the examined meal on an expert's final conclusion (Eli-Chukwu & Ngozi Clara, 2019).

In the food sector, AI has become a contemporary technology as a consequence of rising food consumption and global population expansion over the previous several decades. The ability of the aforementioned intelligent systems to perform a variety of tasks in the food business, including food quality assessment, control tools, food categorization, and prediction, has increased demand for them. As a consequence, this article analyses the advantages, limitations, and formulations of those multiple applications in order to serve as a guide for selecting the most effective methods for boosting future breakthroughs in artificial intelligence and the food sector. An electronic nose, electronic tongue, computer vision system, and near-infrared spectroscopy are advocated for both industrial and consumer advantage (Mavani et al.,2021).

For years, AI has been used in the food sector to classify, categorise, forecast, and regulate parameters, as well as for quality control and food safety. In the food business, expert systems, fuzzy logic, ANNs, adaptive neuro-fuzzy inference systems (ANFIS), and machine learning are all widely used approaches. Prior to the deployment of AI, food studies were conducted to educate the public about food knowledge and to enhance the end outcomes associated with food qualities and production. The AI technique offers a number of advantages, and its use in the food business has been growing for decades (Mavani et al., 2021). MeatReg is a web-based application that automates the process of establishing the optimal machine learning method for comparing data from multiple analytical techniques in order to anticipate the counts of microorganisms that might ruin meat independent of the packaging strategy used. As a result, suggestions were provided about which analytical platforms and machine learning methodologies are optimal for predicting each species of the bacterium (Estelles-Lopez et al., 2017).

The goal of this research is to describe machine learning techniques used in the food processing sector for assessing the manufacturing process's quality and associated metrics, such as raw material carbon footprint and energy resources. The classification results of k-Nearest Neighbors, Neural Networks, C4.5, Random Forest, and Support Vector Machines are compared. The use of the Random Forest approach produced the best results for categorising two goods' processes. The categorization will assist us in identifying and obtaining the most suited manufacturing processes and production parameters (Milczarski et al., 2020).

Food production is a complicated process fraught with uncertainty (e.g., stochastic yield and demand, variations in raw material and ingredient prices), resulting in discrepancies between planned and actual output. The purpose of this project is to develop data analytics-based methods for quantifying the severity of these discrepancies, enabling companies to maximise revenues while reducing their environmental effect and aiding with food waste management (Garre et al., 2020).

The study opens with a discussion of the fundamental theoretical ideas behind ANNs, fuzzy logic, and intelligent agents. Artificial biomimetic technology and numerous conventional drying technologies are described in artificial biomimetic technology (Eliazàt, n.d.).

SVMs were used to give graphic tequila authenticity based on mineral content. Support vector machines were used to construct an adequate classification model. The best classification result is achieved using a linear SVM model with 100% prediction capacity (Ceballos-Magaña et al., 2012). The human senses of smell, taste, hearing, and touch may all be used to judge the quality of food and beverages. Sensory assessment is also used to evaluate sparkling wines and fizzy drinks. FIZZeyeRobot is a robotic pourer that normalises the variability of foam and bubble growth when pouring into a vessel (Condé et al., 2017).

Color is a visually important characteristic of dried fruits and vegetables because it influences customer choice and acceptance. For the prediction of colour changes in mushroom slices throughout the drying process, a novel model based on extreme machine-learning algorithms incorporating Bayesian techniques (BELM) has been created (Liu et al., 2020). The effects of drying factors on the colour variations of apple slices were explored utilising a hot air drying system in conjunction with a real-time computer vision system. Apple slices' colour features and moisture content were also associated with drying factors and drying time using a multilayer perceptron (MLP) artificial neural network (ANN). With a correlation value better than 0.92, the MLP ANN correctly mimicked the colour and moisture changes of apple slices. As a consequence, the computer vision system with ANN may be utilised to analyse and control real-time changes in food colour and moisture content during drying in a non-invasive, low-cost, and uncomplicated way (Kumar et al., 2020).

The obesity pandemic is sweeping the world, and conducting periodic countrywide surveys to estimate the obesity rate is expensive. On the other hand, country-level food sales data may be acquired from a number of sources on a regular basis for a low cost proposed approach has been shown to be reliable in terms of absolute prediction error as well as the fraction of nations where obesity prevalence may be accurately predicted (Dogan et al., 2021).

As the implications of AI on society, politics, and business concerns increase, the food and technology sectors, as well as regulators, will adjust. The purpose of this research is to demonstrate how an industry-wide software solution was used to overcome several challenges and ethical issues in order to allow the connectivity and openness of food safety data across the supply chain. The food and technology sectors, as well as regulators, must work to define and address society, policy, and business (Friedlander et al., 2020). The objective is to establish consistent and repeatable processes for monitoring product quality. To automate operations, expert systems should include human expert knowledge in the quality control process (Goyache et al., 2001).

The purpose of this project is to employ artificial intelligence to design an integrated system for anticipating product development success and choosing a sound market-product strategy in food companies. The purpose of this study is to design a decision-taking and decision-making system for managers in the food and beverage sector in the area of new product development and strategy. As a consequence, this paper reviews the early literature on the creation of new products and the elements that influence their success, as well as market-product strategy, adaptive neural fuzzy inference, and fuzzy inference systems. Rather than depending on expert opinion, this evaluation offered a new way for the future team rules fuzzy inference system in the data-driven method (Ben Ayed et al., 2021). When social distance is taken into account for health reasons, AI may be a critical ally at all levels of the agri-food system, including agriculture, the food business, and large-scale distribution. It focuses on reinventing "how" firms are operated, especially operational procedures, as well as a full redesign of the whole business model in light of potentially unpredictable economic situations such as the one resulting from the COVID-19 pandemic (Di Vaio et al., 2020).

Due to the daily fluctuations in client demand, many firms predict daily food sales. This information may be gleaned from a customer's prior purchasing transaction history. Forecasting sales is becoming more important in major food companies, such as supermarket networks. Oracle's Retail Demand Forecasting (Tsoumakas & Grigorios, 2019) is an example of a system that is more specialised.

Our goal is to adhere to high-standard processes in the food business by applying innovative technologies such as AI and big data. This article highlights studies on artificial intelligence and big data analytics in the food sector, including ML, ANNs, and other techniques. This method entails analysing data patterns and modifying workflows in order to provide accurate, trustworthy output that requires less human resources, is more efficient, and assists the user in forecasting future scenarios over time (Sharma et al., 2021). This research will discuss data analytics-based solutions for projecting the severity of these discrepancies, with the goal of increasing corporate profitability while decreasing environmental impact through food waste management (Garre et al., 2020). This article defines and categorises the production line difficulties addressed by machine learning, specifies the industrial domains targeted, explores the machine learning techniques used, and explains the independent and dependent variables included in the models (Kang et al., 2020).

In terms of statistical certainty, demographic estimates corroborate the FAO's conclusion about overpopulation growth (Guh, 2003), which is shown in Fig. 5A as a global heat map. Figure 5C illustrates greenhouse gas emissions on a continental scale, demonstrating that rising modernity linearizes a rise in environmental threats while Figure 5D illustrates a continental portfolio's value-added proportion of GDP.

*Figure 5. Statistic demonstrating the food crisis (A) world population increase heat map, (B) poor nutrition in Asia, (C) agricultural greenhouse gas emissions and (D) GDP value-added share (2013) (Godfray et al.,2010)*

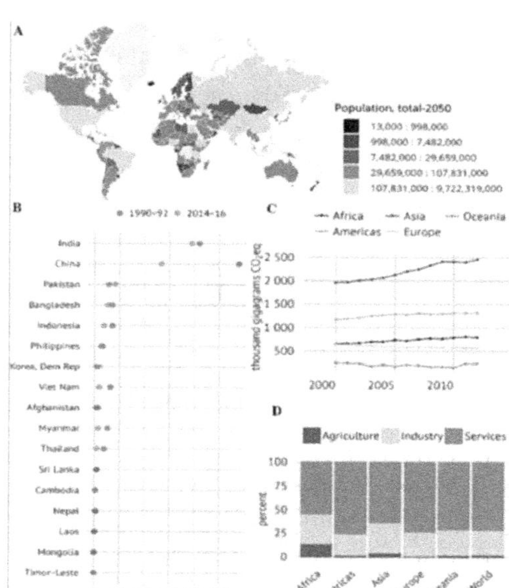

Due to the unpredictable nature of current manufacturing processes, make-to-order procedures are frequently used. Having the correct inventory on hand might help avoid delays. However, current inventory planning methods do not completely account for scheduling risk (Figure 6).

Because activity durations are simulated using probability distributions, critical routes and significant activities must be identified in stochastic scheduling. As a result, a route's criticality analysis must incorporate the risk of becoming critical. Accordingly, Markov Activity Networks is a set of algorithms that simulate the execution of a group of activities with exponentially distributed durations using Markov models. These methods use a transient evaluation of the underlying continuous time Markov chain to establish route and activity criticality (Radke et al., 2013).

*Figure 6. General risk management process*
*(Schwarz et al., 2015)*

## CONTROL OF STATISTICAL PROCESSES

The food business has long been seen as an important part of the economy. Cost-cutting and food-safety benefits may be gained by controlling statistical processes. More sustainable food production is now possible because of the use of Sector 4.0 technologies like deep learning and computer vision in the food industry (Kakani et al., 2012). An effective method for reducing variability and increasing predictability is Statistical Process Control (SPC). SPC approaches may be used in the food business because of the high level of process variability. The capacity of the control chart to distinguish between data that is out of control and data that is naturally variable (data that is under control) is one of its greatest assets. One of the most important additions to regular control charting is control chart pattern detection. An expert system with rules and an artificial neural network is the most common AI technology employed in SPC applications. Automated SPC seems to benefit greatly from the combination of ANN, ES, and the process cost modeling approach (Guh, 2003).

## AGRONOMIC ENGINEERING (AGRITECH)

AgriTech is a term used to convey the idea of moving away from traditional agricultural practices and toward more efficient methods based on modern technology. This field of study is concerned with the creation and operation of a farm in order to earn a livelihood, while also dealing with global commerce, customer requirements, agricultural regulations, and environmental challenges. A Farm Management Information System is a computer system that gathers and analyses data, stores and distributes it in the form of information that is necessary to carry out the farm's operational tasks," It will be tough to manage an FMS described in figure 7. Some automation of the management system is necessitated by the huge number of services, users, and end devices, as well as the compatibility with the communication network architecture. In order to implement this automation, the system must be aware of its surroundings and be able to react autonomously in the event of an emergency (Spanaki et al., 2021) & (Kaloxylos et al., 2012) . In order to boost productivity, reduce costs, integrate systems, and promote sustainable agriculture and food production methods, artificial intelligence (AI) and data science are being investigated.

*Figure 7. Food management system*
*(Kaloxylos et al., 2012)*

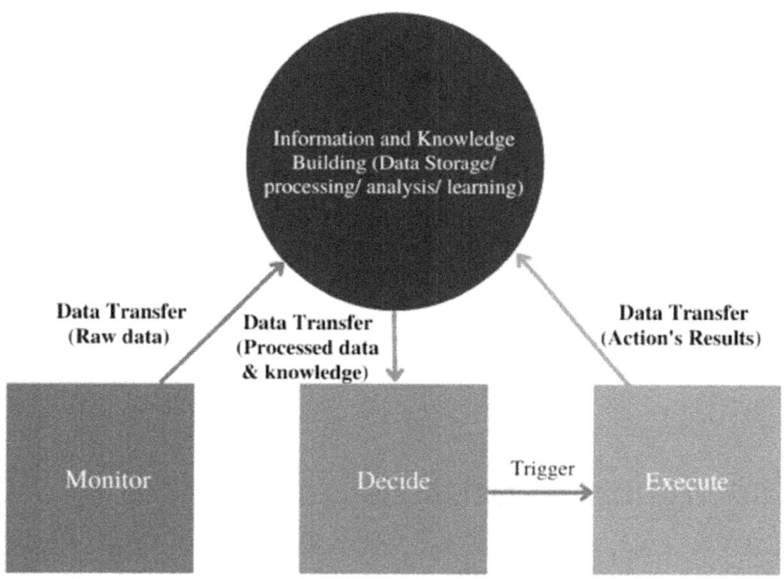


Smart Agriculture aims to make farming more efficient by using technology such as geographic mapping, sensors, machine-to-machine communication, data analytics, and smart information platforms represented in figure 8. To feed more people while squandering less resources, conventional agricultural processes must become more intelligent, productive, and sustainable via the use of Information and Communications Technology (ICT). Smart Agriculture refers to the use of disruptive ICT technologies to improve food production efficiency by boosting crop output, decreasing waste, and improving market access. While many Smart Agriculture technologies are game changers, not all are commercially viable now. According to Accenture study for Vodafone's Connected Agriculture report in 2019, most Smart

Agriculture applications are now only partially adopted, if at all, and mostly by big farms in industrialized nations. The great majority of smallholders still do not have access to these technologies. However, we anticipate that in a decade, technology will be more deeply embedded in global food supply networks. We need widespread access to high-speed internet and inexpensive smart devices. Although these technologies are not yet accessible everywhere, they are predicted to be almost widespread by 2030. Farmers will be able to boost resource efficiency, productivity, and resilience via the use of ICT, as well as minimize food waste along the supply chain. While these technologies are fundamental, Figure 8 identifies disruptive technologies in four critical areas that will drive change, ranging from increasing productivity and resource efficiency to strengthening shock resilience and lowering global food waste (Smarter2030.gesi.org, 2015).

*Figure 8. Food – Future of smart agriculture: Technology vision for 2030*

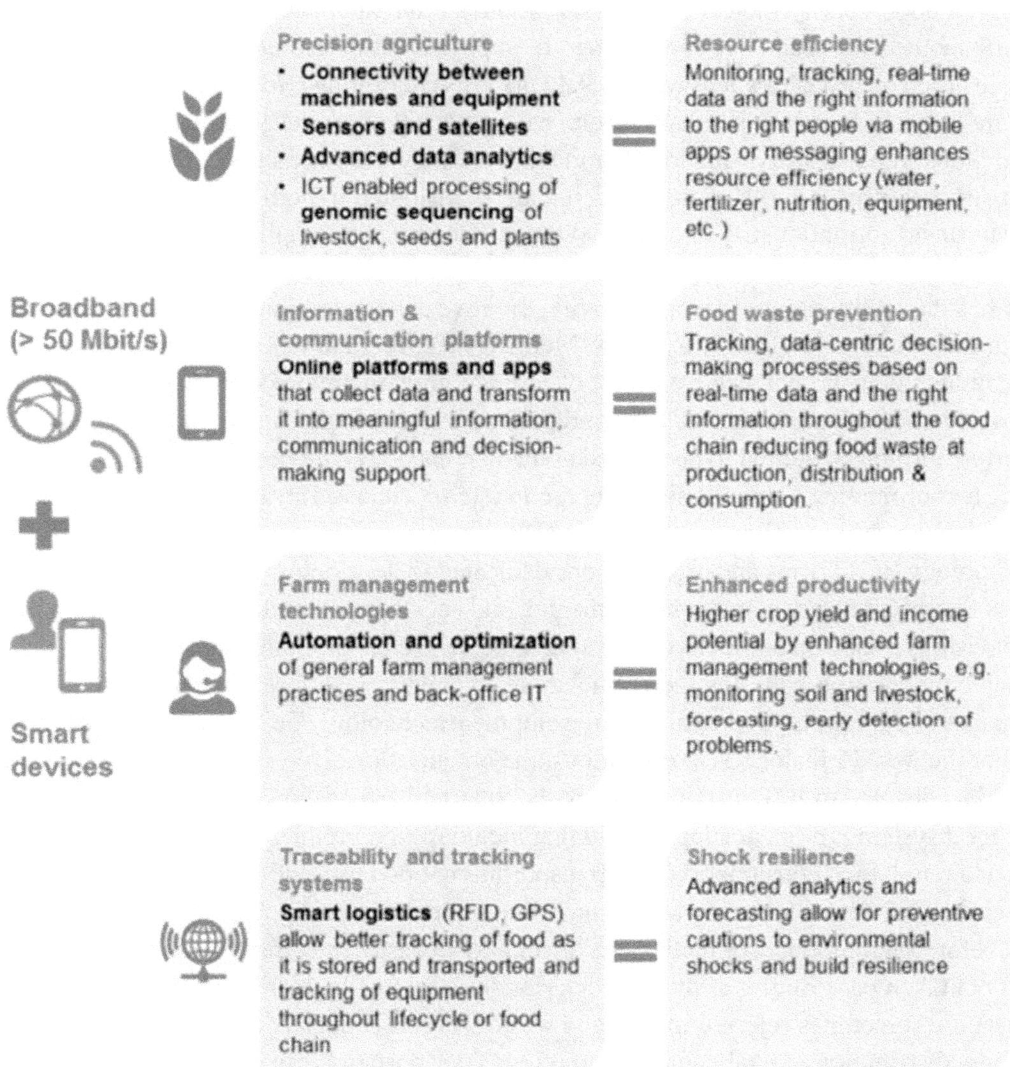

This paper presents the difficulties, possibilities, and viable solutions for the post-COVID-19 period, which is focused on intensively preserving the agri-food supply chain while providing the growing demand for green agreement ideas (Rowan et al., 2020). Paludiculture, the advancement of wet peatland innovation, has the ability to aggressively maintain and integrate agri-food and green technology, hence facilitating the COVID-19 pandemic transition. The establishment of new sustainable multi-actor innovation centers seems to be assisting, connecting, and empowering firms in their recovery and pivoting away from the COVID-19 epidemic. AI has the potential to alter all facets of the agri-food system, including agriculture, the food business, and large-scale distribution. It focuses on reimagining "how" firms operate, including operational procedures, as well as a thorough redesign of the whole business model to account for unforeseeable economic situations such as the COVID-19 pandemic (Rowan et al., 2020).

In order to optimise agricultural inputs and returns based on supply and demand, robotic data collection, decision-making, and corrective action is required. Constant agricultural stress affects global food security. This research uses a probabilistic approach based on human-centric AI to analyse data from the Global Food Security Index (GFSI). The simplicity and applicability of this intuitive probabilistic reasoning method for predictive forecasting are important. It makes use of a user-friendly AI-enabled technique based on probabilistic Bayesian Network (BN) reasoning (How et al., 2020). Extreme weather events are growing more frequent and severe, causing concern for developing nations' agricultural sectors and global food security. Due to a rise in the frequency and intensity of severe weather events such as droughts, heatwaves, and flooding, which may be attributed to both human-induced climate change and the intrinsic nonstationarity of climate systems. This creates complications across several economic sectors and has significant consequences for global food security (Biffis et al., 2017).

ICT has the ability to increase productivity, improve connection, minimize food waste, and perhaps overhaul the whole food chain (Castillo & Meliif, 1970). Agriculture and food systems are industries that are in high demand but also have a lot of room to grow in a more sustainable manner. Meanwhile, as the world's food system evolves toward digitalization and higher industrialization and automation, ICTs have enormous potential to be utilised at many points throughout the food chain on a global scale, enabling and promoting a more sustainable food system (Yuan & Luyao, 2019).

The Global e-Sustainability Initiative (GeSI) is a strategic alliance of information and communication technology (ICT) firms and organisations dedicated to developing and supporting technologies and practices that promote economic, environmental, and social sustainability. GeSI, which was founded in 2001, envisions a sustainable society via responsible, ICT-enabled change. GeSI encourages worldwide and open collaboration educates the public about its members' efforts to enhance their sustainability performance and supports new sustainable development technology. GeSI's membership comprises more than 30 of the world's major ICT corporations; the organisation also works with a variety of worldwide stakeholders devoted to ICT sustainability goals. Smarter2030.gesi.org, 2015)

The food system is a linear supply chain that includes food production, food processing, food distribution, and retail, food consumption, and transportation. Food loss and food waste occur throughout the supply chain, from manufacturing to consumption, as depicted in Fig. 9. Food losses are "food spilt or spoiled before final products or retail stage", while food wastes are "food suitable for human consumption but not taken because it is allowed to deteriorate or thrown by merchants or consumers". Food loss and waste are sometimes referred to as "food waste". Figure 9 depicts the processes of manufacturing, processing, distribution & retail, and consumption. Transportation can be considered as a link, not considered as a process while researching ICT solutions or analysing sustainability (Yuan & Luyao, 2019).

*Figure 9. Illustration of the food system studied*
*(Yuan & Luyao, 2019)*

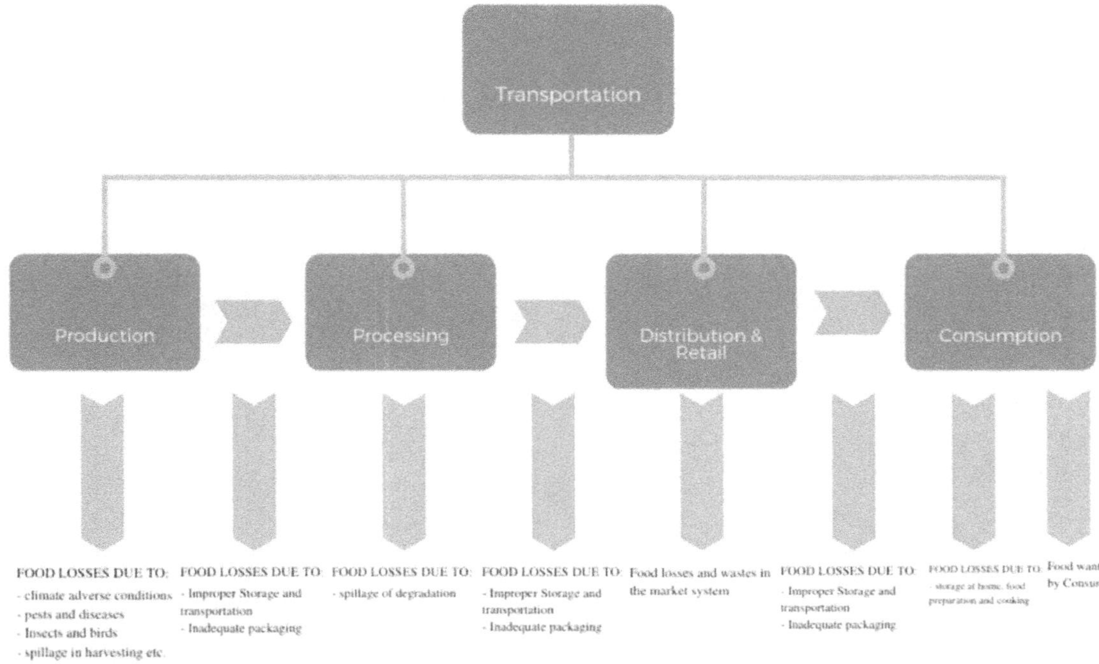


## ICT-ENABLED FOOD SYSTEM SOLUTIONS

Information and communication technology (ICT) is the integration of information technology (IT) and communication technology, with a focus on unified communications that allow users to access, store, send, receive, and alter data. ICT may have to enable impacts on sustainability throughout the usage phase. ICT-enabled solutions throughout the food chain are investigated at the level of ICT services to determine what and/or how ICT may be used. There are numerous highly promising areas for Information and communication technologies across the food chain to increase communication and system efficiency (Yuan & Luyao, 2019). The application of ICT in the Food Industry is represented in figure 10.

By making food chains more transparent and giving real-time information on particular goods, ICT has the potential to significantly cut food waste throughout the supply chain. Less food waste during distribution, transit, and consumption equals more food on the market, perhaps better nutritional results, and lower emissions owing to waste averted (Smarter2030.gesi.org, 2015).

*Figure 10. Applications of ICT in the food industry*

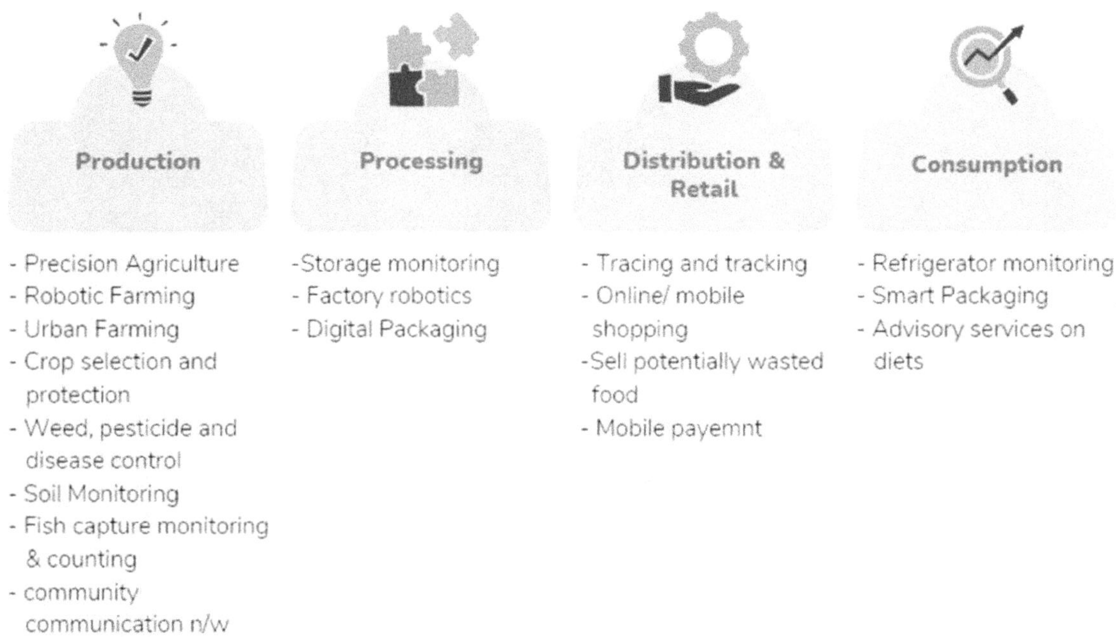


Information and communication technology (ICT) may assist increase productivity and minimise food waste. Smart Agriculture will increase yields by 30%, eliminate 20% of food waste, and provide \$1.9 trillion in economic benefits. Simultaneously, Smart Agriculture has the potential to cut water use by 250 trillion liters while also reducing CO2 emissions by 2.0Gt CO2e (Smarter2030.gesi.org, 2015).

AI has the potential to make a positive influence. According to a GeSI (2015) report, the promising effects of AI-enabled technology solutions in agriculture production could result in environmental benefits such as a 30% increase in crop yields, a reduction of over 300 billion liters of water consumption, and a reduction of oil consumption by 25 million barrels per year by 2030. These results include an estimate of ICT usage's carbon impact, which is ten times less than the averted emissions. Food security challenges include increasing the availability of high-quality food, improving distribution and access, increasing use, and strengthening resilience. Reducing GHG emissions, unsustainable water withdrawals, biodiversity loss, chemical use/pollution, and so on are challenges in terms of environmental sustainability (Yuan & Luyao, 2019).

## COMPARATIVE ANALYSIS

In this paper, we have examined several studies conducted by various researchers using various characteristics of fruits and vegetables in terms of classification model accuracy and performance, including Random Forest (RF), Support Vector Machines (SVM), K-Nearest Neighbors (KNN), Linear Discriminant Analysis (LDA), etc. Table 2 depicts the quality analysis of fruits and vegetables using different combinations of classifiers used by individual researchers. The most important aspect of evaluating food quality is classification, which contributes to a structure in which artificial simulations of human think-

ing are used to aid humans in making complex judgments speedily, correctly, and consistently. Image capturing devices used in food applications include cameras, ultrasound, magnetic resonance imaging (MRI), electrical tomography, and computed tomography (CT). Characteristics of a dataset of a variety of fruits and vegetables acquired by researchers with the help of various image capturing devices are shown in Table 1.
Table 1. shows the characteristics of images acquired by researchers

*Table 1. Comparison of classification approaches for fruit and vegetable quality assessment*

| Reference(s) | Input Image | Feature Extraction | Classifier | Accuracy |
|---|---|---|---|---|
| (Pydipati et al., 2006) | Citrus | Color co-occurrence methods | Generalized Squared Distance | 95.00% |
| (Zhu et al., 2007) | Golden apples | Gabor feature vectors | PCA | 90.60% |
| (Kim et al., 2007) | Grapes | Intensity texture feature | Discriminant Analysis | 96.00% |
| (Rocha et al., 2010) | Fruits & Vegetables | GCH + CCV + BIC + Unser | Multiclass SVM | 97.00% |
| Arivazhagan et al., 2010) | Fruits & Vegetables | Co-occurrence features such as contrast, energy,local homogeneity, cluster shade and cluster | Minimum distance classifier | 86.00% |
| (Faria et al., 2012) | Fruits & Vegetables | Prominence | Classifier Fusion | 98.80%±0.9 |
| (Danti et al., 2012) | Fruits & Vegetables | Mean and range of Hue and Saturation | BPNN Classifier | 96.40% |
| Suresha et al., 2012) | Fruits & Vegetables | Texture features | Decision-tree Classifier | 95.00% |
| (Dubey et al., 2016) | Normal apples | GCH, LBP, CLBP, GCH, CCV | Multiclass SVM | 95.94% for CLBP |
| (Pujari et al., 2013) | Fruits & Vegetables | Texture features | BPNN Classifier | 84.65% |
| (Ashok et al., 2013) | Normal apples | Mean Boundary Gradient | Probabilities NN | 88.33% |
| (Rokunuzzaman et al., 2013) | Tomato | Color Feature | Rule based+NN | 84% for rule based & 87.50% for NN |
| Zhang et al., 2014) | Any | Color, Texture and Shape | FSCBC+FNN | 89.10% |
| Wang et al., 2015) | Any | K- fold Stratified | WE+PCA+BBO+FNN | 89.50% |
| (Zhang et al., 2016) | Any | Color, Texture and Shape | BBO+FNN | 89.11% |
| (Nandi et al., 2016) | Mangoes | Shape | Support vector Regression | 87.00% |

Fruit and vegetable photos may be defined by a collection of attributes such as colour, size, shape, and texture using image processing techniques. These characteristics are utilised to create a training set. A classification algorithm is employed to extract the knowledge base required to make a decision in an unknown case. Figure 11 demonstrates the accuracy of various categorization approaches used for fruit and vegetable quality assessments.

*Figure 11. Effectiveness of classification algorithms for fruit and vegetable quality analysis*

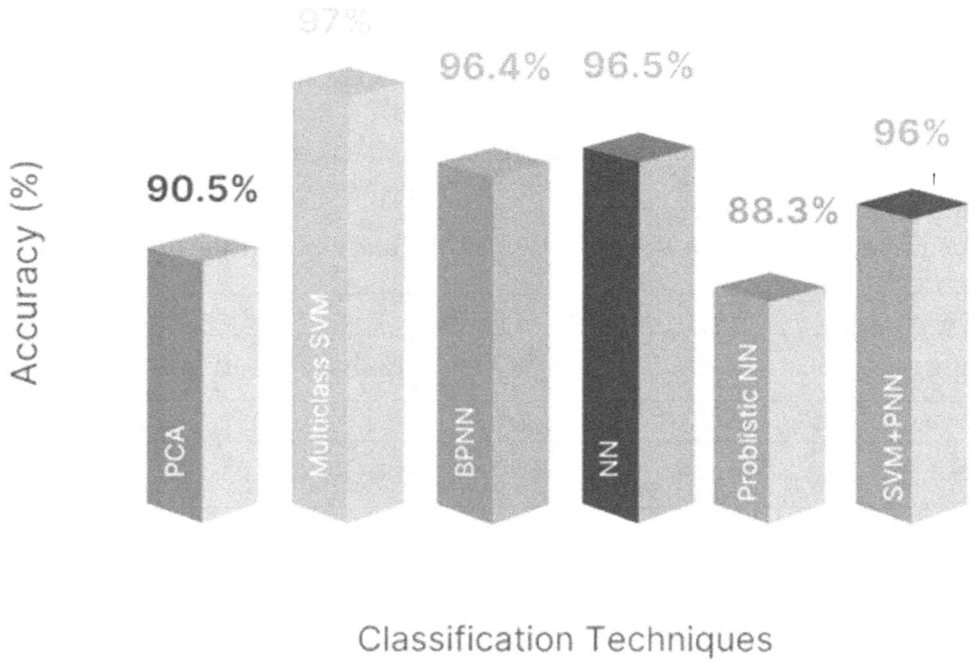


## CONCLUSION

This paper discussed the importance and application of evolving technologies such as Artificial Intelligence and Machine Learning in the food sector. Various problems and challenges are resolved by advanced features such as using robots and automatics processes for sorting, food packaging, food hygiene, and customized products. It enables production optimization and waste reduction that leads to increased revenue. A further role of Information and Communication Technology (ICT) based food system provides better opportunities in this field.

The food processing and handling industry is being transformed by artificial intelligence. It is expected to have a significant impact on the food processing and handling (FP&H) business in the future years. For these firms, AI will help them increase revenue by speeding up the manufacturing process, saving maintenance time, minimising the chance of failure by automating practically every activity, and offering an outstanding customer experience by predicting their preferences. Many FP&H organizations will be seeking AI solutions in the near future to enable them to not only stay competitive but also manage the industry.

## REFERENCES


Ahmad, A., & Dey, L. (2007). A k-mean clustering algorithm for mixed numeric and categorical data. *Data & Knowledge Engineering*, *63*(2), 503–527. doi:10.1016/j.datak.2007.03.016

Alexandratos, N. (Ed.). (1995). *World agriculture: towards 2010: an FAO study. FAO.*

Arivazhagan, S. (2010). Fruit recognition using color and texture features. *Journal of Emerging Trends in Computing and Information Sciences*, *1*(2), 90–94.

Ashok, V., & Vinod, D. S. (2014). Automatic quality evaluation of fruits using probabilistic neural network approach. In *2014 International Conference on Contemporary Computing and Informatics (IC3I)* (pp. 308-311). IEEE.

Ben Ayed, R., & Hanana, M. (2021). Artificial Intelligence to Improve the Food and Agriculture Sector. *Journal of Food Quality*.

Biffis, E., & Chavez, E. (2017). Satellite data and machine learning for weather risk management and food security. *Risk Analysis*, *37*(8), 1508–1521.

Carey, J. (2017). News Feature: On the brink of a recycling revolution? *Proceedings of the National Academy of Sciences of the United States of America*, *114*(4), 612–616. doi:10.1073/pnas.1620655114 PMID:28119494

Castillo & Meliif. (1970). Automated quality control in the food industry combining artificial intelligence techniques with fractal theory. *WIT Transactions on Information and Communication Technologies, 10*.

Ceballos-Magaña, S. G., Jurado, J. M., Muñiz-Valencia, R., Alcázar, A., de Pablos, F., & Martín, M. J. (2012). Geographical authentication of tequila according to its mineral content by means of support vector machines. *Food Analytical Methods*, *5*(2), 260–265. doi:10.100712161-011-9233-1

Chidinma-Mary-Agbai. (2020). Application of artificial intelligence (AI) in food industry. *GSC Biological and Pharmaceutical Sciences*, *13*(01), 171–178. doi:10.30574/gscbps.2020.13.1.0320

Condé, B. C., Fuentes, S., Caron, M., Di Xiao, R. C., & Howell, K. S. (2017). Development of a robotic and computer vision method to assess foam quality in sparkling wines. *Food Control*, *71*, 383–392. doi:10.1016/j.foodcont.2016.07.020

Danti, A. (2012). Mean and range color features based identification of common Indian leafy vegetables. *International Journal of Signal Processing, Image Processing and Pattern Recognition*, *5*(3), 151–160.

Deng, X., Cao, S., & Horn, A. L. (2021). Emerging applications of machine learning in food safety. *Annual Review of Food Science and Technology*, *12*(1), 513–538. doi:10.1146/annurev-food-071720-024112 PMID:33472015

Di Vaio, A., Boccia, F., Landriani, L., & Palladino, R. (2020). Artificial intelligence in the agri-food system: Rethinking sustainable business models in the COVID-19 scenario. *Sustainability*, *12*(12), 4851.

Dogan, A., & Birant, D. (2021). Dogan, Alican, and DeryaBirant. "Machine learning and data mining in manufacturing. *Expert Systems with Applications*, *166*, 114060. doi:10.1016/j.eswa.2020.114060

Dubey, S. R., & Jalal, A. S. (2016). Apple disease classification using color, texture and shape features from images. *Signal, Image and Video Processing*, *10*(5), 819–826.

Eli-Chukwu, N. C. (2019). Applications of artificial intelligence in agriculture: A review. *Engineering, Technology & Applied Science Research*, *9*(4), 4377–4383. doi:10.48084/etasr.2756

Eliazàt, A. (2020). *4 Ways AI is revolutionizing the food industry*. https://autome.me/4-ways-ai-is-revolutionizing-the-food-industry

Estelles-Lopez. (2017). An automated ranking platform for machine learning regression models for meat spoilage prediction using multi-spectral imaging and metabolic profiling. *Food Research International*, *99*, 206–215.

Faria, Alex dos Santos, Rocha, & Torres. (2012). Automatic classifier fusion for produce recognition. In *2012 25th SIBGRAPI Conference on Graphics, Patterns and Images* (pp. 252-259). IEEE.

Food Waste. (n.d.). https://www.usda.gov/foodwaste/faqs/

Friedlander, A., & Zoellner, C. (2020). Artificial Intelligence Opportunities to Improve Food Safety at Retail. *Food Protection Trends*, *40*(4), 272–278.

Garre, A., Ruiz, M. C., & Hontoria, E. (2020). Application of Machine Learning to support production planning of a food industry in the context of waste generation under uncertainty. *Operations Research Perspectives*, *7*, 100147. doi:10.1016/j.orp.2020.100147

Garre, A., Ruiz, M. C., & Hontoria, E. (2020). Application of Machine Learning to support production planning of a food industry in the context of waste generation under uncertainty. *Operations Research Perspectives*, *7*, 100147.

Göbel, C., Langen, N., Blumenthal, A., Teitscheid, P., & Ritter, G. (2015). Cutting food waste through cooperation along the food supply chain. *Sustainability*, *7*(2), 1429–1445. doi:10.3390u7021429

Godfray, H. (2010). Food security: The challenge of feeding 9 billion people. *Science, 327*(5967), 812-818.

Goyache, F., Bahamonde, A., Alonso, J., Lopez, S., del Coz, J. J., Quevedo, J. R., Ranilla, J., Luaces, O., Alvarez, I., Royo, L. J., & Diez, J. (2001). The usefulness of artificial intelligence techniques to assess subjective quality of products in the food industry. *Trends in Food Science & Technology*, *12*(10), 370–381. doi:10.1016/S0924-2244(02)00010-9

Guh, R.-S. (2003). Integrating artificial intelligence into on-line statistical process control. *Quality and Reliability Engineering International*, *19*(1), 1–20. doi:10.1002/qre.510

How, M.-L., Chan, Y. J., & Cheah, S.-M. (2020). Predictive insights for improving the resilience of global food security using artificial intelligence. *Sustainability*, *12*(15), 6272.

Jhawar, J. (2016). Orange sorting by applying pattern recognition on colour image. *Procedia Computer Science*, *78*, 691–697.

Kakani, V., Nguyen, V. H., Kumar, B. P., Kim, H., & Pasupuleti, V. R. (2020). A critical review on computer vision and artificial intelligence in food industry. *Journal of Agriculture and Food Research*, *2*, 100033.

Kaloxylos, Eigenmann, & Teye, Politopoulou, Wolfert, Shrank, & Dillinger. (2012). Farm management systems and the Future Internet era. *Computers and Electronics in Agriculture*, *89*, 130–144.

Kang, Z. (2020). CagatayCatal, and BedirTekinerdogan. "Machine learning applications in production lines: A systematic literature review. *Computers & Industrial Engineering*, *149*, 106773.

Kim, Burks, Qin, & Bulanon. (2009). Classification of grapefruit peel diseases using color texture feature analysis. *International Journal of Agricultural and Biological Engineering*, *2*(3), 41–50.

Kumar, I. (2020). Opportunities of Artificial Intelligence and Machine Learning in the Food Industry. *Journal of Cleaner Production*, 122574.

Kumar, Y., Kaur, K., & Singh, G. (2020). Machine Learning Aspects and its Applications Towards Different Research Areas. *2020 International Conference on Computation, Automation and Knowledge Management (ICCAKM)*, 150-156. 10.1109/ICCAKM46823.2020.9051502

Liu, Z.-L., Nan, F., Zheng, X., Zielinska, M., Duan, X., Deng, L.-Z., Wang, J., Wu, W., Gao, Z.-J., & Xiao, H.-W. (2020). Color prediction of mushroom slices during drying using Bayesian extreme learning machine. *Drying Technology*, *38*(14), 1869–1881. doi:10.1080/07373937.2019.1675077

Mavani, N. R. (2021). Application of Artificial Intelligence in Food Industry—A Guideline. *Food Engineering Reviews*, 1–42.

Milczarski, P. (2020). Machine Learning Application in Energy Consumption Calculation and Assessment in Food Processing Industry. In *International Conference on Artificial Intelligence and Soft Computing* (pp. 369-379). Springer.

Minaei-Bidgoli, B., Parvin, H., Alinejad-Rokny, H., Alizadeh, H., & Punch, W. F. (2014). Effects of resampling method and adaptation on clustering ensemble efficacy. *Artificial Intelligence Review*, *41*(1), 27–48. doi:10.100710462-011-9295-x

Misra, N. N. (2020). IoT, big data and artificial intelligence in agriculture and food industry. *IEEE Internet of Things Journal*.

Naik, S., & Patel, B. (2017). Thermal imaging with fuzzy classifier for maturity and size based non-destructive mango (MangiferaIndica L.) grading. In *2017 International Conference on Emerging Trends & Innovation in ICT (ICEI)* (pp. 15-20). IEEE.

Nandi, C. S. (2016). A machine vision technique for grading of harvested mangoes based on maturity and quality. *IEEE Sensors Journal*, *16*(16), 6387–6396.

Parvin, H., Alinejad-Rokny, H., Minaei-Bidgoli, B., & Parvin, S. (2013). A new classifier ensemble methodology based on subspace learning. *Journal of Experimental & Theoretical Artificial Intelligence*, *25*(2), 227–250. doi:10.1080/0952813X.2012.715683

Parvin, H., MirnabiBaboli, M., & Alinejad-Rokny, H. (2015). Proposing a classifier ensemble framework based on classifier selection and decision tree. *Engineering Applications of Artificial Intelligence*, *37*, 34–42. doi:10.1016/j.engappai.2014.08.005

Pereira, L. F. S. (2018). Predicting the ripening of papaya fruit with digital imaging and random forests. *Computers and Electronics in Agriculture*, *145*, 76–82.

Pujari, Devdas, Yakkundimath, & Byadgi. (2013). Grading and classification of anthracnose fungal disease of fruits based on statistical texture features. *International Journal of Advanced Science and Technology*, *52*(1), 121–132.

Pydipati, R., Burks, T. F., & Lee, W. S. (2006). Identification of citrus disease using color texture features and discriminant analysis. *Computers and Electronics in Agriculture*, *52*(1-2), 49–59.

Radke, A. M. (2013). A risk management-based evaluation of inventory allocations for make-to-order production. *CIRP Annals*, *62*(1), 459–462.

Rocha, Hauagge, & Wainer, & Goldenstein. (2010). Automatic fruit and vegetable classification from images. *Computers and Electronics in Agriculture*, *70* (1), 96-104.

Rokach, L. (2010). Ensemble-based classifiers. *Artificial Intelligence Review*, *33*(1-2), 1–39. doi:10.100710462-009-9124-7

Rokunuzzaman, M., & Jayasuriya, H. P. W. (2013). Development of a low cost machine vision system for sorting of tomatoes. *Agricultural Engineering International: CIGR Journal*, *15*(1).

Rong, D., Rao, X., & Ying, Y. (2017). Computer vision detection of surface defect on oranges by means of a sliding comparison window local segmentation algorithm. *Computers and Electronics in Agriculture*, *137*, 59–68.

Rowan, N. J., & Galanakis, C. M. (2020). Unlocking challenges and opportunities presented by COVID-19 pandemic for cross-cutting disruption in agri-food and green deal innovations: Quo Vadis? *The Science of the Total Environment*, 141362.

Schwarz, Jürgen, & Sánchez. (2015). *29 Implementation of artificial intelligence into risk management decision-making processes in construction projects.* Universität der BundeswehrMünchen, Institutfür-Baubetrieb.

Sharma, S., Gahlawat, V. K., Rahul, K., Mor, R. S., & Malik, M. (2021). Sustainable Innovations in the Food Industry through Artificial Intelligence and Big Data Analytics. *Logistics*, *5*(4), 66.

Si, Y. (2017). Potato tuber length-width ratio assessment using image analysis. *American Journal of Potato Research*, *94*(1), 88–93.

Spanaki, K. (2021). Disruptive technologies in agricultural operations: A systematic review of AI-driven AgriTech research. *Annals of Operations Research*, 1–34.

Strategy, A. (2015). *SMARTer2030: ICT solutions for 21st century challenges. The Global eSustainability Initiative (GeSI).* Tech. Rep. Brussels.

Suresha, & Kumar, & Kumar. (2012). Texture features and decision trees based vegetables classification. *International Journal of Computer Applications*, *975*, 8878.

Surya Prabha, D., & Satheesh Kumar, J. (2015). Assessment of banana fruit maturity by image processing technique. *Journal of Food Science and Technology*, *52*(3), 1316–1327.

The future of Food and Agriculture: Trends and challenges. (2017). *FAO*. https://www.fao.org/3/i6583e/i6583e.pdf /

Tsoumakas, G. (2019). A survey of machine learning techniques for food sales prediction. *Artificial Intelligence Review*, *52*(1), 441–447.

Wang, S., & Zhang, Y. (2015). Fruit classification by wavelet-entropy and feedforward neural network trained by fitness-scaled chaotic ABC and biogeography-based optimization. *Entropy (Basel, Switzerland)*, *17*(8), 5711–5728.

Yuan, L. (2019). *Sustainability Aspects of ICT in Agriculture and Food Systems.* Academic Press.

Zhang, A., Lipton, Z. C., Li, M., & Smola, A. J. (2021). *Dive into deep learning*. arXiv preprint arXiv:2106.11342.

Zhang, Y., Phillips, P., & Wang, S. (2016). Fruit classification by biogeography-based optimization and feedforward neural network. *Expert Systems: International Journal of Knowledge Engineering and Neural Networks*, *33*(3), 239–253.

Zhang, Y., & Wang, S. (2014). Fruit classification using computer vision and feedforward neural network. *Journal of Food Engineering*, *143*, 167–177.

Zhu, B., & Jiang, L. (2007). Gabor feature-based apple quality inspection using kernel principal component analysis. *Journal of Food Engineering*, *81*(4), 741–749.

# Chapter 12
# Applications of Machine Learning in Food Safety

**Rakesh Mohan Pujahari**
*ABES Institute of Technology, India*

**Rijwan Khan**
https://orcid.org/0000-0003-3354-3047
*ABES Institute of Technology, India*

## ABSTRACT

*Food safety has a major correlation with health related to the public. Machine learning can be a great help for large volume and emerging data sets to enhance the safety of the food supply and minimise the impact of food safety incidents. Pathogen genomes which are food borne and unique data streams, transactional, including text, and trade data, have ample emerging applications initiated by a machine learning approach, like prediction of antibiotic resistance, source related pathogens, and detection of food borne outbreak and also assessment of risk. In this chapter, a gentle introduction of machine learning in the pretext of food safety and a detailed overview of various developments and applications has been enumerated.*

## INTRODUCTION

Public health continues to be jeopardised by food safety. Machine learning has the ability to improve food safety and lessen the effect of food safety accidents by exploiting massive, developing data sets. Foodborne pathogen genomes and novel data streams, such as text, transactional, and trade data, have seen new applications enabled by machine learning, including antibiotic resistance prediction, pathogen source attribution, and foodborne outbreak detection and risk assessment. We give a light introduction to machine learning in the context of food safety, as well as an overview of recent advancements and applications, in this article. Many of these applications are still in their early stages, thus general and domain-specific problems and obstacles connected with machine learning are being identified and addressed.

DOI: 10.4018/978-1-6684-5141-0.ch012

Foodborne infections continue to pose a significant and long-term threat to public health. Foodborne illness affects 1 in every 6 citizens (or 48 million individuals) each year, resulting in 128,000 admits in hospitals and 3,000 casulties (Scallan et al. 2011) worldwide. The survey of fit and healthy persons world campaign released its vision document in 2010, which included food related safety as a priority (Koh 2010). As per surveillance data from the Foodborne Diseases Active Surveillance Network (FoodNet), not any one of the vision's targets for eradicating almost six major pathogens which are foodborne till 2020 had reached as close as 2019. (Table 1).

Population growth, urbanisation, and globalisation have all occurred, driving and feeding into macro societal changes (Doyle et al. 2015, Phillips 2006). Large number of modulations and advancements related to business of food and supply-chains, comparable to other sectors and industries, have created large chunk of data in nearby years. To increase the safety of the food supply, a variety bunch of data have been investigated in novel methods and at various stages related to the farm-to-table concept. For example, meteorological and terrain data have been checked for guessing contamination of pathogen on farms produce (Strawn et al. 2013), and auditing which is paperless and keeping of record, initiated 1.4 million of periodic measurements which is monthly monitoring of cooking temperature level of rotisserie related to commitment food safety of chickens for in the retail setting (Strawn et al. 2013). (Yiannas 2015).

*Table 1. Healthy people 2020 objectives and 2019 preliminary data*

| Pathogen[a] | Healthy People 2020 objective[b] | 2019 preliminary data[c] |
|---|---|---|
| *Campylobacter* | 8.5[d] | 19.5 |
| *Salmonella* | 11.4 | 17.1 |
| Shiga toxin–producing *Escherichia coli* | 0.6 | 6.3 |
| *Listeria* | 0.2 | 0.3 |
| *Vibrio* | 0.2 | 0.9 |
| *Yersinia* | 0.3 | 1.4 |

Pathogen Detection at the National Centre for Biotechnology Information. The widespread application of WGS in microbiology related to public health safety has spawned the data-driven field of epidemiology which is genomic in nature (Deng et al. 2016).

The developments which are fresh and recent in the field of data science approach and food related safety which have sparked debate over the Big Data domain (Marvin et al. 2017), a term not generally linked to the safety of food. ML has been looked to be viable method for data volume analysis in the case of food and safety to tackle the analytical hurdles posed by the deluge of data.

This has been intended to present a full overview of the emerging subject by explaining principles of the approach, analysing recent and remarkable advances, and highlighting problems and potential dangers, looking at the quick rise of ML related developments in the food related safety. Machine learning has been utilised in several fields of agriculture and food studies, like quality evaluation and food processing, as a data analytics tool which is general-purpose (Du & Sun 2006).

## 2. FUNDAMENTALS OF MACHINE LEARNING

### 2.1. Machine Learning(ML): A Brief History

The phrase "machine learning" was invented by Arthur Samuel (1959, p. 211) to describe method how a computer system can learn to act and perform well in game related to checkers in a process that "would be defined as involving the process of learning if done by human beings or animals." Samuel's concept was later broadened to include any branch which enables learning of computers without having to be explicitly programmed. The use of method like least-squares to characterise orbits related to planetary motion on the basis of measurements (data) as old as early eighteenth century; geodesists and astronomers used method like least-squares to aid sailors navigate the seas (Stigler 1986). After WWII, visionaries like Alan Turing (Turing 1950) drew up blueprints for current machine learning theory and technologies.

Between the 1960s and the 1990s, few most frequently applied models and algorithms, like neighbours, neural networks, and random forests, were developed. The rapid growth of machine learning helped a lot to solve and analyse from large volume of data sizes, growing computer power, which is increasing rapidly, and new modeling of old tools after Big Data became popular in both general public and the scientific community in the 1990s and 2000s, leading to a slew of new findings.

### 2.2. Machine Learning and its Relevance

Machine learning, a subsidiary of artificial intelligence, related to typical problem-solving using algorithm in that it does not aim to programme an entire collection of specific rules and instructions.

A system of machine learning processes itself from examples and generalises to new case studies which is based on how similar they are to already quoted in various instances of ML which modifies a module and its focus include its factors because of maximum and then it forecasts by utilising new (test) data (which is learning and model-based).

Machine learning is appealing for certain sorts of tasks because of its data-driven and rule-agnostic qualities. Firstly, it(ML) may develop a suitable approximation for various problems that are not easy to define theoretically (Shardanand & Maes 1995) or that lack detailed solution strategies.

Secondly, some situations are too cumbersome for already present approaches to solve, requiring a prohibitively extensive number of rules to be conceptualised through programmes. As an example, in the context of the game of Go, a brute-force search to determine the best move which is impossible. Thirdly, some jobs, like finding new spam in emails and in the context of social media, need the adaptation of hard-coded rules to fresh data. Finally, ML (machine learning) may addresses and verifies heuristics on large-volume of challenges, like DeepMind's AlphaGo's Go tactics and strategies, which were invented and discovered again (Baker & Fan 2017).

### 2.3. Main Types of Methods of ML

Systems of machine learning might get supervision or assistance during training. Learning might be classified as unsupervised, semi-supervised and supervised or reconstructed depending on the degree of supervision offered. The data related to training which is provided to the learning process is labelled with the output which is desired or the ground reality in common regulated learning related tasks like regression as well as classification. Many pet photos for training must be gathered and tagged with the various categories like dogs, cats or in no case to build a cat/dog classifier. Many examples of homes are gathered to construct a model based on regression, each comprising both attributes and a particular label: its particular price, in order to build up a regressor which aims at a numeric data, such as prices related to houses, by using a collection of characteristics (e.g., neighbourhood, size, year made).

Unsupervised learning involves leaving training data unlabelled and relying on the algorithm to uncover hidden patterns. Customer groups can be identified without an a priori determined categorization by using behavioral/transactional data. Anomaly and novelty detection is another major application. A learning system that is shown primarily regular network data, for example, can start detecting cyber-attacks.

Labelling of data is time consuming and not always available for huge data sets.

There are usually only a few labelled samples amid a large number of unlabelled cases. Algorithms which are Half-supervised can exaggerate on unlabelled data load of contribution by relations of feature-target as a blend of supervised and unsupervised learning, frequently using the notion that close samples which are supposed to have similar labels (Zhu et al. 2003).

Unlike supervised learning, which depends on a system of learning(agent) to identify the suitable path or strategy(policy) in each circumstance, the multiple time of undergoing the learning depends on the module to get the best way (policy) in each case. Designated agent learns using a trial-and-error model in which the acts it takes are rewarded or penalised, with the aim of maximising the reward with due time. From learning of robots up till their movement (Haarnoja et al. 2018) to the AlphaGo algorithm defeating global Go champion (Silver et al. 2017), reinforcement is used extensively in robotics and gaming.

## 2.4. Algorithms and their Examples

Algorithms related to ML of various levels of sophistication have been created to adjust issues of various stages of complexity. Figure 1 summarises basically four prominent and basic learning methods.

K-means is an unsupervised technique for dynamically grouping comparable data into clusters. It prototypes clusters using geometric centres of observations (centroids), and a particular observation is designated to a groups if it is nearer to its centre as compared to nearby centre of triangle (Figure 1a).

Typical SVM which is nothing but a support-vector machine represents observations of distinct classes as points and utilises planes to best divide them. A fresh observation is designated and mapped into space, and classification related to it is anticipated based on which side of the plane it falls on.

*Figure 1.*

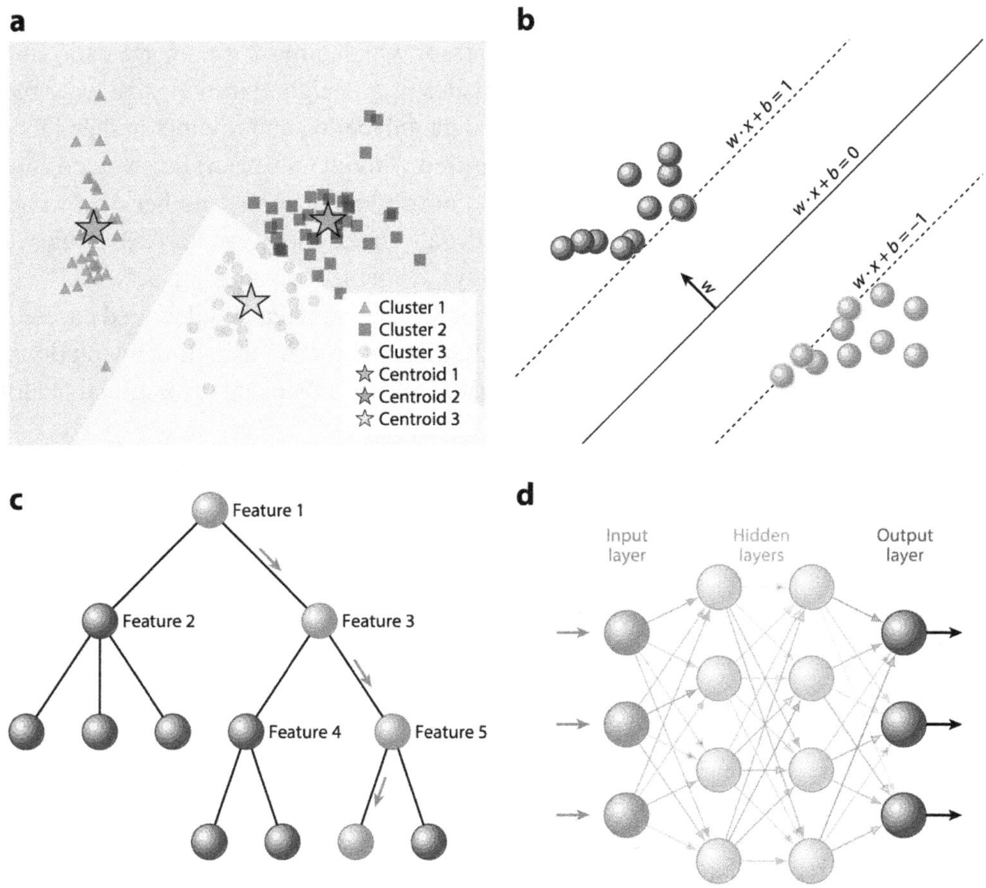


Examples of machine learning models. (a) A decision boundary plot of k-means clustering with three clusters, with new samples being grouped to a cluster by the colored region it lands on. (b) A line dividing two classes in a support-vector machine with a certain margin. w contains the trainable parameters, and x stands for the vector representation of a sample. (c) A decision tree with five features. A sample is categorised into a certain class which is designated as red arrow. (d) A neural network with two hidden layers; the arrow stands for a connection between units, with transparency indicating the connection strength. The Python source code to generate panel a was adapted from https://scikit-learn.org/stable/auto_examples/cluster/plot_kmeans_digits. html under a BSD license. Panel b was adapted from https://commons.wikimedia.org/wiki/File:Svm_ max_sep_hyperplane_with_margin.png, under a

SVM is highly effective when dealing with data that is well split into multiple classes (Figure 1b).

Through the interaction of multiple attributes, the typical decision related trees (Figure 1c) initiates to break various points into different types of categories again and again. A newly designated sample which follows certain branches indicated by the features until it lands on a particular leaf with the expected class. It is feasible to randomise this strategy by averaging the output of several separately designated module of random forest trees and to develop an intensive quantitative factor by introducing an objective function (gradient boosting). Ensemble learning is used in both of these strategies.

An artificial neural network (ANN) is made up of layers of interconnected artificial neurons called processing units that replicate biological brain networks (Figure 1d).

These subsets also known as nodes which receive data and further goes for processing which comprises a weights and input linear combination, as well as an activation related function which is nonlinear, and signals at the output up to the successive strata of layers of nodes. An input layer, which may be one or more intermediary layers known as a hidden layer, and one output layer make up each ANN. They work together to turn the primary inputs into regression or classification results. Deep neural networks, which are at the heart of deep learning, are ANNs with numerous hidden layers.

## 2.5. Challenges, Pitfalls and Good Practices

All the machine learning models are heavily reliant on in which way properly the data are assimilated, since in computer science domain maxim "garbage in, garbage out" overcomes. On data which is raw, a variety of feature engineering approaches can be used to pick necessary features and deal with data which is missing (e.g., data augmentation/imputation). In machine learning, it's typical to incorporate a lot of features at initially in order to capture as many underlying patterns and relationships as possible. However, as the number of features increases beyond a limit, the efficiency related to the learnt model degrades, a phenomenon known as "the curse of dimensionality" (Friedman 1997). To reduce the amount of features, mathematically established dimension reduction approaches such as principal component analysis can be utilised.

It's also possible to pick features by assessing the role of subsets of characteristics on the particular model efficiency (Kohavi & John 1997). Data leaking is a common blunder in data preparation that can culminate to a over analysis of the model. The word "leakage" indicates to when the test data assembly are not divided accurately, allowing the system model to see information from the sample set while being trained.

Data bias is the second important issue. When social network data is used to anticipate opinion related to public, for example, the demographical and behavioural anomalies between the researched and target populations should be considered. Another bias that affects supervised models is overfitting/underfitting, often known as the bias–variance trade-off dilemma.

Affected predictions won't be able to include all the design patterns in the sample data with the equal precision as the training data. If the model's structure is made complicated, for example, may improve precision in modelling relationships in data related to training. At the same time, overfitting is a danger, as overestimating on the testing data may become fluctuating and have a lot of volatility. Overfitting can be managed using regularisation approaches.

When a model fails to perform as predicted, it's tempting to switch to a different model or technique. It's possible, however, the model hasn't seen an ample amount of data. Natural language discontinuity is an example of how multiple learning algorithms, even preliminary one, can show performance nearly at per well when given data in adequate volume (Banko 2001).

Last but not least, one should not only "listen (just) to the data" (Nisbet et al. 2009, p. 739). The performance of a single metric model, like categorisation accuracy, is insufficient to determine its relevance. Highly essential training sets, the model's unpredictability, or model overfitting can all lead to erroneous discoveries. As a result, understanding how a model gets at its conclusions is crucial, but it can also be difficult. Some hidden and inherent models may simultaneously lessen variance in case of boosting ability of prediction, albeit in unbiased manner (Hastie et al. 2009) and human-understandable mechanisms of prediction related to parameter. There has been a case made in favour of models that are intrinsically interpretable, particularly when using ML in important decision taking (Rudin 2019).

## 3. ML APPLICATIONS OF GENOMIC DATA

Machine learning shows promise in genetics, genomics, and medicine for discoveries which are biological in nature and predictions from enormous genomic data sets. Sequence annotation is the process of teaching machine learning algorithms to recognise DNA sequences patterns and elements (Libbrecht & Noble 2015). ML algorithms are used to identify genomic signatures or biomarkers to aid in illness decision-making on the basis of clinical, diagnosis and medication development and discovery (He et al. 2019, Vamathevan et al. 2019). Due to the smaller genomes related to the pathogens, one may presume that machine learning (ML) pathogen genomes analysis are foodborne can be materialised by repeating it again and again current methodology, and thus simple.

However, ML is overwhelmingly employed to capture the fast-developing foodborne pathogen genomes resources and the related metadata, opportunities which are domain-specific and constraints emerge. Predictions related to antimicrobial resistance (AMR) and attribution related to genomic source of particular diseases have been the focus of such applications, which are still in their infancy.

### 3.1. Prediction of Antimicrobial Resistance

The typical method of determining antimicrobial resistance or susceptibility is to use tests which are phenotypic, which quantify the antibacterial agent growth inhibition pure culture bacteria of the population.

The broth related dilution, which incorporates a variety of doses of antibiotic and establishes the least concentration (MIC) drug to dormant or limit the development of a unique method to isolate bacteria, is a standard technique for AST (antimicrobial susceptibility testing). Resistant (less chance of a favourable output), susceptible (high chance of favourable conclusion), and occasionally level at the middle (Humphries et al. 2019, Turnidge & Paterson 2007) clinical parameters are used to segregate AST results into types based on the likelihood of a favourable treatment result.

It's difficult to standardise phenotypic AST among laboratories, and it takes a long time to get clinically useful data.

Excessive recommendations of the agents of etiologic nature which are immune to antibiotics that have the capability to alter efficient treatment mechanism and parts to resistant strains spreads in hospitals. During clinical arrangements, rapid and accurate AST can intimate accurately decision-taking clinically and enhance the management of antibiotic, sometimes additional recommendation of antibiotics in case of which resistant nature of etiologic agents which can slow the process of effective treatment and make contribution for resistant strains spread across hospital. WGS has become commonplace in the public health monitoring of AMR. Regardless of phenotypic resistance, NARMS has begun performing WGS on all anti-typhoidal Salmonella from clinical sources (NARMS 2015).

AST has been extensively researched as a supplement to or replacement for the AST phenotype (Ellington et al. 2017, Hendriksen et al. 2019) which is WGS-related. The first focus was on a particular approach which is rule-based that required prior knowledge of AMR related determinants.

Typically, pathogen genomes are queried for a sequenced panel of AMR genes to produce binary categorization as resistant to associated medicines. In spite of its success in identifying AMR related to unique or genes which are uncatalogued in various pathogens which are foodborne (Feldgarden et al. 2019, McDermott et al. 2016, Zankari et al. 2013), this method is restricted to familiar variants showing resistance and is not able to demarcate AMR designated by genes which are uncatalogued. Hence, AMR is difficult to detect since it is polygenically defined by numerous genes or caused by small changes.

ML is used with these hopes of succeeding a few of the above quoted constraints and using the eruption of volume of data related to WGS in the fight against the AMR (Table 2) recording and making algorithm in case of ML.

### 3.1.1. Classification Categorically

Most doctors prefer categorical analysis of AST data, which is inculcated to categorisation by ML, which is among the most common uses of monitored learning (Turnidge & Paterson 2007). In a study of 78 Enterobacteriaceae related isolates, which includes Escherichia coli, Pesesky et al. (2016) compared ML and rule dependent categorization of AMR. The Resfams database was used to curate a list of known resistance proteins for both types of classification (Gibson et al. 2015). The rules-based strategy used individually adapted rules for each class, whereas the method of machine learning used a unique logistic model of regression comprising 6 classes of 12 antibiotics. For 90% of isolates studied, the two classifiers performed similarly, agreeing with phenotypic AST.

Abbreviations: AMR, antimicrobial resistance; LR, logistic regression; MIC, minimum inhibitory concentration; SCM, Set Covering Machine; SNP, single-nucleotide polymorphism; SVM, support vector machine; WGS, whole-genome sequencing.

Previous studies established the feasibility of training models of machine learning for AMR analysis in variety of clinically important bacteria (Macesic et al. 2017), like Mycobacterium tuberculosis (Niehaus et al. 2014), Staphylococcus aureus (Davis et al. 2016), Streptococcus pneumonia (Davis et al. 2016, Drouin et al. 2016, Li et al. 2016), and Neisseria gonorrhoeae (Eyre et al. 2017). Studies related to foodborne pathogens quickly ensued, taking advantage of the established infrastructure of AMR.

Most doctors prefer categorical interpretation of AST data, which is recommended for classification by machine learning, most common use of ML is supervised learning (Turnidge & Paterson 2007). In a study of 78 number of Enterobacteriaceae isolates, which includes Escherichia coli, Pesesky et al. (2016) compared ML and rule dependent categorization of AMR. The Resfams database was used to curate a list of known resistance proteins for both types of classification (Gibson et al. 2015). The rules-based strategy used individually adapted rules for each class, whereas the method of machine learning used a unique regression model based on logistic approximately for six various classes having twelve number of antibiotics. For 90% of the isolates studied, the two classifiers performed similarly, agreeing with phenotypic AST.

Due to the short set of training and a bias inclined to highly resistant isolates, in no case classifier was designated more accurate to be used as a initial determination by clinically done criteria. The study explained how the process could be improved in the future. It was discovered that the absence or below normal presentation in feature selection of some resistance determinants, their contribution to AMR for effective learning, particularly when rare and unique genes which are AMR-causing were implicated.

*Table 2. Selected studies on antimicrobial resistance prediction using WGS and machine learning*

| Organism | Machine learning model | Prediction type | Size of training set | Features | Number of drugs | Reference |
|---|---|---|---|---|---|---|
| *Salmonella enterica* | XGBoost | MIC determination | 5,278 | *k*-mer | 15 | Nguyen et al. 2019 |
| *S. enterica* | LR, SCM | AMR classification | 97 | AMR genes (LR), *k*-mer (SCM) | 7 | Maguire et al. 2019 |
| *Klebsiella pneumonia* | XGBoost, AdaBoost, bagging, random forest, SVM, extremely randomized trees | MIC determination | 1,668 | *k*-mer | 20 | Nguyen et al. 2018 |
| *Neisseria gonorrhoeae* | Multivariate linear regression | MIC determination | 670–681 | SNPs, deletions, genes | 5 | Eyre et al. 2017 |
| *Streptococcus pneumonia* | Mode MIC, random forest, elastic net | MIC determination | 2,528 | AMR genes | 1 | Li et al. 2016 |
| *Clostridium difficile, Mycobacterium tuberculosis, Pseudomonas aeruginosa, S. pneumonia* | SCM | AMR classification | 111–556 | *k*-mer | 3–5 for each organism | Drouin et al. 2016 |
| *Escherichia coli, Enterobacter aerogenes, Enterobacter cloacae, K. pneumonia* | LR | Susceptibility classification | 78 | AMR genes | 12 | Pesesky et al. 2016 |
| *Acinetobacter baumannii, Staphylococcus aureus, S. pneumoniae, M. tuberculosis* | AdaBoost | AMR classification | 99–1,350 | *k*-mer | 1–5 for each organism | Davis et al. 2016 |
| *M. tuberculosis* | LR, SVM | AMR classification | 652 | SNPs | 4 | Niehaus et al. 2014 |

Drouin et al. (2016) devised a reference-free method for classifying and identifying AMR biomarkers in 4 multiple variants of bacteria, enlisting Clostridium difficile, a foodborne disease.

The model which was based on already familiar AMR variants and impartial through any separate sets of AMR genes, using a display of training of k-mer related characteristics of genome which are input (every genome is characterised by length of k). Hence the designated system of model might possibly find novel AMR biomarkers as learning related to it was not constrained by present wisdom of AMR techniques.

Here, the study overcame the curse of dimensionality by employing k-mers. A k-mer output is related to a genome which is very frequently and extremely dimension related, comprising of various k-mers that heavily outweigh the designated genomes they put forward, on the basis of k-mer genome size. The enormous multidimensional feature space in machine learning which makes the model vulnerable to theft and overlifting.

An overfit model classification of AMR utilising k-mers deceives by unnecessarily relying on the basis of details, such as anonymous and untimely connections among AMR and k-mers based characteristics for the training module data degrading the ability when applicable on a fresh set of test data. Hence the designated set covering machine called as SCM method was extended up to the whole characteristics related place, which included every k-mers starting as minimum as 12 upto 123 million (k = 31).

According to reports, the Set Covering Machine model must not overmatch as well as overperformed the other models which reduce the feature space through future choice.

Then, even though there were considerably more features than cases, the authors done an analysis theoretically on the above limit of the various difficulties concerning to model of SCM which is related to k-mer was not subjected to match accurately (genomes).

The model of SCM was shown to depend on few k-mers between already verified ML models to provide strong predictions of AMR after assessing the impact of specific k-mer characteristics in AMR specifications. The SCM related model was able to demarcate out AMR biomarkers across genomes and converge on essential k-mers of biological significance to AMR because of this advantage, which was complemented with the wide coverage of the whole features.

It has been speculated that the ability to create small number of markers that can make it easier for domain experts to interpret machine learning results and translate them into most basic clinical related diagnostics like chain reaction of polymerase, as interpreting and translating ML findings is a barrier to practical applications.

A group of approximately ninety seven Salmonella enterica specimen of non-typhoidal characteristics are collected as derived from farms of poultry broiler, were then subjected to the reference-free SCM model (Maguire et al. 2019). Similarly, for seven medications evaluated, the model has been able to identify and learn major factors to drive the AMR.

A logistic classifier of regression is utilised and initiated genes of AMR as characteristics as well as a straight forward match of the to analyse genes AMR in the cases related to the designated genomes were also tested in this investigation. The logistic and SCM regression classifiers correctly classified AMR to all seven antibiotics with a precision of more than 0.9, beating the AMR gene-matching technique, which dramatically predicted in exaggerated manner AMR (accuracy upto 0 to 0.5). Above mentioned findings imply that direct AMR gene profiling has limitations, such as inaccurate AMR task to the paralogs of the AMR related genes and the non-ability to discover intra-genic AMR isolates like promoter and regulatory regions.

This chapter emphasised the difficulty of label of imbalance, where resistant isolates were underestimted by several medicines (labels), as well as the not able to forecast MIC due to the limited to volume of the set.

### 3.1.2. Concentration Prediction

MIC measurement, as opposed to categorical classification, allows for more detailed characterisation and exact monitoring of AMR. MIC and its calculations is problematic related to sets of training which are very short for heavy modelling related to regression analysis, as proposed in the above mentioned research (Drouin et al. 2016, Maguire et al. 2019, Pesesky et al. 2016).

Levels of MIC were predicted by Li et al. (2016) and Eyre et al. (2017) using models with over 2,500 S. genomes which are pneumoniae and over 600 N. genomes which are gonorrhoeae, respectively. Both research result leads to curated AMR outcomes as characteristics, demonstrating that machine learning can accurately MIC prediction using well calibrated AMR genes and their mutations.

In a most thorough ML research on AMR, Nguyen et al. (2019) used the k-mer dependent and MIC prediction in nontyphoidal Salmonella, as indicated by sample size and the cases of medicines tested (Table 2). The related training sets modules include 5,728 samples of genomes of Salmonella with related MICs and about fifteen collected antibiotics, over a 15-year period (2002-2016). The study utilised in the algoritm of extreme gradient boost (XGBoost) (Chen & Guestrin 2016). This is a popular applications in between ML users (https://www.kaggle.com/competitions) and it is a leading model in the multiple Kaggle related issues (Niehaus 2016). The approach is important for a designated and scalable version of decision trees which is gradient based, which uses many decision trees related to weak learners for getting better results (ensemble learning).

Earlier, the XGBoost based model beat rest prominent ML based models in predicting MIC in case of specimen Klebsiella pneumoniae, even when assimilated polymorphisms or AMR genes were not included (Nguyen et al. 2018). The regressors related to XGBoost, predicted MICs with a average precision of 0.95 in 1 twice dilution step. The prediction models met FDA guidelines for systems which are automated for all 15, ME rates variants of antibiotics and seven out of the fifteen antibiotics of VME, as calibrated by important errors like false resistant results and MEs, and very important errors like false susceptible results and VMEs, that are routinely employed in evaluation of AST. In this case it has been developed and monitored separate validation sets (using 10 percent of the data) which were non-overlapping with the training and test related sets to avoid overfitting.

The research also focussed on both previously unanswered concerns about utilising genomes related to bacteria in the form of training. Initially, the training set volume is sometimes decided randomly; it is uncertain how large it material should be more relevant for effective learning. The authors demonstrated that models of MIC prediction may be created with less than 500 genomes genetically modified and provide better than 0.90 accuracy by subsampling the training related set while increasing differences in genetic concern in between subset. This conclusion has ramifications for developing realistic models concerning ML, as training huge sets of genomes are calculative and intensive, needing 1.5 TB capacity random access memory for study's entire 4,500 genomes.Secondly, it's unclear how much the diversity and spatial specimen of the training set influence biological trait learning. Models related to Training with currently revolving strains is apparently beneficial for tracking changing features in microbial diseases to notify current epidemiology. MIC prediction which are tested in subsequent samples by infering from earlier isolates, which gives the rapid growth of AMR in near years. Different partitioning of training sets from previous various years and test sets from years later yielded stable predictions, indicating the model's long-term value.

## 3.2. Attribution of Foodborne Pathogens and its Genomic Source

As 95 percent of foodborne illnesses are random, non-outbreak cases with difficult-to-determine due to food exposures and contamination of sources. It's very importnt to comprehend illness related to foodborne epidemiology and establish techniques to check and reduce such maladies because as foodborne infections is mainly unknown because of lack of information. Salmonella and E. coli, for example, bacteria which are zoonotic enteric with livestock and wild animals as their principal reservoirs. Unlike

AMR, where several genetic factors have been found and defined, zoonotic host related specifications and tropism have a limited mechanistic knowledge.

### 3.2.1. Potential prediction and Host Specifications

Lupolova et al. (2016) employed a categorisation of SVM to forecast E. coli O157 carrier specifications. Using pan-genome content to train 185 genomes both from cow and human infections, approximately human being 85 percent in number and approximately 91 percent of cattle which were accurately categorised among each hosts isolation. In this case, neither clustering study utilising scaling multidimensionally based on content which is pan-genomic nor phylogenetic analysis by utilising core genome single-nucleotide related polymorphisms revealed such a host dichotomy (SNPs). These findings were interpreted as the machine learning approach's unique ability to extract specific data from genomes of E. coli O157, implying distinct zoonotic potentials of the virus, according to the investigators. The SVM classifier's host-specificity prediction was subsequently extended to Salmonella (Lupolova et al. 2017). Forecasting of sixty-seven to ninety percent of Salmonella Typhimurium specimen of human being, swine, avian, and bovine related sources, a prediction congruent with the isolation host was made. Despite the fact that Salmonella Typhimurium has a wide zoonotic range of host, only a limited percentage of the specimen studied showed high level of probability ratings for various hosts. In E. coli, an equally important observation was found, and these findings were interpreted as demarcated host constraint, with just a small percentage demonstrating ability to colonise several hosts.

Strains of Salmonella have been linked to infection of invasive extraintestinal and which causes loss of genes as well as gene functions (pseudogene creation) which are not essential for each host to survive (Thomson et al. 2008). Wheeler et al. (2018) built a typical classifier of random forest to locate intruding as well as populations which are host-adapted of S. enterica, which includes prototypes which are put forwarded as well as lineages which are emerging from those, using mutations which are atypical, indicating a functional change in protein. Furthermore, the most informative genes for classification were discovered to suggest metabolic pathway breakdown, which is a prevalent motif in host adaptability.

### 3.2.2. Zoonotic-Source and its Contribution

Zhang et al. (2019) used genomic data to apply a classifier on zoonotic-source contribution of random forest of Salmonella Typhimurium. Between 1949 and 2014, the classifier was developed using over 1,200 genomes gathered from cases related to human and zoonotic sources s (PulseNet, NARMS, and GenomeTrakr). The classifier accurately ascribed sample isolates from seven with respect to eight significant zoonotic outbreaks of Salmonella Typhimurium between 1994 and 2013. It was calculated to be having 83 percent accuracy in forecasting poultry, swine, bovine, and wild bird origin. Remarkably, the analysis aimed at predictions of machine learning by interpreting the results biologically.

Phylogenetic research identified livestock lineages with consistent mutation rates, allowing for the conclusion of a recent but significant link with animal production systems. As animal-associated lineages split from generalist populations, genomic analysis revealed an increased piling of lineage related pseudogenes as signs of possible host adaptability. In several animal isolates, metabolic profiling revealed probable metabolic acclimation, indicating host adaptation. Multiple essential features have been experimentally involved in bacterial contact with animal hosts, and the study found that utilising a set of approximately 50 critical genetic traits from the whole characteristics accommodate more than 3,000 SNPs, genes and indels, significant source prediction could be made.

When the below mentioned mechanism is complex or poorly understood, these findings indicate a basis for using machine learning to solve problems of food safety.

Furthermore, some of the learnt characteristics may provide information that can be used to further investigate the mechanism.

Since then, various studies with an emphasis on technique have been conducted on the possibilities of genetic source attribution by machine learning. Lupolova et al. (2019) conducted an assessment which is technical and comparison of various ML developed models for host forcasting as well as methodologies(statistical) for characteristics selection. Guillier et al. (2020) created a software process to simplify zoonotic contribution of source of Salmonella Typhimurium using subsidiary genes at a time as characteristics.

## 3.3. Machine Learning and its uses in Food Safety with Genome Data and their Challenges and Potential Pitfalls

AST and the source attributions are two methods of phenotypic inference from genomic data that are similar but not the same. Machine learning inference may be taken as a subsection of microbial as well as genome-wide assimilated case studies when it is used to find genetic variants causally linked to specific phenotypes (mGWAS). Adapted from human genetics GWAS methodologies, mGWAS encounter unique obstacles and drawbacks, such as linkage genome-wide related imbalance and very important population aggregate, like separate clonal groups and lineages (Eyre et al. 2017, San et al. 2019). These features which are genetic origin as well as demographic features leads to the discovery of correlational but not causal genotype–phenotype relationships.).

Such obstacles and hazards have only recently been considered in the utilization of ML in safety of food related genomic studies, in one way during selection of feature (Lupolova et al. 2019) or after approval of results (Drouin et al. 2016). AMR and other phenotypes studies could be easily examined in lab. Since AMR is frequently caused using an unique individual or a group of genes, confirming the functionality of related AMR related bio-markers is quite simple. Source association, on the other hand, is a complicated phenomena shaped by a complex interplay of bacterial, host, and environmental factors. It can be difficult to conduct an experimental assembly for host specifications and exact validation of bio-marker related functions (for example, due to a lack of appropriate animal models).

The use of different host(surrogate) markers in predicting complicated phenotypes like association of source, adds to the complexity. Phylogenetic related data markers of lineage, which is space wise limited to a particular source, for example, could correctly identify the particular source while having no role which is active in the source relationship. Related surrogate and their markers might be used for attribution in relation to source, ususally when specific cause related to host or may be environment related and concerned tropism is complex to determine. Although they no relevance in AMR calculation and future studies since verification is done functionally of various AMR determinants is often expected. Surrogate sub-markers are allocated for microbiological source and its follow up, since a long time (Scott et al. 2002).

In ML, the design of training set has a significant impact on the investigation's conclusion.

Due of differences in the design of genome training sets, SalmonellaTyphimurium source attribution has resulted in conflicting outcomes (Wheeler 2019). Machine learning based predictions of host specifications, Lupolova et al. (2017) revealed that above accuracy of 90% in designating human hosts for genomes and an unexpectedly high occurrence of strains related to host which is restricted, perceived adapted human populations which are included. These findings contradict accepted epidemiology, which states that animal reservoirs are the source of most human Salmonella Typhimurium infections. Zhang et al. (2019) commended for identifying humans as a source class and reservoir of Salmonella Typhimurium and contested the assumption of generally applying strains of human being which are adapted, assuming the prevalence of zoonotic transmission.

It has been claimed that the Lupolova study's human host prediction accuracy was exaggerated human related isolates which are closely related by the training sample, with 85 percent having human isolate as nerby phylogenetic neighbour (Zhang et al. 2019). Zhang et al. (2019) demonstrated that human related isolates are not distinguishable as compared to the genomic level at livestock isolates level, disputing the survival of human host which are separate and related impressions in Salmonella Typhimurium samples and indicating amalgamated cause of human infections in a series of zoonotic sources by using a set of various set of human isolates to minimise biased training samples due to epidemiological and redundancy which are phylogenetic.

## 4. USE OF NOVEL DATA STREAMS(NDS) IN MACHINE LEARNING APPLICATIONS

It focuses on numerous resources of novel date streams (Althouse et al. 2015) which might utilise to support and verify food safety related research and practise when combined with ML or related data either in scientific or computational methodologies. (Althouse et al. 2015, Bansal et al. 2016, Oldroyd et al. 2018, Timmins et al. 2018). continuous, data collection which are automated, passive, which provides enhanced breadth, detail, and scalability of observations that data systems set up at primary level particularly for food safety of NDS. Importantly, these vast digital data sources include important structure (many features) that might be extracted using extensive analysis and processing, making machine learning approaches ideal for analysis.

### 4.1. Text Data

Transactional, text and trade data are three key NDS data sources that are examined here in conjunction with machine learning analysis. Text related data has been utilised in conjunction with techniques of machine learning and is the subject of the most attention. Although promising, trade data and transactional have only a few early uses.

#### 4.1.1. Data Types

Public Post Data which are user-generated and web generated data are two types of NDS text data which are combined with ML algorithms in food related safety uses. Posts on media which are on social sites like Twitter sites (Devinney et al. 2018, Harris et al. 2017, Harrison et al. 2014, Kuehn et al. 2014, Sadilek et al. 2017) as well as Facebook sites or consumer review sites which are crowdsourced like Yelp (Effland et al. 2018, Nsoesie et al. 2014, Schomberg et al. 2016) and also online sites like Amazon (Maharana et al).

Post data has a variety of properties that can be mined and analysed in food safety related uses. A post's content in the shape of text may comprise normal writing of language, as well as a title and hashtags (user keywords). Text form data may be examined to assess the mood of postings or content (good, negative, or neutral). Temporality geotags that designate the exact source of the post or where it has been correlated, have been utilised to relate a user having a indicated source of uncleanliness (e.g., restaurant), may be the pricing and the rating of a exact position of metadata which are non-text that might arrive with a designated post and may put to analysis.

Articles related to news media and the websites of both academic and professional organisations are examples of web data sources. Surveillance systems for safety event related food (e.g., recalls, spread) that collect, rank and monitor, required material and across the important keywords various websites and spatial locations have been built using online data (Chen et al. 2016, Kate et al. 2014). Other sources include detailed assessments of consumer formed data for illness surveillance which are foodborne (Oldroyd et al. 2018) and natural utilisation of food research related text data (Tao et al. 2020).

### 4.1.2. Methods and Approach

To convert raw text which is high dimensional food safety information input, a multistep analytic approach is necessary (Figure 2). Data-gathering techniques may be used first, followed by data pre-processing cleaning (such as grammar mistakes, and metadata noise removal and removal of filler word) and decrease of volume of data (screening), prior to ML applications can be used to learn and data prediction (Tao et al. 2020, Zhai&Massung 2016). Keyword demarcation is frequently used as an initial step in extracting or filtering data which might contain information in relation to food safety. Separate list of specialised terms such as disease, throw up, vomit, barf, food poisoning, and nausea, is frequently generated a priori.

*Figure 2. Flowchart of text data by ML analysis*

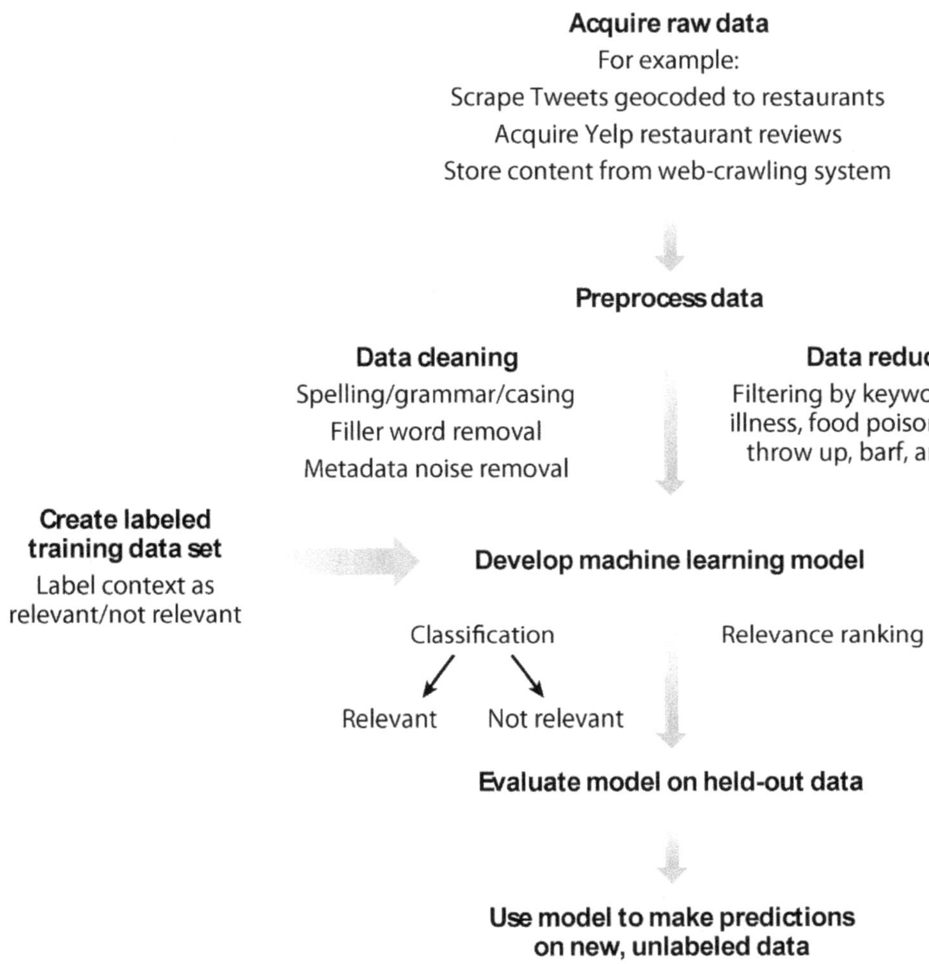


The classification of post or website data as relevant or not relevant to food safety occurrences is a popular machine learning task. Decision trees, SVMs, naive Bayes, and neural networks are some of the classification techniques employed in these situations (Oldroyd et al. 2018). Machine learning algorithms are used to find and rank relevant content where a basic keyword search would be insufficient to construct a monitoring system that searches and sorts information across various websites (Kate et al. 2014). Cosine similarity (Drury & Roche 2019); a scoring system based on factors such as geographic region, time period, and food-related terms (Chen et al. 2016); and text classification algorithms, such as the ranking support vector machine (Rank-SVM) have all been used in this job (Kate et al. 2014).

## 4.2. Transactional Data

NDS is a type of transactional data that is used to investigate foodborne disease outbreaks. Patient interviews are used in traditional epidemiological investigations to uncover commonalities among case patients, as following which microbiological experiments on specimen as suspected (Smith et al. 2015,

World Health Organ. 2008). Data has been shown to enhance, complement, and even supplement traditional investigative methodologies in producing ideas related to the causal vehicle of food at the outset of a calculation, pinpoint the source of contamination, hotels serving food, or somewhere else in the distribution and supply-chain. Applications which are transactional in nature mainly have inculcated analysing particular consumer related data, such as loyalty card, credit card, printed copy of barcode scanner and card related history of employee, gathered related to known case-studies of infected persons or marketing and business points in that case food is contaminated, is suspected to be bought. Since 2006, these data sets have aided in the identification of more than 20 epidemic suspects (Moller et al. 2018). Standard techniques(statistical) were used to analyse purchase records to find commonalities, such as ratios to compare patient case histories with few groups of shoppers. These are useful since data records are frequently obtained on an individual basis.

Machine learning techniques could be used to analyse aggregated sales data, such as spatially aggregated or store-based retail sales/loyalty card data. The application of this data in epidemic surveillance and outbreak investigations has been demonstrated in a few situations, assisting in the identification of the causal food carrying vehicle. Kaufman et al. (2014) devised a method for determining food item which is contaminated that triggered an epidemic by correlating outbreak, reported in areas like markets (Hu et al. 2016). Food articles with sales design and pattern that are similar to the spread related distribution are thought to be the effective food.

## 4.3. Trade Data

The trade data has lately been discovered to offer unique applications in case of food safety risk related assessment, which has historically been kept or maintained for statistical analysis or company operations. In the context of NDS, we define trade data as generally freely available record details, statistics of aggregated records of flows characterising the sequence of first production, then consumption and at last movement of food using supply chains among various regions countries both inside and outside supply system areas. Federal trade statistics, as well as production, consumption data, international trade, are examples of data sources. These data sources allow for the correlating the supply-chain, permitting for complex, typically risk assessments of food safety.

Because precise, finely grained data on agricultural commodity supply chain networks is typically unavailable, trade structures can be modelled. Many factors influence supply-chain networks, including production sites, population hubs, and storage and transportation facilities (Balster & Friedrich 2019, Venkatramanan et al. 2017). Models of food-flow network architectures have been developed using innovative applications related to data related to factors, in association with sophisticated modelling approaches and algorithms of machine learning.

Through a recurring, optimization process which is mass-balanced, the flows from start to end are approximated using special gravity models calculated with data across trade-flows, cross-regional, from a transport analysis nationally. A comparable effort to food-flow modelling with area module utilising machine learning was created to predict food flows over approximately 3,142 nations and across seven commodity categories (Lin et al. 2019). To train the model, a centrally regularised machine learning approach module of gamma regression and involving logistic is used to guarantee that the aggregated networks features match structural properties of monitored empirical food-flow networks.

These modelled supply network have initiated opportunities related to food safety related to analysis of risk when combined with other modelling methodologies. The method has been tested on recent spread of diseases which are foodborne, and it is extended to address the issues of highlighting the food related vehicle spread using a statistical based assignment including graded clustering (A. Horn, M. Fuhrmann, T. Schlaich, A. Balster, A. Kaesbohrer, M. Filter, H. Friedrich, unpublished results).

### 4.4.1. Novel Data Streams and Data Access

Due to absence of specific foodborne disease–related phrases is a common text data problem. Words like nausea, sick, and diarrhoea can be used to describe a wide range of diseases and conditions, generating false positives. ML algorithms created for sarcasm identification, founded in sentiment analysis as well as pattern recognition, are being applied to such challenges with promising results (Bouazizi & Ohtsuki 2016, Oldroyd et al. 2018, Pinheiro et al. 2015).

Various strategies discussed are required the use of ML in conjunction with mathematical or predictive mechanistic related modelling, which adds basis and rationale to a problem-solving strategy. Food safety questions can be investigated using predictive modelling methods like models which are based on agent methodology (Zoellner et al. 2019), for supply chain (Garre et al. 2019, Horn & Friedrich 2019, Manitz et al. 2014), and rest simulation-based or probabilistic system modules with scanty insource data. In various instances that does not need calculation but sometimes requires effictive IT systems modules for capturing, designating and gathering real-time data, like the mapping, data input, and visualisation mechanisms used time to time.

As NDS and its employed techniques discussed here can enrich or substitute already present data and analytical tools to solve issues of food safety, but not a final fool proof of investigation.

## REFERENCES

Aiello, L. M., Schifanella, R., Quercia, D., & Prete, L. D. (2019). Large-scale and high-resolution analysis of food purchases and health outcomes. *EPJ Data Science*, *8*(1), 14. doi:10.1140/epjds13688-019-0191-y

Allard, M. W., Strain, E., Melka, D., Bunning, K., Musser, S. M., Brown, E. W., & Timme, R. (2016). Practical value of food pathogen traceability through building a whole-genome sequencing network and database. *Journal of Clinical Microbiology*, *54*(8), 1975–1983. doi:10.1128/JCM.00081-16 PMID:27008877

Altenburger, K. M., & Ho, D. E. (2019). When algorithms import private bias into public enforcement: The promise and limitations of statistical debiasing solutions. *Journal of Institutional and Theoretical Economics*, *175*(1), 98–122. doi:10.1628/jite-2019-0001

Althouse, B. M., Scarpino, S. V., Meyers, L. A., Ayers, J. W., Bargsten, M., Baumbach, J., Brownstein, J. S., Castro, L., Clapham, H., Cummings, D. A. T., Del Valle, S., Eubank, S., Fairchild, G., Finelli, L., Generous, N., George, D., Harper, D. R., Hébert-Dufresne, L., Johansson, M. A., ... Wesolowski, A. (2015). Enhancing disease surveillance with novel data streams: Challenges and opportunities. *EPJ Data Science*, *4*(1), 17. doi:10.1140/epjds13688-015-0054-0 PMID:27990325

Baker, L., & Fan, H. (2017). Innovations of AlphaGo. *DeepMind Blog*. https://deepmind.com/blog/article/innovations-alphago

Balster, A., & Friedrich, H. (2019). Dynamic freight flow modelling for risk evaluation in food supply. *Transp. Res.E*, *121*, 4–22. doi:10.1016/j.tre.2018.03.002

Banko, M. (2001). Scaling to very very large corpora for natural language disambiguation. In *ACL '01: Proceedings of the 39th Annual Meeting on Association for Computational Linguistics* (pp. 26–33). Stroudsberg, PA: Assoc. Comput. Linguist. 10.3115/1073012.1073017

Bansal, S., Chowell, G., Simonsen, L., Vespignani, A., & Viboud, C. (2016). Big data for infectious disease surveillance and modeling. *The Journal of Infectious Diseases*, *214*(suppl 4), S375–S79. doi:10.1093/infdis/jiw400 PMID:28830113

Bouazizi, M., & Ohtsuki, T. (2016). A pattern-based approach for sarcasm detection on Twitter. *Morbidity and Mortality Weekly Report*, *46*, 258–261.

Chen, S., Huang, D., Nong, W., & Kwan, H. S. (2016). Development of a food safety information database for Greater China. *Food Control*, *65*, 54–62. doi:10.1016/j.foodcont.2016.01.002

Chen, T., & Guestrin, C. (2016). XGBoost: a scalable tree boosting system. In *KDD '16: Proceedings of the 22nd ACMSIGKDD International Conference on Knowledge Discovery and Data Mining* (pp. 785–94). New York: Assoc. Comput. Mach. 10.1145/2939672.2939785

Connelly, R., Playford, C. J., Gayle, V., & Dibben, C. (2016). The role of administrative data in the big data revolution in social science research. *Social Science Research*, *59*, 1–12. doi:10.1016/j.ssresearch.2016.04.015 PMID:27480367

Dallman, T., Inns, T., Jombart, T., Ashton, P., Loman, N., Chatt, C., Messelhaeusser, U., Rabsch, W., Simon, S., Nikisins, S., Bernard, H., le Hello, S., Jourdan da-Silva, N., Kornschober, C., Mossong, J., Hawkey, P., de Pinna, E., Grant, K., & Cleary, P. (2016). Phylogenetic structure of European Salmonella Enteritidis outbreak correlates with national and international egg distribution network. *Microbial Genomics*, *2*(8), e000070. doi:10.1099/mgen.0.000070 PMID:28348865

Davis, J. J., Boisvert, S., Brettin, T., Kenyon, R. W., Mao, C., Olson, R., Overbeek, R., Santerre, J., Shukla, M., Wattam, A. R., Will, R., Xia, F., & Stevens, R. (2016). Antimicrobial resistance prediction in PATRIC and RAST. *Scientific Reports*, *6*(1), 27930. doi:10.1038rep27930 PMID:27297683

Deng, X., den Bakker, H. C., & Hendriksen, R. S. (2016). Genomic epidemiology: Whole-genome-sequencing-powered surveillance and outbreak investigation of foodborne bacterial pathogens. *Annual Review of Food Science and Technology*, *7*(1), 353–374. doi:10.1146/annurev-food-041715-033259 PMID:26772415

Devinney, K., Bekbay, A., Effland, T., Gravano, L., Howell, D., Hsu, D., O'Hallorhan, D., Reddy, V., Stavinsky, F., Waechter, H. N., & Gutelius, B. (2018). Evaluating Twitter for foodborne illness outbreak detection in New York City. *Online Journal of Public Health Informatics*, *10*(1), e120. doi:10.5210/ojphi.v10i1.8894

Doyle, M. P., Erickson, M. C., Alali, W., Cannon, J., Deng, X., Ortega, Y., Smith, M. A., & Zhao, T. (2015). The food industry's current and future role in preventing microbial foodborne illness within the United States. *Clinical Infectious Diseases*, *61*(2), 252–259. doi:10.1093/cid/civ253 PMID:25824814

Drouin, A., Giguère, S., Déraspe, M., Marchand, M., Tyers, M., Loo, V. G., Bourgault, A.-M., Laviolette, F., & Corbeil, J. (2016). Predictive computational phenotyping and biomarker discovery using reference-free genome comparisons. *BMC Genomics*, *17*(1), 754. doi:10.118612864-016-2889-6 PMID:27671088

Drury, B., & Roche, M. (2019). A survey of the applications of text mining for agriculture. *Computers and Electronics in Agriculture*, *163*, 104864. doi:10.1016/j.compag.2019.104864

Du, C., & Sun, D. (2006). Learning techniques used in computer vision for food quality evaluation: A review. *Journal of Food Engineering*, *72*(1), 39–55. doi:10.1016/j.jfoodeng.2004.11.017

Effland, T., Lawson, A., Balter, S., Devinney, K., Reddy, V., Waechter, H. N., Gravano, L., & Hsu, D. (2018). Discovering foodborne illness in online restaurant reviews. *Journal of the American Medical Informatics Association: JAMIA*, *25*(12), 1586–1592. doi:10.1093/jamia/ocx093 PMID:29329402

Ellington, M. J., Ekelund, O., Aarestrup, F. M., Canton, R., Doumith, M., Giske, C., Grundman, H., Hasman, H., Holden, M. T. G., Hopkins, K. L., Iredell, J., Kahlmeter, G., Köser, C. U., MacGowan, A., Mevius, D., Mulvey, M., Naas, T., Peto, T., Rolain, J.-M., ... Woodford, N. (2017). The role of whole genome sequencing in antimicrobial susceptibility testing of bacteria: Report from the EUCAST Subcommittee. *Clinical Microbiology and Infection*, *23*(1), 2–22. doi:10.1016/j.cmi.2016.11.012 PMID:27890457

Eyre, D. W., De Silva, D., Cole, K., Peters, J., Cole, M. J., Grad, Y. H., Demczuk, W., Martin, I., Mulvey, M. R., Crook, D. W., Walker, A. S., Peto, T. E. A., & Paul, J. (2017). WGS to predict antibiotic MICs for Neisseria gonorrhoeae. *The Journal of Antimicrobial Chemotherapy*, *72*(7), 1937–1947. doi:10.1093/jac/dkx067 PMID:28333355

Feldgarden, M., Brover, V., Haft, D. H., Prasad, A. B., & Slotta, D. J. (2019). Validating the AMRFinder Tool and Resistance Gene Database by using antimicrobial resistance genotype-phenotype correlations in a collection of isolates. *Antimicrob. Agents Chemother.*, *63*(11).

Friedman, J. H. (1997). On bias, variance, 0/1—Loss, and the curse-of-dimensionality. *Data Mining and Knowledge Discovery*, *1*(1), 55–77. doi:10.1023/A:1009778005914

Garre, A., Fernandez, P. S., Brereton, P., Elliott, C., & Mojtahed, V. (2019). The use of trade data to predict the source and spread of food safety outbreaks: An innovative mathematical modelling approach. *Food Research International*, *123*, 712–721. doi:10.1016/j.foodres.2019.06.007 PMID:31285021

Gates, M. C., Holmstrom, L. K., Biggers, K. E., & Beckham, T. R. (2015). Integrating novel data streams to support biosurveillance in commercial livestock production systems in developed countries: Challenges and opportunities. *Frontiers in Public Health*, *3*, 74. doi:10.3389/fpubh.2015.00074 PMID:25973416

Gertler, M., Czogiel, I., Stark, K., & Wilking, H. (2017). Assessment of recall error in self-reported food consumption histories among adults—particularly delay of interviews decrease completeness of food histories— Germany, 2013. *PLoS One*, *12*(6), e0179121. doi:10.1371/journal.pone.0179121 PMID:28640839

Gibson, M. K., Forsberg, K. J., & Dantas, G. (2015). Improved annotation of antibiotic resistance determinants reveals microbial resistomes cluster by ecology. *The ISME Journal*, *9*(1), 207–216. doi:10.1038/ismej.2014.106 PMID:25003965

Greis, N. P., & Nogueira, M. L. (2017). A data-driven approach to food safety surveillance and response. In Food Protection and Security: Preventing and Mitigating Contamination during Food Processing and Production. doi:10.1016/B978-1-78242-251-8.00005-9

Guillier, L., Gourmelon, M., Lozach, S., Cadel-Six, S., Vignaud, M. L., Munck, N., Hald, T., & Palma, F. (2020). AB_SA: Accessory genes-based source attribution - tracing the source of Salmonella enterica Typhimurium environmental strains. *Microbial Genomics*, *6*(7), mgen000366. doi:10.1099/mgen.0.000366 PMID:32320376

Gupta, A., Nelson, J. M., Barrett, T. J., Tauxe, R. V., Rossiter, S. P., Friedman, C. R., Joyce, K. W., Smith, K. E., Jones, T. F., Hawkins, M. A., Shiferaw, B., Beebe, J. L., Vugia, D. J., Rabatsky-Ehr, T., Benson, J. A., Root, T. P., & Angulo, F. J. (2004). Antimicrobial resistance among Campylobacter strains, United States, 1997–2001. *Emerging Infectious Diseases*, *10*(6), 1102–1109. doi:10.3201/eid1006.030635 PMID:15207064

Haarnoja, T., Ha, S., Zhou, A., & Tan, J. (2018). *Learning to walk via deep reinforcement learning*. arXiv:1812.11103 [cs.LG]

Halevy, A., Norvig, P., & Pereira, F. (2009). The unreasonable effectiveness of data. *IEEE Intelligent Systems*, *24*(2), 8–12. doi:10.1109/MIS.2009.36

Hall, A. J., Wikswo, M. E., Manikonda, K., Roberts, V. A., Yoder, J. S., & Gould, L. H. (2013). Acute gastroenteritis surveillance through the National Outbreak Reporting System, United States. *Emerging Infectious Diseases*, *19*(8), 1305–1309. doi:10.3201/eid1908.130482 PMID:23876187

Harris, J. K., Hawkins, J. B., Nguyen, L., Nsoesie, E. O., Tuli, G., Mansour, R., & Brownstein, J. S. (2017). Using Twitter to identify and respond to food poisoning: The Food Safety STL Project. *Journal of Public Health Management and Practice*, *23*(6), 577–580. doi:10.1097/PHH.0000000000000516 PMID:28166175

Harris, J. K., Mansour, R., Choucair, B., Olson, J., & Nissen, C. (2014). Health department use of social media to identify foodborne illness—Chicago, Illinois, 2013–2014. *MMWR. Morbidity and Mortality Weekly Report*, *63*, 681–685. PMID:25121710

Harrison, C., Jorder, M., Stern, H., Stavinsky, F., & Reddy, V. (2014). Using online reviews by restaurant patrons to identify unreported cases of foodborne illness—New York City, 2012–2013. *MMWR. Morbidity and Mortality Weekly Report*, *63*, 441–445. PMID:24848215

Hastie, T., Tibshirani, R., & Friedman, J. (2009). *The Elements of Statistical Learning: Data Mining, Inference, and Prediction*. Springer. doi:10.1007/978-0-387-84858-7

He, J., Baxter, S. L., Xu, J., Xu, J., Zhou, X., & Zhang, K. (2019). The practical implementation of artificial intelligence technologies in medicine. *Nature Medicine*, *25*(1), 30–36. doi:10.103841591-018-0307-0 PMID:30617336

People, H. (2010). *Healthy People 2020 Food Safety Objectives*. HHS. https://www. healthypeople.gov/2020/topics-objectives/topic/food-safety/objectives

Hendriksen, R. S., Bortolaia, V., Tate, H., Tyson, G. H., Aarestrup, F. M., & McDermott, P. F. (2019). Using genomics to track global antimicrobial resistance. *Frontiers in Public Health*, *7*, 242. doi:10.3389/fpubh.2019.00242 PMID:31552211

Hochreiter, S., & Schmidhuber, J. (1997). Long short-term memory. *Neural Computation*, *9*(8), 1735–1780. doi:10.1162/neco.1997.9.8.1735 PMID:9377276

Horn, A., & Friedrich, H. (2019). Locating the source of large-scale outbreaks of foodborne disease. *Journal of the Royal Society, Interface*, *16*(151), 20180624. doi:10.1098/rsif.2018.0624 PMID:30958197

Hu, K., Renly, S., Edlund, S., Davis, M., & Kaufman, J. (2016). Amodeling framework to accelerate food-borne outbreak investigations. *Food Control*, *59*, 53–58. doi:10.1016/j.foodcont.2015.05.017

Humphries, R. M., Abbott, A. N., & Hindler, J. A. (2019). Understanding and addressing CLSI breakpoint Revisions: A primer for clinical laboratories. *Journal of Clinical Microbiology*, *57*(6), e00203–e00219. doi:10.1128/JCM.00203-19 PMID:30971460

Huyghe, E., Verstraeten, J., & Van Kerckhove, A. (2017). Clicks as a healthy alternative to bricks: How online grocery shopping reduces vice purchases. *JMR, Journal of Marketing Research*, *54*(1), 61–74. doi:10.1509/jmr.14.0490

Kate, K., Chaudhari, S., Prapanca, A., & Kalaganam, J. (2014). FoodSIS: a text mining system to improve the state of food safety in singapore. In *KDD '14: Proceedings of the 20th ACM SIGKDD International Conference on Knowledge Discovery and Data Mining*, (pp. 1709–18). New York: Assoc. Comput. Mach. 10.1145/2623330.2623369

Kaufman, J., Lessler, J., Harry, A., Edlund, S., Hu, K., Douglas, J., Thoens, C., Appel, B., Käsbohrer, A., & Filter, M. (2014). A likelihood-based approach to identifying contaminated food products using sales data: Performance and challenges. *PLoS Computational Biology*, *10*(7), e1003692. doi:10.1371/journal.pcbi.1003692 PMID:24992565

Koh, H. K. (2010). A 2020 vision for healthy people. *The New England Journal of Medicine*, *362*(18), 1653–1656. doi:10.1056/NEJMp1001601 PMID:20445177

Kohavi, R., & John, G. H. (1997). Wrappers for feature subset selection. *Artificial Intelligence*, *97*(1-2), 273–324. doi:10.1016/S0004-3702(97)00043-X

Krizhevsky, A., Sutskever, I., & Hinton, G. E. (2012). ImageNet classification with deep convolutional neural networks. In Advances in Neural Information Processing Systems 25 (NIPS 2012). Neural Inf. Proc. Syst. Found.

Krumme, C., Llorente, A., Cebrian, M., Pentland, A. S., & Moro, E. (2013). The predictability of consumer visitation patterns. *Journal of the American Medical Association*, *312*, 117–118.

Lazer, D., Pentland, A., Adamic, L., Aral, S., Barabási, A.-L., Brewer, D., Christakis, N., Contractor, N., Fowler, J., Gutmann, M., Jebara, T., King, G., Macy, M., Roy, D., & Van Alstyne, M. (2009). Computational social science. *Science*, *323*(5915), 721–723. doi:10.1126cience.1167742 PMID:19197046

LeCun, Y., Boser, B., Denker, J. S., Henderson, D., Howard, R. E., Hubbard, W., & Jackel, L. D. (1989). Backpropagation applied to handwritten zip code recognition. *Neural Computation*, *1*(4), 541–551. doi:10.1162/neco.1989.1.4.541

LeviR.RenegarN.SpringsS.ZamanT. (2019). Supply chain network analytics guiding food regulatory operational policy. SSRN. doi:10.2139/ssrn.3374620

Levow, G. (2006). Unsupervised and semi-supervised learning of tone and pitch accent. In *Proceedings of the Human Language Technology Conference of the North American Chapter of the Association for Computational Linguistics*, (pp. 224–31). New York: Assoc. Comput. Linguist. 10.3115/1220835.1220864

Li, Y., Metcalf, B. J., Chochua, S., Li, Z., Gertz, R. E. Jr, Walker, H., Hawkins, P. A., Tran, T., Whitney, C. G., McGee, L., & Beall, B. W. (2016). Penicillin-binding protein transpeptidase signatures for tracking and predicting β-lactam resistance levels in Streptococcus pneumoniae. *mBio*, *7*(3), e00756–e16. doi:10.1128/mBio.00756-16 PMID:27302760

Libbrecht, M. W., & Noble, W. S. (2015). Machine learning applications in genetics and genomics. *Nature Reviews. Genetics*, *16*(6), 321–332. doi:10.1038/nrg3920 PMID:25948244

Lin, X., Ruess, P., Marston, L., & Konar, M. (2019). Food flows between counties in the United States. *Environmental Research Letters*, *14*(8), 084011. doi:10.1088/1748-9326/ab29ae

Lu, X., Horn, A., Su, J., & Jiang, J. (2019). A universal measure for network traceability. *Omega*, *87*, 191–204. doi:10.1016/j.omega.2018.09.004

Lupolova, N., Dallman, T. J., Holden, N. J., & Gally, D. L. (2017). Patchy promiscuity: Machine learning applied to predict the host specificity of Salmonella enterica and Escherichia coli. *Microbial Genomics*, *3*(10), e000135. doi:10.1099/mgen.0.000135 PMID:29177093

Lupolova, N., Dallman, T. J., Matthews, L., Bono, J. L., & Gally, D. L. (2016). Support vector machine applied to predict the zoonotic potential of E. coli O157 cattle isolates. *Proceedings of the National Academy of Sciences of the United States of America*, *113*(40), 11312–11317. doi:10.1073/pnas.1606567113 PMID:27647883

Lupolova, N., Lycett, S. J., & Gally, D. L. (2019). A guide to machine learning for bacterial host attribution using genome sequence data. *Microbial Genomics*, *5*(12), e000317. doi:10.1099/mgen.0.000317 PMID:31778355

Macesic, N., Polubriaginof, F., & Tatonetti, N. P. (2017). Machine learning: Novel bioinformatics approaches for combating antimicrobial resistance. *Current Opinion in Infectious Diseases*, *30*(6), 511–517. doi:10.1097/QCO.0000000000000406 PMID:28914640

Maguire, F., Rehman, M. A., Carrillo, C., Diarra, M. S., & Beiko, R. G. (2019). Identification of primary antimicrobial resistance drivers in agricultural nontyphoidal Salmonella enterica serovars by using machine learning. *mSystems*, *4*(4), e00211–e00219. doi:10.1128/mSystems.00211-19 PMID:31387929

Maharana, A., Cai, K., Hellerstein, J., Hswen, Y., Munsell, M., Staneva, V., Verma, M., Vint, C., Wijaya, D., & Nsoesie, E. O. (2019). Detecting reports of unsafe foods in consumer product reviews. *JAMIA Open*, *2*(3), 330–338. doi:10.1093/jamiaopen/ooz030 PMID:31984365

Manitz, J., Kneib, T., Schlather, M., Helbing, D., & Brockmann, D. (2014). Origin detection during food-borne disease outbreaks - a case study of the 2011 EHEC/HUS outbreak in Germany. *PLoS Currents*, *6*. Advance online publication. doi:10.1371/currents.outbreaks.f3fdeb08c5b9de7c09ed9cbcef5f01f2 PMID:24818065

Marvin, H. J., Janssen, E. M., Bouzembrak, Y., Hendriksen, P. J., & Staats, M. (2017). Big data in food safety: An overview. *Critical Reviews in Food Science and Nutrition*, *57*(11), 2286–2295. doi:10.1080/10408398.2016.1257481 PMID:27819478

McDermott, P. F., Tyson, G. H., Kabera, C., Chen, Y., Li, C., Folster, J. P., Ayers, S. L., Lam, C., Tate, H. P., & Zhao, S. (2016). Whole-genome sequencing for detecting antimicrobial resistance in nontyphoidal Salmonella. *Antimicrobial Agents and Chemotherapy*, *60*(9), 5515–5520. doi:10.1128/AAC.01030-16 PMID:27381390

Møller, F. T., Mølbak, K., & Ethelberg, S. (2018). Analysis of consumer food purchase data used for outbreak investigations, a review. *Eurosurveillance*, *23*(24), 1700503. doi:10.2807/1560-7917.ES.2018.23.24.1700503 PMID:29921346

Munck, N., Leekitcharoenphon, P., Litrup, E., Kaas, R., Meinen, A., Guillier, L., Tang, Y., Malorny, B., Palma, F., Borowiak, M., Gourmelon, M., Simon, S., Banerji, S., Petrovska, L., Dallman, T. J., & Hald, T. (2020a). Four European Salmonella Typhimurium datasets collected to develop WGS-based source attribution methods. *Scientific Data*, *7*(1), 75. doi:10.103841597-020-0417-7 PMID:32127544

Munck, N., Njage, P. M. K., Leekitcharoenphon, P., Litrup, E., & Hald, T. (2020b). Application of whole-genome sequences and machine learning in source attribution of Salmonella Typhimurium. *Risk Analysis*, *40*(9), 1693–1705. doi:10.1111/risa.13510 PMID:32515055

Nguyen, M., Brettin, T., Long, S. W., Musser, J. M., Olsen, R. J., Olson, R., Shukla, M., Stevens, R. L., Xia, F., Yoo, H., & Davis, J. J. (2018). Developing an in silico minimum inhibitory concentration panel test for Klebsiella pneumoniae. *Scientific Reports*, *8*(1), 421. doi:10.103841598-017-18972-w PMID:29323230

Nguyen, M., Long, S. W., McDermott, P. F., Olsen, R. J., Olson, R., Stevens, R. L., Tyson, G. H., Zhao, S., & Davis, J. J. (2019). Using machine learning to predict antimicrobial MICs and associated genomic features for nontyphoidal Salmonella. *Journal of Clinical Microbiology*, *57*(2), e01260–e18. doi:10.1128/JCM.01260-18 PMID:30333126

Niehaus, D. (2016). *Tree Boosting with XGBoost—Why Does XGBoost Win "Every" Machine Learning Competition?* Nor. Univ. Sci. Technol.

Niehaus, K. E., Walker, T. M., Crook, D. W., Peto, T. E. A., & Clifton, D. A. (2014). Machine learning for the prediction of antibacterial susceptibility in Mycobacterium tuberculosis. In *IEEE-EMBS International Conference on Biomedical Health and Informatics* (pp. 618–21). Piscataway, NJ: IEEE. 10.1109/BHI.2014.6864440

Nisbet, R., Miner, G., & Elder, J. (2009). *Handbook of Statistical Analysis and Data Mining Applications*. Academic.

Norström, M., Kristoffersen, A. B., Görlach, F. S., Nygård, K., & Hopp, P. (2015). An adjusted likelihood ratio approach analysing distribution of food products to assist the investigation of foodborne outbreaks. *PLoS One*, *10*(8), e0134344. doi:10.1371/journal.pone.0134344 PMID:26237468

Nsoesie, E. O., Kluberg, S. A., & Brownstein, J. S. (2014). Online reports of foodborne illness capture foods implicated in official foodborne outbreak reports. *Preventive Medicine*, *67*, 264–269. doi:10.1016/j.ypmed.2014.08.003 PMID:25124281

Oldroyd, R. A., Morris, M. A., & Birkin, M. (2018). Identifying methods for monitoring foodborne illness: Review of existing public health surveillance techniques. *JMIR Public Health and Surveillance*, *4*(2), e57. doi:10.2196/publichealth.8218 PMID:29875090

Oliver, N., Lepri, B., Sterly, H., Lambiotte, R., Deletaille, S., De Nadai, M., Letouzé, E., Salah, A. A., Benjamins, R., Cattuto, C., Colizza, V., de Cordes, N., Fraiberger, S. P., Koebe, T., Lehmann, S., Murillo, J., Pentland, A., Pham, P. N., Pivetta, F., ... Vinck, P. (2020). Mobile phone data for informing public health actions across the COVID-19 pandemic life cycle. *Science Advances*, *6*(23), eabc0764. doi:10.1126ciadv.abc0764 PMID:32548274

Pang, H., Mokhtari, A., Chen, Y., Oryang, D., Ingram, D. T., Sharma, M., Millner, P. D., & Van Doren, J. M. (2020). A predictive model for survival of Escherichia coli O157:H7 and generic E. coli in soil amended with untreated animal manure. *Risk Analysis*, *40*(7), 1367–1382. doi:10.1111/risa.13491 PMID:32378782

Pesesky, M. W., Hussain, T., Wallace, M., Patel, S., Andleeb, S., Burnham, C.-A. D., & Dantas, G. (2016). Evaluation ofmachine learning and rulesbased approaches for predicting antimicrobial resistance profiles in gram-negative bacilli from whole genome sequence data. *Frontiers in Microbiology*, *7*, 1887. doi:10.3389/fmicb.2016.01887 PMID:27965630

Phillips, L. (2006). Food and globalization. *Annual Review of Anthropology*, *35*(1), 37–57. doi:10.1146/annurev.anthro.35.081705.123214

Pinheiro, V., Pontes, R., & Furtado, V. (2015). A #hashtagtokenizer for Social Media Messages. *Int. J. Comput. Linguistics Appl.*, *6*, 141–158.

Quade, P., & Nsoesie, E. O. (2017). A platform for crowdsourced foodborne illness surveillance: Description of users and reports. *JMIR Public Health and Surveillance*, *3*(3), e42. doi:10.2196/publichealth.7076 PMID:28679492

Rortais, A., Belyaeva, J., Gemo, M., van der Groot, E., & Linge, J. P. (2010). MedISys: An early-warning system for the detection of (re-)emerging food- and feed-borne hazards. *Food Research International*, *43*(5), 1553–1556. doi:10.1016/j.foodres.2010.04.009

Rudin, C. (2019). Stop explaining black box machine learning models for high stakes decisions and use interpretable models instead. *Nature Machine Intelligence*, *1*(5), 206–215. doi:10.103842256-019-0048-x PMID:35603010

# Chapter 13
# Artificial Intelligence for Improving Food Quality

**Avinash Kumar Sharma**
*Department of CSE, ABES Institute of Technology, AKTU, Ghaziabad, India*

**Shivani Singh**
*Department of CSE, ABES Institute of Technology, AKTU, Ghaziabad, India*

**Shivam Kumar**
*Department of CSE, ABES Institute of Technology, AKTU, Ghaziabad, India*

## ABSTRACT

*There is no doubt about the fact that artificial intelligence (AI) has stepped into improving quality of food throughout the world. Artificial intelligence makes it possible to be done for machine to profit as a material or substance of fact examining facts from different data sources and make conclusions, and can perform most humanly tasks with more than the limits of act without any fault. Due to the interference of humans in the production and packaging of food product, there is lack in maintaining the demand supply chain and also in safety of food to prevail over these issues of food industry. Automation is the best possible solution. It wraps up an enormous amount of streaming data, often noted as "big data," which brings fresh and new opportunities to watch agricultural and food processes. Besides sensors, big data from social media is becoming important for the food industry. During this review, the authors present a summary of artificial intelligence (AI) and its role in shaping further of agri-food systems.*

## INTRODUCTION

The approximation of food production will increase throughout the world by 65-115% by 2050 to cater the expeditiously growing population (Alexandratos,2010; Schierhorn, 2016). So the feasibility of agricultural field is key to ensure hunger eradication and guarantee food security for ever growing population. A proper formatted and detectable system has became a need for the quality control in food sector because of the appearance of the several food safety and incidents in food industry like as Jordanian market scandal that proved the food contain concerning level of carcinogenic substance and bovine spon-

DOI: 10.4018/978-1-6684-5141-0.ch013

giform encephalopathy. Few more challenges going to occur in next coming years like weather, climatic change condition along with maintainable water management because of water scarcity(Donepudi,2014). Due to all these reason an establishment of strategic agricultural sustainability is needed. Therefore, the requirement of innovative change is more important than any thing in current memory(Fargose,2015). Artificial intelligence has capability to address to such difficulties leading worldwide (Misra,2020).

The study's main goal is to develop artificial intelligence and procedures for gaining access to food quality and safety efforts in the food industry (Fesaghandis,2018). To expect desirable solutions helping farmer to adopt sustainable practices like iot, ai cloud computing is a key choice ((Fesaghandis,2018; Schierhorn,2016). The subset of AI that is machine and deep learning algorithms are extensively used(Khalid,2017). The moto of this review is to present the application of AI in agri-food sector.

The focus of this audit is on AI applications that correspond to the four pillars of the food security (accessibility, availability, usage, and strength (Buschulte,2015). Artificial intelligence advancements are being used to address each of the pillars of food security listed above. Ai technology in the food and beverage market was valued at $3.06 billion in 2020, according to Mordor Intelligence, and is predicted to reach $30 billion by 2026, with a CAGR of over 45.70% between 2021 and 2026. (Domingo,2008; Hamzah,2018).

The current article examines how AI breakthroughs might benefit the global farming and food industries, as well as the ways in which AI can handle some of the most pressing issues.

The safety of the food is one of the foundations of food quality(Kenefick,1999). A important element of AI is reducing the presence of diseases and detecting poisons in food production. The processing section is another pillar. In this section, AI can increase output and decrease waste by replacing employees on the line whose sole responsibility is to detect objects that are inappropriate for processing.

The supply chain is the other pillar of food quality. The food quality is directly influenced by the supply chain. Meal delivery on schedule and maintaining food quality are critical, particularly in this remote day. UberEats, a ride-hailing company's four-year-old meal delivery wing, serves more than half of the United States' population, providing deliveries for around 100,000 eateries. It also works in 300 cities and 300 other places around the world. The company is incorporating AI to create restaurant and menu item recommendations, optimise delivery, and investigate the usage of drones. Michelangelo, a machine learning platform, is used by the firms for a variety of services. In Michelangelo, they have a number of models operating, including meal delivery time forecasts, search rankings, search autocomplete, and restaurant rankings.

There's no disputing that AI has risen in popularity in recent years, with many companies investing heavily in researching the technology's possibilities in the industry(Kulkarni,2015). When AI is used in the food industry, it improves the quality of the food in terms of processing, supply chain, and safety.

The use of robots helps control cross-contamination in food processing plants. Any human labour has the risk of cross-contamination. Workers assigned to packaged food can easily transfer pathogens from product to product or from one area of the facility to another. Because jobs in food processing facilities are usually side-by-side, once specific cells are contaminated, the contaminants can easily spread from one worker to another.

Robots that are fixed in place and handle all aspects of a specific packaging job can help locate potential contamination, making it easier for processors to minimize cross-contamination and ensure food safety. If not cleaned properly, robots can still cause cross-contamination, but an additional set of robots can also solve this problem. For example, a robot supplier in the food processing industry has developed a robot that can clean entire work cell.

These robots work in pairs, start at the end of each operating cycle, and use high-power water to flush the work cell, the packaging robots used there, and themselves.

## COLLABORATIVE ROBOTICS (COBOTS)

One major recent innovation in robots has a new focus on tech that is collaborative. These new robots, unlike conventional robotics, aren't always built to fully automate a particular task. All things being equal, they are worked to cooperate and work cooperatively close by people where essential. Artificial intelligence-based machine vision technology helps them navigate factory floors safely or assist in tasks like assembly and machine tending. Safety features like force limiters and padded joints help prevent injuries that can occur while working in close proximity to conventional robots. These features also enable them to work in tight spaces without the use of safety cages that conventional robots sometimes require. In factories and food processing plants, they can provide assistance and speed up existing workflows (Kuula,2009; Fujita,2010). For example, an article in Asia Pacific Food Industry cites one case study from a Swedish food processor, Orkla Foods. The company integrated cobots into the production line packaging vanilla cream, freeing up the human workers who had been responsible for the task. Before the cobots were introduced, workers had to bag and manually pack the vanilla cream into cartons. (Marstio,2018)

Cobots are often lightweight and easy to reprogram on-the-fly, allowing workers to quickly move them from task to task as needed. In many cases, an entire fleet of cobots can be repositioned and re-programmed in half a day, allowing a business to reconfigure its robots to handle entirely new tasks without additional capital investment. This flexibility can also make cobots a better fit for personalized products than other systems. As product specifications change, a cobot can be easily programmed and reprogrammed to handle the differences. The use of these robots can also help prevent cross contamination, like more conventional robotics. New technology—like machine vision and collaborative robotics technology—is helping to expand the use cases of robots in the food processing industry. These robots can often improve product quality more effectively than process changes alone, and may help manage a labour gap that could persist well into the future(Cannella,2015).

### AI in Improving Quality of Foods

It would be wonderful if the planter could plant higher quality food under ideal growth conditions. Firms have started looking into how they may use this element of AI techniques to help farmers. Artificial intelligence can give farmers continuous experiences from their fields, permitting them to distinguish regions that need water system, preparation, or pesticide treatment. It seems to be nice if artificial intelligence can directly help farmers in guessing the type of crop grown in that suitable condition.

### Chatbot for Farmers

Chatbots are nothing more than a type of virtual assistant that allows robots to connect with humans. We can identify natural language using artificial intelligence-based chatbots and machine learning technologies, allowing us to communicate with users in a more personalized manner. This facility is primarily used for retail, tourism, media, and agricultural. They supply farmers with advice and other proposals by assisting them in resolving outstanding challenges. There are only a few requirements: literacy is required, it delivers a natural and familiar experience, and it does not require users to learn any latest innovative concepts or interactive methods. Experts can readily update or customise the communication system's educational foundation.

## Application of AI with its Drawbacks and Solutions

AI technology is transforming a variety of industries, including government departments, medical, advertising, and finance, to mention a few (Y,2016). By understanding the complexities of machine learning and artificial intelligence, the food and beverage industry has begun to use the technology's predictive analysis. With each passing day, the food sector evolves, and this has been reflected in Artificial intelligence is being used by an increasing number of food firms to increase revenue and productivity(Perdomo,2016). Cost, cultural shifts, professional skills requirements, transparency difficulties, and single-mindedness are all challenges in using artificial intelligence technology(Stroshine,2015). Reduce the expenses of equipment repair and maintenance (Piotter,2020; Ritenour,2011). It is self-evident that if a corporation uses machines, a significant portion of its resources are spent on cleaning, processing, and maintaining those devices(Piotter,2020). According to ongoing research initiatives, a technology known as self optimizing-clear-in-place can enhance cleaning time and dramatically minimise cleaning resources, especially water (Khalid,2017). During the pre-rinse and detergent phases of the cleaning process, a self-optimizing clean-in-place (SOCIP) system will use an artificially intelligent, multi-sensor inspection system to monitor and assess exactly how much food residue and microbiological wastes are left inside equipments(Niphadka,2011). This will ultimately lead to improved employee efficiency and human resource management as there is a much better system inplace to provide early detection of any fault (Shafie,2018).

## AI's significance in the food industry

*Figure 1.*

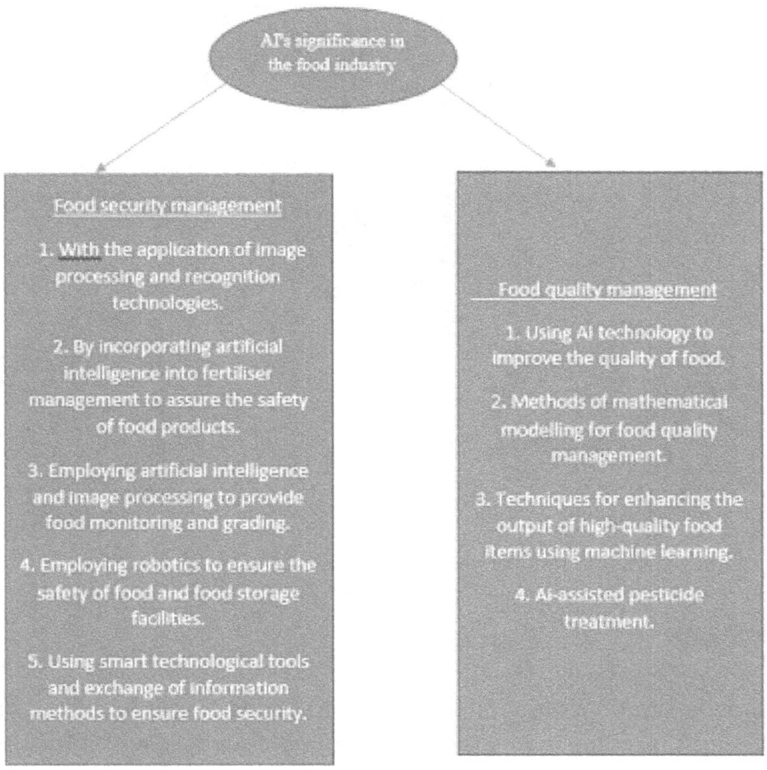


## Artificial Intelligence in Food Processing Industry

The writers will emphasize the impact and use of many context of ai in the food processing sector, such as object recognition, analysis of data, deep learning, machine learning, and robots, in this segment. With food processing, the food handling sector is also crucial, and AI plays a significant role in completing the full processing unit operation. As illustrated in Figure 2 there are some major applications chosen from the food processing and handling business.

*Figure 2. Major application of AI in food industry*

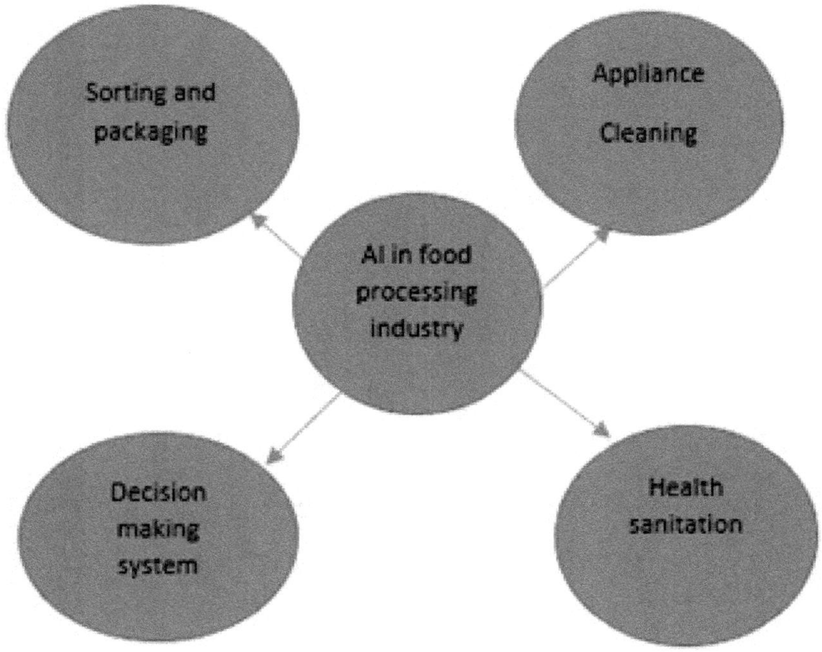


## Sorting and Packaging

In the food handling industry, the most complex and time-taking task is bundling and ordering of food items in proper sequence that is the work of assembling unit. So a vision of AI-based frameworks with the goal that can reduce the error, and the creation pace of the business is quickly expanded. Due to the irregular shapes, colors, and sizes of veggies, fruits, and other packaged foods, developing AI-based technologies is a difficult challenge. A significant amount of information is necessary for constructing an AI-based sequencing and packaging system so that the system may be adequately trained and accomplish the duties efficiently (Tripathi,2020; Oktarina,2019). Several scholars and research organizations are working behind this idea. One of the inventions in this fields TOMRA which work on the sorting algorithm very effectively. The addition of such kinds of inventions in food industry will help quicker creation rate, excellent yielding, and work cost cutting.

The AI developed decision making system consists of several procedures and methods i.e., high-resolution cameras, laser-technology-based systems, X-ray-based systems, IR spectroscopy, drones and 3D printing. These devices and advancements are utilized at the input, inspect every component of the food items. Traditional systems can only distinguish between pleasant and unpleasant products based by their look. Its been discovered that employing TOMRA, the separating and sorting challenges can be enhanced by 5–10% in the situation of potatoes alone. (Onishchuk,2020; Bünger,2021).

*Figure 3. Work flow of AI in sorting of food*

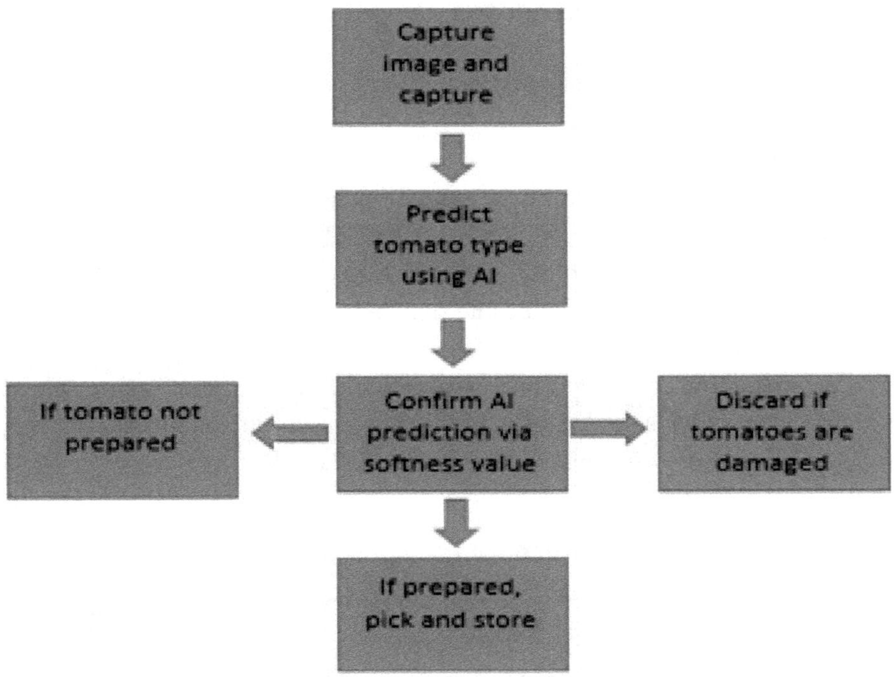


## Health Sanitation

It has been found that some countries like USA has initiated food processing unit sanitation artificial intelligence based systems also takes care of such guidelines .KanKan, a Chinese technology startup, signed a major partnership last year to provide an AI-powered solution for boosting personal hygiene among restaurant employees. The system employs cameras to keep an eye on the employees. It makes advantage of facial and object identification techniques of machine learning to ensure whether the workers are wearing mask and following the food safety rules or not, if it finds violation of law it captures the image on the screen for further review. The accuracy of this technology is more than 96%. The entire system found very useful, so it has been planned to expand it to the organization as much as possible (Philbrick,2020).

## Decision-Making System for Customers

AI has stepped in the part where human being were totally involved i.e. decision making part where assuming interference of machines and robots were not even thought. Now, though, It has been discovered that AI is guiding clients in selecting new essences as well as assisting food processing companies in producing distinct flavouring amalgamations (Dehghanian,2020; Cheng,2021). In 2018, Kellogg introduced Bear-Naked-Custom, a product that let users to create their own customised granola using more than 50 ingredients. It keeps track of each person's flavour preferences, client taste preferences, and other details. This type of data is critical for introducing a new product onto the market (Dehghanian,2020;

Cheng,2021).Now not only mechanical task but artificially intelligent machines can now make decisions on their own and can think logically as like human.

## Equipment Cleaning and Maintenance

When it comes to the food processing sector, it's critical to make sure that the equipment is properly cleaned and maintained. Due to the inability of present technology to identify how unclean food and beverage processing equipment is, everyday cleaning can take up to five hours to reduce food safety hazards(Bhullar,2020). Cleanliness accounts for 30% of energy and water consumption. Excessive chemical use can cost manufacturers a lot of money and the environment a lot of harm. Such jobs are easily handled by artificial intelligence(Demirci,2020). Clean-in-place (CIP) equipment aids in the control, monitoring, and documentation of hygienic processes cleaning procedures. Clean-in-place, technologies have grown more efficient and cost effective, not only help assure product safety but may also provide efficiencies to cleaning processes (Demirci,2020; Piotter,2021).

*Figure 4. Use of automatic robots for cleaning*

## Development of New Products

Deep learning and statistical analytics are used by AI technology to assess customer taste preferences and predict on how they'll react to novel flavours. Companies can divide the data into demographic groupings to aid in the development of new items that suit to the tastes of their target market (Bottani E, 2019). Manufacturers might use these to predict which products will be effective before they hit the shelves. Coca-Cola has set up self-service soft drink machines at a number of restaurants and other places, enabling customers to customise their drinks. Consumers may conceivably produce hundreds of different cocktails by just changing the basic drinks. Hundreds of additional drink stations, each serving a different type of beverages daily. Coca-Cola is employing artificial intelligence to examine a large amount of customer request data. CHERRY SPRITE was the first product to originate from this data, since ai technology anticipated that customers would produce a significant amount of cherry-flavored Sprite on their own and it would market well(Kumar,2021).

## Data Analysis at Food Industry

A huge number of well-known companies as well as food outlets abound in the food-based economy. This sector is losing popularity as a place to start a new firm due to increased competition(Yang,2020). In the food industry, the only way to keep ahead of the competition is to use technology, particularly data science. The information on data analysis in the food business is shown in Figure 5.

*Figure 5. Data analysis at food industry*

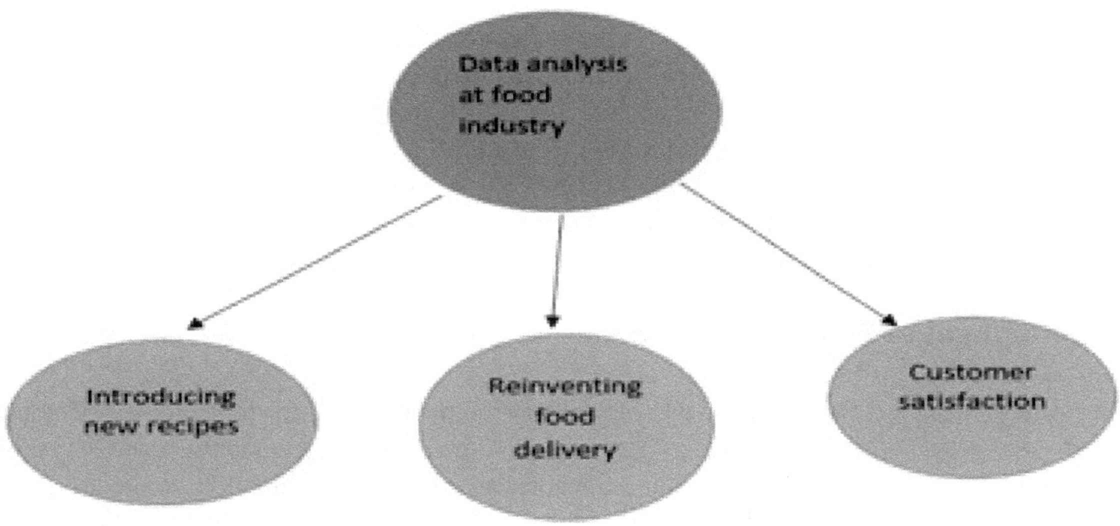


## Customer Satisfaction

Gobble's founder, Ooshma Garg, expressed the opinion that the food industry may be described as a digital firm. For the rest of the globe, it was a debatable assumption, but there is some truth to it(Feng,2020; Vorobyev,2021). In today's technology-driven sectors, data science has become a must for upgrading and managing their different business methods(Prokopov,2020) Gobble is an excellent example of a business that totally relies on data science to forecast supply and demand for its clients. It provides ten-minute meal kits to its customers and has thousands of regular customers who order from a variety of menus. It gathers information such as purchasing history, customer behaviour, feedback, and food preferences over time.

## Tinkering with New Ingredients

By combining the ingredients, a same dish can be prepared in a variety of ways. Furthermore, those items can be cooked in a variety of different ways, opening up a world of unlimited culinary options. Numerous recipes are accessible online, and they contain a large dataset that allows anyone from a novice to a specialist to explore ingredients in a variety of cuisines. The similarities and differences among the various cuisines can be determined by the researchers. North American and Western European cuisines, for example, are entirely dependent on products that have the same flavour components as East Asian and Southern European cuisines. Finally, researchers can determine which feed additives have a good flavour and identify a popular cuisine in a few provinces. This fundamental understanding also enables artificial-intelligence-based algorithms to suggest alternative types of component mixtures to chefs, resulting in a menu expansion and increased earnings for the food sector.

## Modifying Food Delivery

Online food delivery services like Swiggy, Zomato, and Uber-eat collect a lot of data about their clients' ordering habits and dish preferences. Data science and artificial intelligence (AI) can be used by food-based professionals to develop simpler, more cost-effective, and faster delivery systems(Semipyatny,2020). Some established industries are benefiting from AI, and there are some legitimate opportunities for market dominance. However, it is still in its infancy in the food sector and, as a result, requires more skilled optimization by food firms so that people may achieve better meals with better customer service.

## For Betterment of Farming Conditions

It would be wonderful if the farmer could plant better food under ideal growth conditions. Companies have already started looking into how they may use this element of intelligent machines to help farmers (Damayanthi,2019). Sentiment is one of the companies attempting to develop a system that will allow the impacts of elements such as light intensity, weather, salinity, and water stress on basil to be carefully monitored. Until now, development has only been accomplished in the lab, with the goal of developing precise "ingredients" for creating perfect foods.

To give a few examples of modern artificial intelligence applications in the food industry, they include producing large reductions in downtime, eliminating consumer discomfort at the time of sale, speeding up manual activities, and improving the worker-overtime ratio (Mujumdar,2017).

## Need of AI for Food Adulteration

Artificial Intelligence (AI) has the potential to be a boon to the food business. It may assist in precision agriculture and a variety of other uses in food production, therefore it plays an important role in supporting our food system. In the food industry, it can also be employed as a quality assurance metric.

AI is transforming how people think about food production, quality, and delivery, and the era of clever mobile apps has played a significant role in this shift. Artificial brains can be used to efficiently develop and analyse food databases. This has the ability to make the food sector healthier and more affordable for both employees and customers.

The technologies employed in the food industry to identify food adulteration are highly costly and complicated. Quality evaluation methods necessitate specific infrastructure. These methods necessitate a lot of physical labour and can be time consuming and inefficient. As a result, AI may be utilised as a platform to create a low-cost automated system for detecting adulteration in fruits, vegetables, and dairy products.

Different new approaches in this area have been widely employed to identify food quality, such as electronic tongues, electronic noses, machine vision, spectroscopy and spectral imaging, and etc. These systems can collect a lot of digital data about food composition and qualities, but it's crucial to analyse that data and extract usable information from it. However, implementing these strategies in real-world applications is a difficult undertaking.

To assess the data's quality and analyse it, AI-based approaches will be employed. Partial least squares, artificial neural network (ANN), support vector machine, random forest, k-nearest neighbour (KNN), and other approaches for dealing with vast amounts of data exist. Exploratory factor analysis (PCA), wavelet transform (WT), independent component correlation algorithm (ICA), scale-invariant feature transform, histogram of directed gradient, and so on can all be used for feature extraction. When dealing with this type of data, these strategies are extremely handy. As a result of examining the current state of food quality degradation, academics have been motivated to pursue research in this field.

## THE COVID-19 PANDEMIC AS WELL AS THE NATION'S FOOD SECURITY

The current COVID-19 pandemic has also brought attention to the significance of food and nutrition security, especially in developing countries. It has put the world economy in a dangerous position, and it has had a significant influence on Sri Lanka at all levels. Almost all industries were adversely impacted, with the exception of some critical utilities like water, power, and fuel, as well as those that provide food (agriculture) and medical aid. The authority's decision to maintain food supply as an important service was critical in ensuring food supply and accessibility.

The doss distribution, on the other hand, took longer to restore to normality. Workplace has shifted to "Work remotely" style under the newly normal case.As a result, the need to rethink agricultural production as a profession, as well as technology requirements, in order to attain food security based on new requirements. In this context, we anticipate that Ai technologies, combined with recent technological advancements, will play a larger role. As the world progresses post-COVID, the use of AI to improve food production is increasing, and demands for speed, efficiency, and sustainability are increasing in lockstep with the world's fast increasing population.

## THE ELECTRONIC NOSE

The electronic nose, often known as E-nose, is a device that works in the same way as the human nose to detect aromas and flavours. It's made up of a series of electronic chemical sensors that can detect both simple and complicated scents (Ponnusamy,2020). E-nose has been utilised in gas sensing applications that need examination of each component or mixture of gases/vapors.

In addition, it is critical in the food business for maintaining product quality. It has been used as an environmental protection tool and for the identification of explosive compounds due to its capacity to identify complex scents(Bera,2021). The core hardware component of E-nose is believed to be an array of non-specific gas sensors, which will interact with a range of other sensors. The sensors in the array are then stimulated, and a characteristic response called as a fingerprint is obtained.

E-key nose's system software is its feature extraction and pattern recognition algorithms, which evaluate the response, elicit key details, and then select one. As a result, the software component of the E-nose is critical to its effectiveness. E-nose is made up of three primary components: a sample delivery system, a detection system, and a computation system. In E-nose, the technique used includes ANN, FL, and pattern recognitions.

*Figure 6.*

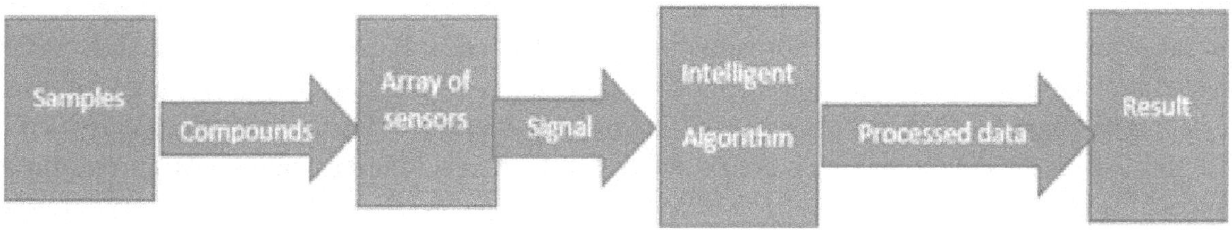


## The Use of Machine Learning in the Restaurant Industry

The use of machine learning and artificial intelligence approaches isn't restricted to creating automated systems, robocrops, soil monitoring, and new product introductions. It can also be used in the restaurant industry for a variety of services. Miso Robotics is a firm established in California that specialises on AI-driven robotic technologies for the kitchen. "Flippy," the company's premier AI kitchen assistant, uses 3D, thermal, and conventional vision to help with grilling, frying, preparing, and plating.

Machine Learning was utilised to teach the robot in this case:

To tell the difference between a raw steak, a prepared steak, and a slice of cheese When and how to flip the steak and when to take it off the grill

We'll go through how colours and forms are utilised to distinguish one thing from another in the identification process. Based on the initial fallacies (burnt or under cooked steak), the algorithm alters internal parameters or sections of its structure and attempts again. The algorithm will eventually figure out that a brown steak is a cooked steak.

Some AI Advancements in Improving Quality of Food in Food Industry

*Table 1.*

| | Technological companies/ technologies | Applications |
|---|---|---|
| 1. | **Kankan** | An artificially intelligent solution to deal with the personal hygiene among the restaurant workers in China. Company uses camera to look after the employee's facial recognition also object recognition software to detect whether workers are following proper good safety law or not. It's accuracy is approximate 96%. |
| 2. | **TOMRA** | This has brought a great change in the field of automation. As TOMRA sorting food develops optical sorting on the basis of sensor with the help of machine learning abilities. This system uses various technological devices such as cameras and near infrared sensors to observe food as normal humans being do. The outcome is less hours spent on manual arranging, more significant returns and better quality. |
| 3. | **SOCIP** | The University of Nottingham is working on a system that employs ultrasonic detection and optical fluorescent scanning to evaluate food leftovers and microbiological waste in a piece of equipment and then improve the cleaning procedure. |
| 4. | **Wellio** | Customers can order groceries online using this technology, which integrates behavioural science and machine intelligence to deliver personalised recipe recommendations. |

## CONCLUSION

This review explored about the numerous data and details to suggest the implementation of AI in dealing with the quality of food. In the present scenario AI has stepped in the food industry and has replaced several human interferences in this field. Day to day AI is improving its older version because of its functionality to amplify hygienics, safety of food, and waste management gadgets. This paper also presented the methodology of AI based robotics in improving and checking the quality of food and how maintenance of robots are taken care, so that it should not affect the quality of food in anyway. because of numerous demanding situations within the food sector and several factors like changing climate, increasing population, packaging, contagious diseases etc. it has became necessity of the hour to digitize food industry.

## REFERENCES


Alexandratos, N. (1995). *World Agriculture: towards 2010: An FAO Study*. Food & Agriculture Org.

Elferink, M., & Schierhorn, F. (2016). Global demand for food is rising. Can we meet it? *Harvard Business Review*, *7*(4), 1–7. https://www.researchgate.net/publication/302466629%0A

Donepudi, P. K. (2014). Technology growth in shipping industry: An overview. *American Journal of Trade and Policy*, *1*(3), 137–142. doi:10.18034/ajtp.v1i3.503

Narvekar, M., & Fargose, P. (2015). Daily weather forecasting using artificial neural network. *International Journal of Computers and Applications*, *121*(22), 9–13. doi:10.5120/21830-5088

Misra, N. N., Dixit, Y., Al-Mallahi, A., Bhullar, M. S., Upadhyay, R., & Martynenko, A. (2020). IoT, big data and artificial intelligence in agriculture and food industry. *IEEE Internet of Things Journal*, *1*, 1.

Soltani-Fesaghandis, G., & Pooya, A. (2018). Design of an artificial intelligence system for predicting success of new product development and selecting proper market-product strategy in the food industry. *The International Food and Agribusiness Management Review*, *21*(7), 847–864. doi:10.22434/IFAMR2017.0033

Iqbal, J., Khan, Z. H., & Khalid, A. (2017). Prospects of robotics in food industry. *Food Science and Technology (Campinas)*, *37*(2), 159–165. doi:10.1590/1678-457x.14616

Filter, M., Appel, B., & Buschulte, A. (2015). Expert systems for food safety. *Current Opinion in Food Science*, *6*, 61–65. doi:10.1016/j.cofs.2016.01.004

López, E. M., García, M., Schuhmacher, M., & Domingo, J. L. (2008). A fuzzy expert system for soil characterization. *Environment International*, *34*(7), 950–958. doi:10.1016/j.envint.2008.02.005 PMID:18378310

Allawi, M. F., Jaafar, O., Ehteram, M., Mohamad Hamzah, F., & El-Shafie, A. (2018). Synchronizing artificial intelligence models for operating the dam and reservoir system. *Water Resources Management*, *32*(10), 3373–3389. doi:10.100711269-018-1996-3

Riely, F., Mock, N., Cogill, B., Bailey, L., & Kenefick, E. (1999). *Food Security Indicators and Framework for Use in the Monitoring and Evaluation of Food Aid Programs, Nutrition Technical Assistance Project*. FANTA.

Athani, Tejeshwar, Patil, Patil, & Kulkarni. (n.d.). *Soil moisture monitoring using IoT enabled Arduino sensors with neural networks for improving soil management for farmers and predict seasonal rainfall for planning future harvest in North Karnataka*. Academic Press.

Arai, Kato, & Fujita. (2010). Assessment of operator stress induced by robot collaboration in assembly. *CIRP Annals – Manufacturing Technology, 59*, 5–8.

Aaltonen, I., Salmi, T., & Marstio, I. (2018). Refining levels of collaboration to support the design and evaluation of human-robot interaction in the manufacturing industry. *Procedia CIRP*, *72*, 93–98. doi:10.1016/j.procir.2018.03.214

Pan, Y. (2016). Heading toward artificial intelligence 2.0. *Engineering*, *2*(4), 409–413. doi:10.1016/J.ENG.2016.04.018

Barrero, O., Rojas, D., Gonzalez, C., & Perdomo, S. (2016). Weed detection in rice fields using aerial images and neural networks. *Proceedings of the the 2016 XXI Symposium on Signal Processing, Images and Artificial Vision (STSIVA),* 1–4. 10.1109/STSIVA.2016.7743317

Barzegar, M., Zare, D., & Stroshine, R. L. (2015). An integrated energy and quality approach to optimization of green peas drying in a hot air infrared-assisted vibratory bed dryer. *Journal of Food Engineering*, *166*, 302–315. doi:10.1016/j.jfoodeng.2015.06.026

Schmidt, R. H., & Piotter, H. M. (2020). The hygienic/sanitary design of food and beverage processing equipment. In *Food Safety Engineering*. Springer. doi:10.1007/978-3-030-42660-6_12

Qin, J., Burks, T. F., Zhao, X., Niphadkar, N., & Ritenour, M. A. (2011). Multispectral detection of citrus canker using hyperspectral band selection. *Transactions of the ASABE*, *54*(6), 2331–2341. doi:10.13031/2013.40643

Iqbal, J., Khan, Z. H., & Khalid, A. (2017). Prospects of robotics in food industry. *Food Science and Technology (Campinas)*, *37*(2), 159–165. doi:10.1590/1678-457x.14616

Tripathi, S., Shukla, S., Attrey, S., Agrawal, A., & Bhadoria, V. S. (2020). *Smart industrial packaging and sorting system. In Strategic System Assurance and Business Analytics*. Springer.

Dewi, T., Risma, P., & Oktarina, Y. (2020). Fruit sorting robot based on color and size for an agricultural product packaging system. *Bulletin of Electrical Engineering and Informatics*, *9*(4), 1438–1445. doi:10.11591/eei.v9i4.2353

Onishchuk, M. O. (2020). *Opto-mechanical sorting of municipal solid waste* (PhD dissertation). BHTY.

Bünger, L. (2021). *Robotic waste sorting*. Worcester Polytechnic Institute. https://www.wpi.edu/Academics/Projects

Rary, E., Anderson, S. M., Philbrick, B. D., Suresh, T., & Burton, J. (2020). Smart sanitation-biosensors as a public health tool in sanitation infrastructure. *International Journal of Environmental Research and Public Health*, *17*(14), 5146. doi:10.3390/ijerph17145146 PMID:32708793

Dehghan-Dehnavi, S., Fotuhi-Firuzabad, M., Moeini-Aghtaie, M., Dehghanian, P., & Wang, F. (2020). Estimating participation abilities of industrial customers in demand response programs: a two-level decision-making tree analysis. *2020 IEEE/IAS 56th Industrial and Commercial Power Systems Technical Conference (I&CPS)*, 1–8. 10.1109/ICPS48389.2020.9176817

Yost, E., & Cheng, Y. (2021). Customers' risk perception and dine-out motivation during a pandemic: Insight for the restaurant industry. *International Journal of Hospitality Management*, *95*, 102889. doi:10.1016/j.ijhm.2021.102889

Misra, N. N., Dixit, Y., Al-Mallahi, A., Bhullar, M. S., Upadhyay, R., & Martynenko, A. (2020). IoT, big data and artificial intelligence in agriculture and food industry. *IEEE Internet of Things Journal*, *1*, 1.

Wang, X., Puri, V. M., & Demirci, A. (2020). *Equipment cleaning, sanitation, and maintenance. In Food Safety Engineering*. Springer.

Wang, X., Puri, V. M., & Demirci, A. (2020). *Equipment cleaning, sanitation, and maintenance. In Food Safety Engineering*. Springer.

Schmidt, R. H., & Piotter, H. M. (2020). The hygienic/sanitary design of food and beverage processing equipment. In *Food Safety Engineering*. Springer. doi:10.1007/978-3-030-42660-6_12

Kumar, A. S., Suresh, G., Lekashri, S., Babu, L. G., & Manikandan, R. (2021). Smart Agriculture System with E–Carbage Using Iot. *International Journal of Modern Agriculture*, *10*(1), 928–931.

Tao, D., Yang, P., & Feng, H. (2020). Utilization of text mining as a big data analysis tool for food science and nutrition. *Comprehensive Reviews in Food Science and Food Safety*, *19*(2), 875–894. doi:10.1111/1541-4337.12540 PMID:33325182

Fainshtein, E., Serova, E., & Vorobyev, P. (2021). Contemporary research and analysis of food industry: case of Russian restaurant business network branch. In *Eurasian Business Perspectives: Proceedings of the 29th Eurasia Business and Economics Society Conference* (pp. 183–199). Springer International Publishing. 10.1007/978-3-030-65085-8_12

Topleva, S. A., & Prokopov, T. V. (2020). Integrated business model for sustainability of small and medium-sized enterprises in the food industry. *British Food Journal*, *122*(5), 1463–1483. doi:10.1108/BFJ-03-2019-0208

Ageikina, I. I., Lazareva, E. G., Mikhailova, I. Y., & Semipyatny, V. K. (2020). Results designing and analysis when introducing new beverage identification criteria. *Food Systems*, *3*(3), 4–7.

Kim, J. J., Kim, I., & Hwang, J. (2021). A change of perceived innovativeness for contactless food delivery services using drones after the outbreak of COVID-19. *International Journal of Hospitality Management*, *93*, 102758.

Withanage, D. P., & Damayanthi, B. W. R. (2019). Factors influencing the youths' interest in. *Agricultural Entrepreneurship in Sri Lanka*, *9*(12), 32–40.

Xu, J.-C., Zhang, M., Mujumdar, A. S., & Adhikari, B. (2017). Recent developments in smart freezing technology applied to fresh foods. *Critical Reviews in Food Science and Nutrition*, *57*(13), 2835–2843.

Marambe, B., & Silva, P. (2020). A sixty-day battle to tackle food security-response of the Sri Lankan government to the COVID-19 pandemic. *Sri Lanka Journal of Food and Agriculture*, *6*(1).

Ponnusamy, P., Roshan Ghias, A., Guo, C., & Sarikaya, R. (2020). Feedback-based self-learning in large-scale conversational ai agents. *Proceedings of the AAAI Conference on Artificial Intelligence*, 34(8), 13180–13187.

Bera. (2021). An application of operational analytics: for predicting sales revenue of restaurant. In *Machine Learning Algorithms for Industrial Applications* (pp. 209–235). Springer.

# Chapter 14
# The Role of Machine Learning and Computer Vision in the Agri-Food Industry

**Aswathy Ravikumar**
https://orcid.org/0000-0003-0897-6991
*Mar Baselios College of Engineering and Technology, India*

## ABSTRACT

*The food industry has been unable to control the demand-supply cycle and has also fallen short on food safety due to human engagement. Food production and distribution activities may be controlled more efficiently while also improving operational competence using an artificial intelligence-based solution. The food industry's future is entirely reliant on drones and also witty and robotic farming, thanks to AI and machine learning. Smart farming (soil monitoring, pest monitoring, and fertilizer management), food processing (production, processing, marketing, and customer feedback), and food safety, among other topics, need a detailed evaluation of machine learning applications in the agrifood business. It is vital to monitor production lines to ensure that the manufacturing process and products fulfill the required quality standards. A plethora of data may now be created throughout the production process due to growing digitization.*

## INTRODUCTION

Artificial intelligence is a term for the building of non-natural human brains capable of learning, planning, sensing, and understanding natural language (Adek & Ula, 2020). Artificial intelligence (AI) is a branch of information technology that focuses on computers that are meant to mimic human behavior. "The scientific and technical knowledge of constructing clever computer programmer in particular," according to AI's father, John McCarthy. (Patil & Banyal, 2019). Agriculture is vital to the economy of the nation later its foods the entire people. Food, or ration, needs for people, and it may be observed as the best consequence of farming, fashioned after farmers allocate many supplies. The goods of the food sector are critical to every country's growth. It also plays an important part in the growth of the

DOI: 10.4018/978-1-6684-5141-0.ch014

countries and global economies. In recent decades, newly created technology such as Machine Learning has produced excellent outcomes in achieving the intended goals. It's critical to look at ML-based smart agriculture and the advanced food business. Such methods encounter shared needs although still bringing good quality items on the schedule. The food sector may manufacture a big number of food goods in less time by using new technology, which would exponentially boost the company's economics (N & Gupta, 2020).An autonomous system is also known as an AI-based system; almost overall country used the aspect of technology. It will resolve to permit worldwide problem inefficiently, computerize Argo business, and alter the products of the food industry. AI-based system line intelligent drones and robots may also play a main role in packing at the lowest costs nowadays. The AI-based system as to the delivery of food goods, it also at dangerous environment time it will complete tasks an in a high-quality manner. The AI-based food industry has essential responsibilities divided into two types: food quality and food security management. In addition to Al, image processing plays a vital role in the prediction of food safety (Khan et al., 2021) . The AI together with image processing plays a vital role in farming activities and the food industry (Sahni et al., 2021; Sharma et al., 2021)

## MACHINE LEARNING

Machine learning's three major goals are supervised machine learning, unsupervised machine learning, and reinforcement learning. A predictive model is constructed via supervised learning by integrating labeled data with prior knowledge of the input and expected output variables. Supervised learning's purpose is to convert the input parameters to the required output variables. Approaches to supervised learning include decision trees, Bayesian networks, and regression analysis. Unsupervised learning methods are used on unlabeled datasets with uncertain dependent and independent variables. Unsupervised machine learning, which is often used for data reduction and exploratory analysis, creates hidden patterns from an unlabeled dataset. In reinforcement learning, the training and testing datasets are combined, and the learner interacts with the environment to obtain data. The learner is rewarded for his interactions with the environment, equating to exploration and exploitation. In opposed to applying previously acquired knowledge, the learner must investigate newly unknown activities to get extra information. As a result, Machine Learning (ML) is a vital component of artificial intelligence, as one of the characteristics of intelligence is the capacity to learn from its environment. Machine learning algorithms, in general, synthesize knowledge from unstructured data in such a manner that the resulting software programs (i.e., expert systems) are capable of performing valuable tasks. The Figure 1 shows the structure of AI, Machine learning and Deep Learning.

*Figure 1. Structure of AI*

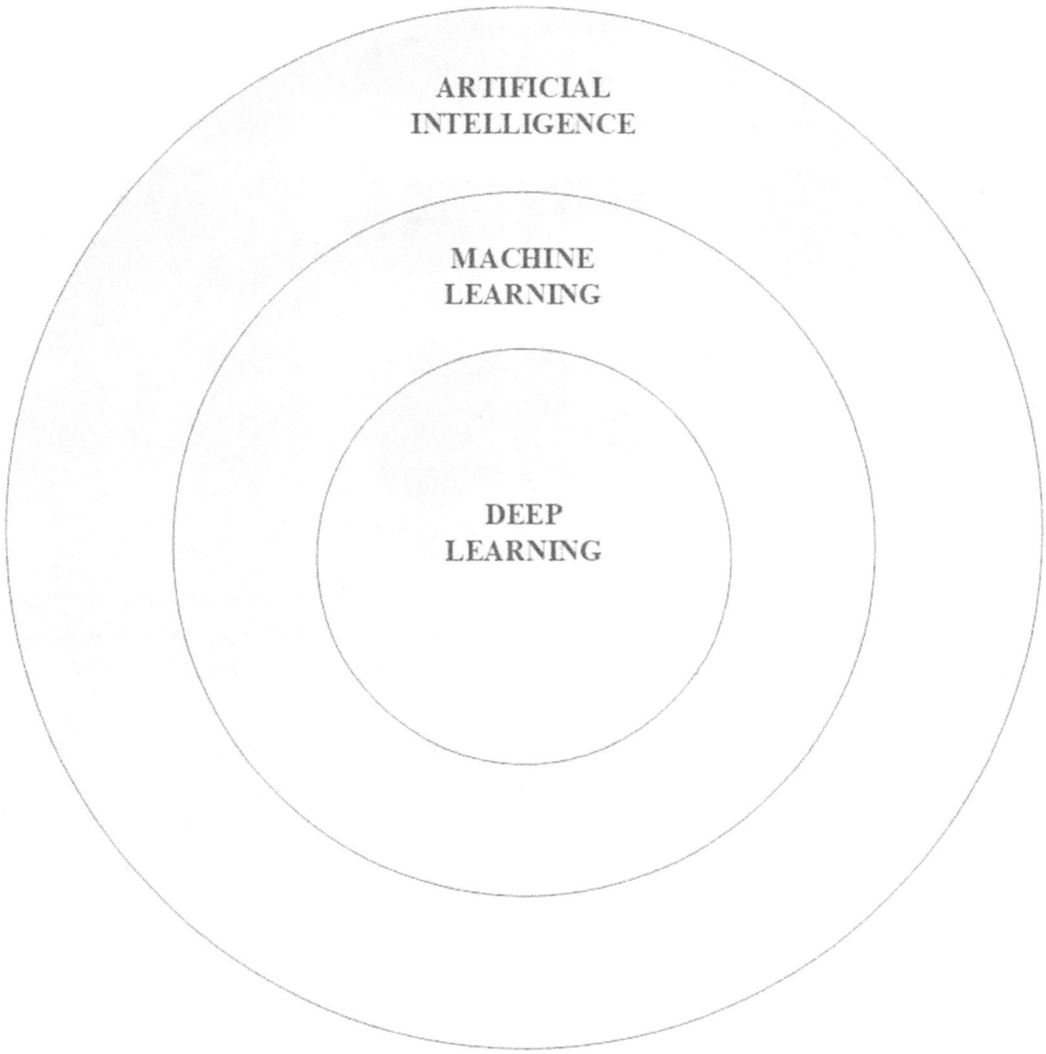


The Figure2 shows the structure of the types of Machine learning and Deep Learning algorithms used in this sector.

*Figure 2. Types of ML algorithms*

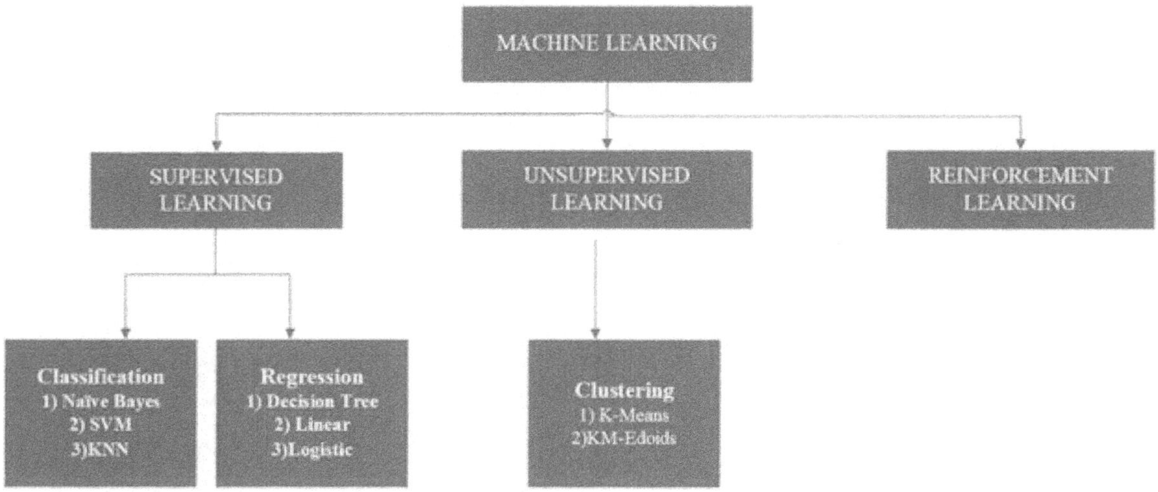


While machine learning approaches are similar in concept to linear regressions, they have substantial distinctions. Machine learning algorithms develop prediction functions that are more complicated than simple linear equations; they are reasonable approximations to much more general variables. However, machine learning algorithms may use category labels as both characteristics (independent variables) and class values, a critical difference. As a consequence, the application area and output possibilities of machine learning algorithms are far greater than those of simple linear least-squares tools. The most often used input for learning functions is a collection of examples of intelligent actions performed in the past by human specialists, referred to as training sets. These examples are defined by a variety of features, including a single attribute termed class of the example. The outputs of these functions are other functions or, more broadly, pieces of information capable of calculating (predicting) the class of unknown events (examples without a known class). Before we can construct training sets for judging the quality of food commodities, we must first articulate the critical features involved in determining product conditions. These relevant features constitute the 'objective' data that we give AI systems. This is a difficult task since the impartiality and correctness of this depiction will affect how the AI algorithms' output is used. We may not always have sufficient data sources to adequately represent our classes, and we are unsure if the data we do have is relevant or simply noise. For instance, while examining cow carcasses, a collection of geometrical shapes may need to be grasped. As a consequence, each geometrical concept must be quantified using a set of mathematical calculations. However, this is just half of the knowledge we must acquire; we must also assign the right score to each item according to our experts. In general, we must connect the kind of behavior we want to evoke with the way each event is shown. Following the establishment of the training sets, we should choose the machine learning algorithm that most interests us to get the best outcomes. Machine learning algorithms come in a variety of flavors. Numerous algorithms may be employed to handle a specific problem. We may opt to choose one algorithm over another depending on the kind of results it generates.

The Figure 3 shows the structure of, Machine learning model used for the agriculture and food applications. The first step is data collection followed by data preprocessing and dimensionality reductions and then application of machine learning algorithms suitable for the application and finally the output prediction and analysis of it.

*Figure 3. Steps in ML model*

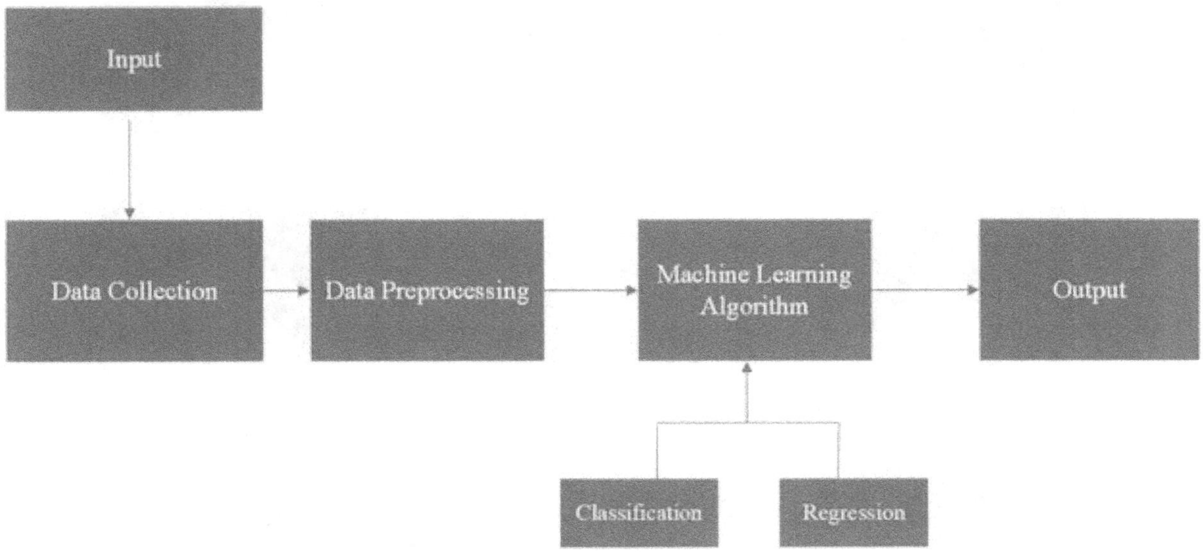


## MACHINE LEARNING IN AGRICULTURE

Technological advancements have assisted agricultural research, particularly via the integration of industrial advances into a sustainable agriculture production system. By electrifying every agricultural process, technology has transformed farming into a viable business. This saves the farmer money and eliminates the need for an intermediary who buys at a discount from producers and sells at a premium to end-users. Recent advances in computational intelligence have enabled the solution of location decision-making difficulties in agroecosystems. The adaptability, promotion, and reduced costs of machine learning enable the evaluation of agricultural systems' complicated relationships between outputs and inputs utilizing analytical approaches that are characterized by non-linearity, time-variable features, and a high number of unknown inputs. This paper covers a range of machines having to learn algorithms for agricultural activities such as crop disease detection, intelligent irrigation, soil classification, monitoring, and tracking. Agriculture is critical to humanity's survival. Agriculture provides a livelihood for a sizable portion of the world's population. Additionally, it offers a plethora of employment opportunities for the local populace. The Figure 4 shows the major application of Machine learning in Agriculture.

*Figure 4. Applications of ML in agriculture*

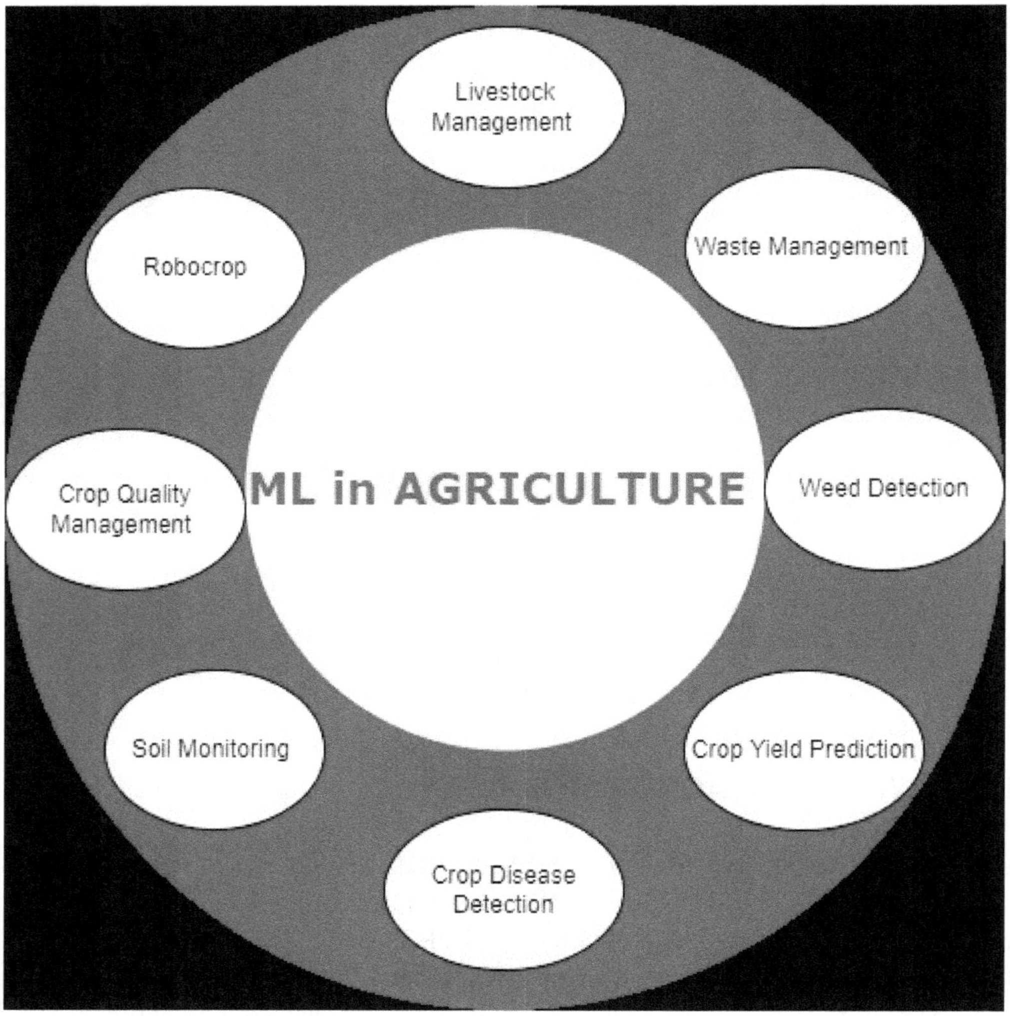


## Soil Monitoring

The food sector is now investigating the benefits of AI-based solutions. Computer vision and machine learning algorithms are critical components of an AI-based platform since they are being used to analyze the series of data and information obtained by AI-based entities to monitor crop and soil health progress. Customers are provided with an awareness of their soil's strengths and weaknesses via the use of computerized tools (Morvan et al., 2008). The fundamental objective of the developed system is to identify defective crops and determine the optimal course of action to ensure healthy crop development. In the Soil Monitoring (SM) scenario, after a farmer sends a sample of soil to the monitoring organization, the customer will get a complete report of the contents of their field soil. Following the analysis, an appropriate judgment was made about bacteria, fungus, and microbial development in general. In Japan, the first AI-based drone was employed for agricultural dusting in 1980. crop health, most corporations are now using farm AI and airborne technologies (Palti & Cohen, 1980). The company's main goal is to reduce expenses and increase agricultural yields. Users will pre-program the drone's path before integrating it with the device. Following that, the computer vision system will take several photographs that will be utilized in the assessment. Now, more advanced technology has ushered in a new age of progress known as the Internet of Things (IoT). The Internet of Things (IoT) plays an essential role in agricultural and soil monitoring decisions (Raut et al., 2018). The display of incoming data, as a result of the analysis, is useful in resource utilization. Identification of system behavior necessitates determining soil trends and making careful decisions to achieve optimum crop production and superior products. agriculture-based Smart agriculture is a term used to describe the Internet of Things. The smart food business is an IoT-based food sector. Smart agricultural explanations are also accompanied by a refined environment, such as high air quality and a well-kept irrigation system.

## Robocrop

The food sector is using modern technology-based equipment to boost productivity as a result of technological advancements. Robocrop (Melander et al., 2015). The large area collected by the input devices and the multiple processing lines allows for excellent typical crop center-line tracking (Machleb et al., 2020). The crop line area is used to compare the final picture to a ground truth gridiron layout. In addition to background control, the pattern-based attribute makes the system incredibly healthy. Because it has several cameras and sensors, it boosts performance and manufacturing rate. Various writers have done remarkable work in the area of harvesting robots, which has significantly enhanced production in recent decades, in robocropping or agricultural automation. Due to advancements and additional advantages such as increased productivity and decreased workforce, systems have become more popular. Weed management near or inside crop rows is made easier and more effective using robotic weeding in the farming sector. Adaptive Robotic Chassis (ARC) (Bucher et al., 2021) is a strawberry flower-specific system that was constructed from a list of existing systems. Strawberry blooms are photographed and processed in that system using an installed camera. Finally, the robot has gotten the required coordinate and is doing the necessary tasks. The robocrop's performance is entirely dependent on the characteristics of the input picture. When the input picture has more prominent characteristics, the results are exceptional.

Weather changes, for example, are tracked and forecasted using learning models. Algorithms based on machine learning play a major part in this. In collaboration with satellites, machine learning algorithms look at agricultural sustainability, weather forecasting, and assessing farms to see whether pests and diseases are present. model excels in delivering high-quality, often updated data or information.

## Crop Yield Prediction

Among the most significant aspects of smart farming is yield forecasting, which is necessary for output mapping, yield estimation, agricultural supply-demand matching, and agronomic practices to increase productivity. ML applications have those reported in (Ramos et al., 2017), which featured a method for identifying coffee fruits on a limb that was fast, minimal, and non-destructive. The technique classifies coffee fruits into 3 groups: harvestable, unharvestable, and fruits with an unknown maturity stage. Additionally, (Su et al., 2017) developed a strategy for forecasting rice development phases using SVM and essential spatial data collected from Chinese meteorological stations.

## Crop Disease Prediction

Pest and disease control in agronomic and thermal conditions is among agriculture's most significant concerns. The most popular way of biological pest control is to evenly spray pesticides over the agricultural area. While effective, this strategy has a hefty environmental and economic cost. Environmental ramifications include residues in agricultural goods, adverse impacts on groundwater contamination, and effects on indigenous wildlife and ecosystems. Based on an image processing technique, the authors of (Ebrahimi et al., 2017) developed a unique approach for parasitic classification and automated detection of thrips in a strawberry plant community for real-time management. When compared to human inspections, automated detection of unhealthy plants increased grain production and was faster.

## Livestock Management

Animal welfare and livestock production are two subcategories of the livestock category. Animal welfare is concerned with the health and well-being of animals, with machine learning being used mostly to monitor animal behavior to identify illnesses early. The paper (Dutta et al., 2014) describes a strategy for identifying cow behaviors by combining machine learning algorithms with data collected by collar sensors equipped with magnetometers and 3 accelerometers. The researchers intended to see if they could forecast oestrus and identify nutritional changes in cattle. (Pegorini et al., 2015) suggested an approach for the automatic identification and classification of chewing behaviors in calves. The term "livestock production" refers to studies conducted to ensure the accurate forecast and estimation of agricultural features to maximize the financial production efficiency system. This subcategory contains four articles, three on cow production and one on chicken egg production. The study of (Craninx et al., 2008) established an approach for forecasting the rumen fermentation pattern using milk fatty acids. The subsequent study (Morales et al., 2016) examined hen production. Specifically, an approach based on the SVM model was created for early detection and warning of issues in commercial egg production. SVM-based techniques were used to develop a system for properly predicting bovine weight trajectories across time (Alonso et al., 2015) Breeders must be able to determine weights properly.

## Water Management

Water management takes time and is necessary for hydrological, climatic change, and agricultural balance to be maintained. Accurately estimating evapotranspiration is a challenging procedure that is crucial for crop production resource management and irrigation system design. Another study (Hansen et al., 2018) developed a computational approach for estimating mean monthly evapotranspiration in arid and semi-arid regions. It was computed between 1951 and 2010 using monthly mean climatic data from 44 meteorological stations. Another article dedicated to machine learning techniques in agricultural irrigation systems (Mehdizadeh et al., 2017) presented two scenarios for calculating daily evaporation using thermal data collected from six weather stations over a lengthy period (i.e., 1961–2014).

## Fruit Recognition

Orange fruit recognition (Singh, Shivam, et al., 2022) is accomplished by the use of various convolutional neural network techniques (CNN). The collection contains a total of 820 example photos for the purpose of image accession. The chosen dataset has been separated into two distinct classes, the first of which comprises photographs of oranges and the second of which contains images of various fruit (apple, banana etc.). Self-created CNNs, LeNet-5, VGG-19, and MobileNetV2 are used for identification; these are prepackaged CNN models with fixed layers. The architecture was developed particularly for the LeNet-5 model, whereas the two models, VGG-19 and MobileNetV2, were imported using preexisting libraries and then customized to meet the authors' requirements.

### Leaf Identification

Numerous approaches and instruments have been developed throughout the years to safeguard plants (Kohli et al., 2022). However, the outcomes were not very efficient. Identifying distinct plant illnesses using their leaves is a difficult challenge for scientists and other researchers who want to save them from extinction. This article describes an automated method for detecting plant leaf disease using supervised classifiers and FCM clustering. To improve the picture, denoising and histogram linearization are used. Following feature extraction, a 13-dimensional feature vector was divided into a training and validation feature set. SVM, PNN, and KNN are used for classification, while PPV, TPR, OCA, and MCA are calculated for each experiment. The obtained result indicates that the greatest OCA is 98.4 percent when the SVM classifier is used in conjunction with the RBF kernel function.

## MACHINE LEARNING IN THE FOOD INDUSTRY

The main applications of Machine learning in the food industry are explained below and shown in Figure 5.

*Figure 5. Applications of ML in the food industry*

## Product Sorting and Packaging

Proper ordering and wrapping of food products are one of the most time-consuming and challenging tasks in the food processing industry. Such a tiresome process may be performed by AI-based systems, minimizing the possibility of mistakes and substantially increasing the industry's output rate. The creation of artificial intelligence-based systems is a difficult challenge owing to the uneven forms, colors, and sizes of food supplies. Numerous research groups developed disparate solutions to accomplish the same aim. One of them is TOMRA, which is very effective in sorting. It significantly increased manufacturing accuracy to 90%. Currently, the majority of product sorting and packing processes are automated. By using such systems, industries achieved many benefits, including increased production rates, higher-quality yields, and labor cost savings. Additionally, each firm discovered that the AI-powered system is more exact.

## Personal Health Sanitation

Several countries across the globe, along with the United States, have developed food processing unit cleaning criteria, it was revealed in the literature. These criteria have also been followed by AI-based systems. A machine learning system is meant to give an anonymous quantity object and face recognition. Individuals who do not adhere to the criteria are monitored using this technique (Rary et al., 2020). If anything matches, it may be handled immediately and without delay. Because the developed system produces excellent results, it has been intended to extend the system to include an increasing number of companies.

## Decision-Making System for Customers

It has been discovered that Machine learning is not only supporting food processing enterprises in producing novel taste combinations but also assisting consumers in selecting creative essences (Dehghan-Dehnavi et al., 2020). Kellog launched BearNaked-Custom in early 2018, allowing customers to customize their granola with over 50 ingredients. By looking after each individual, it maintains track of personal preferences, customer preferences, and much more. This kind of information is crucial for the effective launch of a new product (Yost & Cheng, 2021).

## Equipment Cleaning and Maintenance

In the food processing sector, proper maintenance of processing equipment is crucial. Such a task is easily achieved using artificial intelligence-based technology (Wang et al., 2020). This is performed with the assistance of several sensors and cameras. One of Whitwell and Martec's goods suffers from the possibility of being decreased to 50%, allowing for increased efficiency and reduced time. Martec is currently striving to defend the notion of an AI-powered cleaning station. Martec employs this strategy by growing accumulated information to develop artificial intelligence systems via the use of ultrasonic distance imaging technologies and visual fluorescent techniques (Schmidt & Piotter, 2020). It quantifies the amount of food and microbiological waste remaining within the machine. The system will come to a stop once the whole of the testing phase's report has been published.

## Showcasing New Products

Introducing new products for any industrial operation is a lengthy procedure. It is completely reliant on specific customer segments, especially in the food business. the information gathered by different consumer decision-making systems is beneficial for the introduction of new goods. The ML-based module processes the acquired data and then makes the appropriate choice for the item (Wardah et al., 2020). By using a machine learning method, problems such as "what precisely are buyers seeking for" have been addressed. Almost all food processing and packaging sectors are now using the potential of machine learning to grow and introduce new goods into the market. Previously, this work was conducted through a test case or a survey. The system's success rate is quite low. Now, the whole landscape has shifted, and AI and machine learning are widely employed for these sorts of activities. Coca-Cola has deployed self-service soft drink kiosks around the United States of America. Customers may create thousands of different cocktails simply by adding slight taste adjustments. Such action has been captured by the computer, and the remaining analysis is carried out using machine learning techniques algorithms. These data may be used to introduce new goods.

## Management of the Demand-Supply Chain

As soon as culinary firms are concerned about food standards, better transparency regarding the path food products follows through the global supply chain is required.AI is used to monitor each step of the process. It oversees all aspects of business, from pricing to inventory management. Additionally, it is responsible for anticipating and tracking the movement of belongings from the point of origin to the point of collection for clients. Symphony Retail, powered by artificial intelligence, enables transportation booking, invoicing, and inventory management. Additionally, it preserves discipline and prevents the accumulation of commodities that eventually wind up in depleted material.

## Customer Satisfaction

Ooshma Garg, Gobble's founder, expressed the view that the food sector may be seen as a technology firm. For the rest of the region, this was a debatable assumption, but there is some truth to it (Fainshtein et al., 2021; Tao et al., 2020). In today's technology-driven sectors, data science has become a requirement for upgrading and maneuvering their different business methods. Gobble is an excellent example of a sector that depends entirely on data science to forecast both supply and demand from its consumers. It provides ten-minute supper kits to its consumers and has hundreds of regular clients with a variety of menu options. It gathers data such as purchasing history, consumer behavior, feedback, and food preferences across a range of periods to ensure that it is prepared to fulfill demand (Topleva & Prokopov, 2020).

## Introducing New Recipes

By mixing the components in a single recipe, multiple variations may be created. Additionally, the same elements may be prepared in a variety of various ways, creating an area of limitless possibilities for creating culinary delicacies. Numerous recipes are accessible online and constitute a massive dataset that enables the investigation of components in a variety of cuisines by both amateurs and experts. This fundamental knowledge also enables artificial intelligence-based algorithms to propose alternative sorts of component combinations to chefs, therefore expanding the menu and profit margins of a food sector.

## Online Restaurant Search Engine Powered by Artificial Intelligence

Consumers frequently make restaurant, coffee shop, or pub selections based on their rating in comparison to competitors. It becomes critical for a restaurant company to learn about both good and bad client experiences to attract new customers and retain current ones. The great majority of customers find restaurants via search engines or other internet media. A customer is advised a favorite diner or café based on geographical proximity and review data. Additionally, AI agents aid customers by informing them about promotions, events, and deals at their favorite eateries.

## Predictions for Revenue Using deep learning

A successful business, such as a food service outlet, is dependent on the quality of the food and services given by the operators. Apart from service and cuisine, forecasting the restaurant's sales production is equally critical to the company. The owner of food chains or establishments must develop a solid business plan for future operations to maximize company development and profit. Multiple fitting algorithms may be employed in artificial intelligence to generate a sales prediction (Sanjana Rao et al., 2021). In the food sector, developing an appropriate fitting algorithm for sales forecasting, whether for five months or fourteen months' forecasting, involves a significant amount of time and flawless work.

Robots are commonly accepted in food processing sectors due to their sterile nature, the reason for this is that grains, condiments, and other food goods do not need refrigeration and are located in the most contaminated region. Previously, such food goods were contaminant-free, but the situation has changed dramatically. For these sorts of issues, AI-based technologies may undoubtedly assist in resolving the issue. They are unable of transmitting diseases in the same way that humans do. Additionally, there are some breakthrough ideas using machine learning in food safety techniques that are expected to get widespread recognition shortly. They are primarily concerned with reducing the prevalence of food-borne infections.

## Food Distribution

Fresh farm products are fragile and time-sensitive; thus, food quality and preservation are critical. To preserve the food's integrity and prevent infection, the time necessary to transport fresh food goods must be as short as possible, and the transportation atmosphere must be secure. As a result, the necessity for a sensible logistical delivery mechanism for fresh farm goods develops, which is a key problem for almost all clients. China pioneered cold storage logistics, and many improvements have been made throughout changes concerning issues and problems, but at some point, it was determined that this was insufficient. Due to the difficulties associated with ensuring the delivery of healthy produce, another solution has been identified: route planning for food product delivery without regard for cold storage administration. Path planning for the distribution of food goods may increase the assurance of high-quality food, customer happiness, preventing food contamination, and lowering logistic costs, among other benefits. The results indicate an efficient technique, namely an artificial bee colony algorithm, for distributing fresh food within a time frame without incurring penalties or compromising food quality (Katiyar et al., 2021).

## Identification of Fresh Produce

One of the most significant issues confronting food processing facilities is the infrequent availability of material. Manual sorting is used in food processing factories to sift and separate vegetables, resulting in decreased efficiency and higher costs. Food processing firms can significantly automate food cataloguing by using cameras, lasers, and machine intelligence to allow more efficient food sorting. For example, by using Artificial Intelligence in conjunction with sensor-based optical sorting technologies, the lengthy and time-consuming procedures associated with sorting fresh fruit may be eliminated, resulting in increased output with improved quality and reduced waste. AI is utilized to improve machine calibration in order to handle several product sizes while minimizing waste and expenses.

## Food Safety

Food is a vital aspect of human existence, which makes food quality a critical factor in its choice for ingestion. To ensure food quality is maintained, it must begin at the point where it may be harmed, namely warehouses. Food safety and the security of its warehouses are key problems since many civilians die of poor quality of food. A robot that can guarantee the safety of both the food and the warehouse might be one option, since maintaining large warehouses is a lengthy operation, and occasionally food within the warehouse goes unreported, becoming contaminated. Additionally, a robot can secure stores against burglars in any scenario that would be impossible for humans. This robot would be inexpensive and efficient, while simultaneously maintaining food safety by preserving the food's integrity and assuring its superior quality.

In Figure 6 the brief overall picture of AI and ML in Agriculture is shown the main four steps the observation, act, interpret and decide.

## Effect on Supply Chain

Currently, organizations involved in supply chain management (SCM) rely heavily on their network of partners and suppliers to ensure that commodities are moved successfully and efficiently (Singh, Rawat, et al., 2022). To boost productivity, the only way to succeed is to use the leading technologies just at right moment. This includes strategies and long-term considerations for controlling and resolving the numerous risks that may arise throughout the execution of procedures. However, by incorporating artificial intelligence into each stage of the supply chain, there is significant opportunity for innovation. These phases include conceptualizing the product from scratch, developing the manufacturing process, and delivering the product to the appropriate customers. Machine learning can assist in improving and developing each of these phases, hence exponentially increasing total efficiency at a low cost. This paper makes a hypothetical reference to the use of AI in supply chain management, reimbursement, and the obstacles associated with AI in the supply chain management business. The manuscript's primary objective is to examine a previously published assignment that may serve as the foundation for future industrial research.

*Figure 6. AI and ML in agriculture*

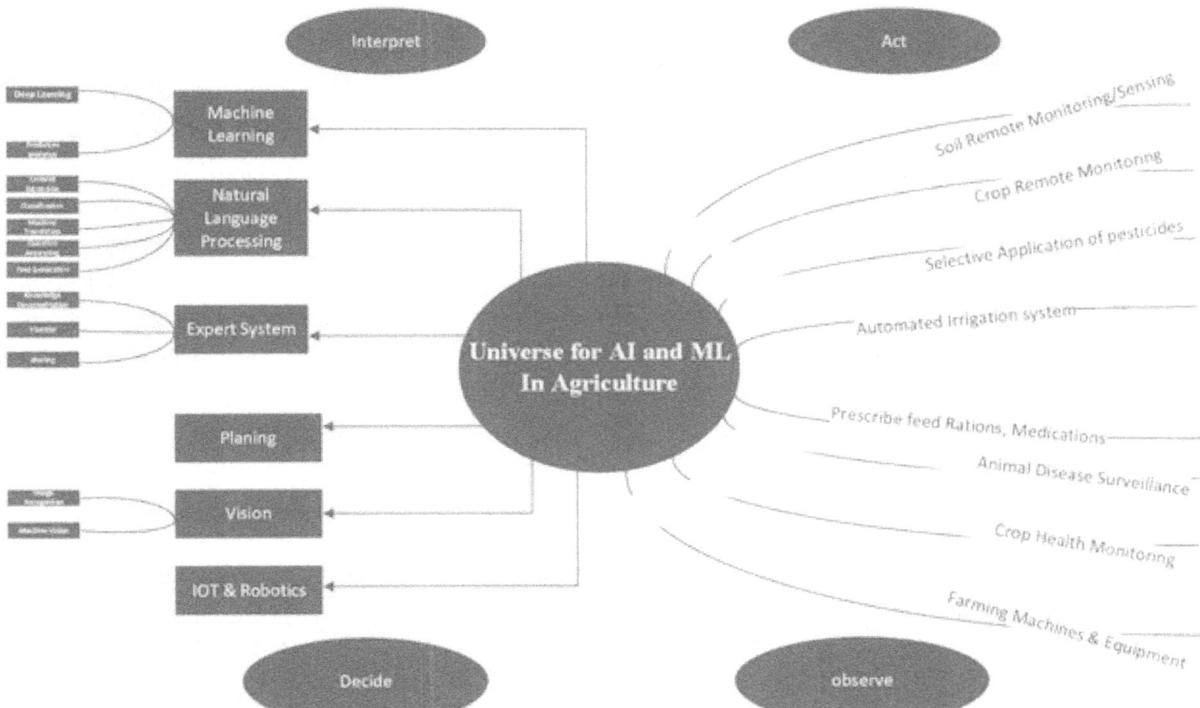


## OPEN RESEARCH PROBLEMS

Although machine learning in agriculture has made significant progress, numerous open challenges persist that have certain common reference points, despite the topic's diversity of subfields. The primary issues arise when sensors are used on farms for a variety of reasons, including high ICT expenses, traditional methods, and a lack of information. Additionally, the majority of available data sources do not reflect real situations, as they are typically generated by a small number of people collecting images or specimens over a short period and from a small geographic area. Additionally, the need for more effective machine learning algorithms and scalable computing architectures has been highlighted, which might result in quick data processing. Another significant outstanding issue is that the great majority of the farmers are not machine learning specialists and hence cannot completely appreciate the underlying patterns discovered by machine learning algorithms. As a result, more intuitive systems should be designed. Simple solutions that are easy to comprehend and use, such as a visualization tool with a consumer interface for the proper display and manipulation of data, would be particularly beneficial. Given farmers' increasing familiarity with smartphones, particular smartphone apps have been presented as a viable solution to the aforementioned difficulty]. Finally, the development of effective machine learning approaches that include expert knowledge from diverse stakeholders, notably in computer science, agriculture, and the commercial sector, should be encouraged as a way of building realistic solutions.

## CONCLUSION

The primary benefits of using AI methods are that they are well-suited to non-linear behaviors, which are prevalent in food processing; they can also be used to record what is learned, and they may be used to identify critical factors impacting the process's performance. All of these qualities of artificial intelligence-based systems enable the generation of trustworthy evaluations similar to verified performance levels in the absence of conventional quantitative methodology. Simultaneously, ideas of process learning may be reinterpreted as procedural guidelines for training future specialists. This is the circumstance were getting electronic representations of the object to be categorized is impossible or unreasonably expensive, and the only way to build a small, sound collection of clear criteria is to acquire an experimental sample. However, the most intriguing use of machine learning is identifying the critical qualities required to solve knowledge-related problems while the underlying process is unknown.

## REFERENCES

Adek, R. T., & Ula, M. (2020). A Survey on The Accuracy of Machine Learning Techniques for Intrusion and Anomaly Detection on Public Data Sets. *2020 International Conference on Data Science, Artificial Intelligence, and Business Analytics (DATABIA)*. 10.1109/DATABIA50434.2020.9190436

Alonso, J., Villa, A., & Bahamonde, A. (2015). Improved estimation of bovine weight trajectories using Support Vector Machine Classification. *Computers and Electronics in Agriculture*, *110*, 36–41. doi:10.1016/j.compag.2014.10.001

Bucher, S., Ikeda, K., Broszus, B., Gutierrez, A., & Low, A. (2021). Adaptive Robotic Chassis (ARC): RoboCrop A Smart Agricultural Robot Toolset. *Interdisciplinary Design Senior Theses*. https://scholarcommons.scu.edu/idp_senior/69

Craninx, M., Fievez, V., Vlaeminck, B., & De Baets, B. (2008). Artificial neural network models of the rumen fermentation pattern in dairy cattle. *Computers and Electronics in Agriculture*, *60*(2), 226–238. doi:10.1016/j.compag.2007.08.005

Dehghan-Dehnavi, S., Fotuhi-Firuzabad, M., Moeini-Aghtaie, M., Dehghanian, P., & Wang, F. (2020). Estimating Participation Abilities of Industrial Customers in Demand Response Programs: A Two-Level Decision-Making Tree Analysis. *2020 IEEE/IAS 56th Industrial and Commercial Power Systems Technical Conference (I CPS)*, 1–8. 10.1109/ICPS48389.2020.9176817

Ebrahimi, M. A., Khoshtaghaza, M. H., Minaei, S., & Jamshidi, B. (2017). Vision-based pest detection based on SVM classification method. *Computers and Electronics in Agriculture*, *137*, 52–58. doi:10.1016/j.compag.2017.03.016

Fainshtein, E., Serova, E., & Vorobyev, P. (2021). Contemporary Research and Analysis of Food Industry: Case of Russian Restaurant Business Network Branch. doi:10.1007/978-3-030-65085-8_12

Hansen, M. F., Smith, M. L., Smith, L. N., Salter, M. G., Baxter, E. M., Farish, M., & Grieve, B. (2018). Towards on-farm pig face recognition using convolutional neural networks. *Computers in Industry*, *98*, 145–152. doi:10.1016/j.compind.2018.02.016

Katiyar, S., Khan, R., & Kumar, S. (2021). Artificial Bee Colony Algorithm for Fresh Food Distribution without Quality Loss by Delivery Route Optimization. *Journal of Food Quality*, *2021*, e4881289. doi:10.1155/2021/4881289

Khan, R., Kumar, S., Dhingra, N., & Bhati, N. (2021). The Use of Different Image Recognition Techniques in Food Safety: A Study. *Journal of Food Quality*, *2021*, e7223164. doi:10.1155/2021/7223164

Kohli, P., Kumar, I., & Vimal, V. (2022). Plant Leaf Disease Identification Using Unsupervised Fuzzy C-Means Clustering and Supervised Classifiers. In R. Agrawal, J. He, E. Shubhakar Pilli, & S. Kumar (Eds.), *Cyber Security in Intelligent Computing and Communications* (pp. 281–293). Springer. doi:10.1007/978-981-16-8012-0_22

Machleb, J., Peteinatos, G. G., Kollenda, B. L., Andújar, D., & Gerhards, R. (2020). Sensor-based mechanical weed control: Present state and prospects. *Computers and Electronics in Agriculture*, *176*, 105638. doi:10.1016/j.compag.2020.105638

Mehdizadeh, S., Behmanesh, J., & Khalili, K. (2017). Using MARS, SVM, GEP and empirical equations for estimation of monthly mean reference evapotranspiration. *Computers and Electronics in Agriculture*, *139*, 103–114. doi:10.1016/j.compag.2017.05.002

Melander, B., Lattanzi, B., & Pannacci, E. (2015). Intelligent versus non-intelligent mechanical intra-row weed control in transplanted onion and cabbage. *Crop Protection (Guildford, Surrey)*, *72*, 1–8. doi:10.1016/j.cropro.2015.02.017

Morales, I. R., Cebrián, D. R., Blanco, E. F., & Sierra, A. P. (2016). Early warning in egg production curves from commercial hens: A SVM approach. *Computers and Electronics in Agriculture*, *121*, 169–179. doi:10.1016/j.compag.2015.12.009

Morvan, X., Saby, N. P. A., Arrouays, D., Le Bas, C., Jones, R. J. A., Verheijen, F. G. A., Bellamy, P. H., Stephens, M., & Kibblewhite, M. G. (2008). Soil monitoring in Europe: A review of existing systems and requirements for harmonisation. *The Science of the Total Environment*, *391*(1), 1–12. doi:10.1016/j.scitotenv.2007.10.046 PMID:18063012

N, T. R., & Gupta, R. (2020). A Survey on Machine Learning Approaches and Its Techniques. *2020 IEEE International Students' Conference on Electrical, Electronics and Computer Science (SCEECS)*, 1–6. doi:10.1109/SCEECS48394.2020.190

Palti, J., & Cohen, Y. (1980). Downy mildew of Cucurbits (Pseudoperonospora Cubensis): The Fungus and its hosts, distribution, epidemiology and control. *Phytoparasitica*, *8*(2), 109–147. doi:10.1007/BF02994506

Patil, G. G., & Banyal, R. K. (2019). Techniques of Deep Learning for Image Recognition. *2019 IEEE 5th International Conference for Convergence in Technology (I2CT)*, 1–5. 10.1109/I2CT45611.2019.9033628

Pegorini, V., Zen Karam, L., Pitta, C. S. R., Cardoso, R., Da Silva, J. C. C., Kalinowski, H. J., Ribeiro, R., Bertotti, F. L., & Assmann, T. S. (2015). In Vivo Pattern Classification of Ingestive Behavior in Ruminants Using FBG Sensors and Machine Learning. *Sensors (Basel)*, *15*(11), 28456–28471. doi:10.3390151128456 PMID:26569250

Ramos, P. J., Prieto, F. A., Montoya, E. C., & Oliveros, C. E. (2017). Automatic fruit count on coffee branches using computer vision. *Computers and Electronics in Agriculture*, *137*, 9–22. doi:10.1016/j.compag.2017.03.010

Rary, E., Anderson, S. M., Philbrick, B. D., Suresh, T., & Burton, J. (2020). Smart Sanitation—Biosensors as a Public Health Tool in Sanitation Infrastructure. *International Journal of Environmental Research and Public Health*, *17*(14), 5146. doi:10.3390/ijerph17145146 PMID:32708793

Raut, R., Varma, H., Mulla, C., & Pawar, V. (2018). Soil Monitoring, Fertigation, and Irrigation System Using IoT for Agricultural Application. In Lecture Notes in Networks and Systems (pp. 67–73). doi:10.1007/978-981-10-5523-2_7

Sahni, V., Srivastava, S., & Khan, R. (2021). Modelling Techniques to Improve the Quality of Food Using Artificial Intelligence. *Journal of Food Quality*, *2021*, e2140010. doi:10.1155/2021/2140010

Sanjana Rao, G. P., Aditya Shastry, K., Sathyashree, S. R., & Sahu, S. (2021). Machine Learning based Restaurant Revenue Prediction. In V. Suma, N. Bouhmala, & H. Wang (Eds.), *Evolutionary Computing and Mobile Sustainable Networks* (pp. 363–371). Springer. doi:10.1007/978-981-15-5258-8_35

Schmidt, R. H., & Piotter, H. M. (2020). The Hygienic/Sanitary Design of Food and Beverage Processing Equipment. In A. Demirci, H. Feng, & K. Krishnamurthy (Eds.), *Food Safety Engineering* (pp. 267–332). Springer International Publishing. doi:10.1007/978-3-030-42660-6_12

Sharma, A., Jain, A., Gupta, P., & Chowdary, V. (2021). *Machine Learning Applications for Precision Agriculture: A Comprehensive Review*. IEEE., doi:10.1002/9781119769323.ch8

Singh, S. P., Shivam, & Kumar, I. (2022). Orange Fruit Recognition Using Neural Networks. In K. N. Das, D. Das, A. K. Ray, & P. N. Suganthan (Eds.), *Proceedings of the International Conference on Computational Intelligence and Sustainable Technologies* (pp. 417–427). Springer. 10.1007/978-981-16-6893-7_38

Singh, S. P., Rawat, J., Mittal, M., Kumar, I., & Bhatt, C. (2022). Application of AI in SCM or Supply Chain 4.0. In S. L. Fernandes & T. K. Sharma (Eds.), *Artificial Intelligence in Industrial Applications: Approaches to Solve the Intrinsic Industrial Optimization Problems* (pp. 51–66). Springer International Publishing. doi:10.1007/978-3-030-85383-9_4

Su, Y., Xu, H., & Yan, L. (2017). Support vector machine-based open crop model (SBOCM): Case of rice production in China. *Saudi Journal of Biological Sciences*, *24*(3), 537–547. doi:10.1016/j.sjbs.2017.01.024 PMID:28386178

Tao, D., Yang, P., & Feng, H. (2020). Utilization of text mining as a big data analysis tool for food science and nutrition. *Comprehensive Reviews in Food Science and Food Safety*, *19*(2), 875–894. doi:10.1111/1541-4337.12540 PMID:33325182

Topleva, S., & Prokopov, T. (2020). Integrated business model for sustainability of small and medium-sized enterprises in the food industry: Creating value added through ecodesign. *British Food Journal*. doi:10.1108/BFJ-03-2019-0208

Wang, X., Puri, V. M., & Demirci, A. (2020). Equipment Cleaning, Sanitation, and Maintenance. In *Food Engineering Series* (pp. 333–353). Springer. doi:10.1007/978-3-030-42660-6_13

Wardah, S., Djatna, T., Marimin, & Yani, M. (2020). New product development in coconut-based agro-industry: Current research progress and challenges. *IOP Conference Series. Earth and Environmental Science*, *472*(1), 012053. doi:10.1088/1755-1315/472/1/012053

Yost, E., & Cheng, Y. (2021). Customers' risk perception and dine-out motivation during a pandemic: Insight for the restaurant industry. *International Journal of Hospitality Management*, *95*, 102889. doi:10.1016/j.ijhm.2021.102889

# Chapter 15
# Artificial Intelligence-Based Food Calories Estimation Methods in Diet Assessment Research

**Naimoonisa Begum**
*Muffakamjah College of Engineering, Hyderabad, India*

**Ankur Goyal**
*Chandigarh University, India*

**Sachin Sharma**
https://orcid.org/0000-0002-6855-6940
*Koneru Lakshmaiah Education Foundation, India*

## ABSTRACT

*The standard of healthy intake of food is the necessity for keeping a balanced diet to prevent the obesity problem and many other health problems in humans. Obesity is increasing at an alarming speed and keeping people's health at risk. Mankind needs to have careful control on their daily intake of calories by choosing healthier foods, which will be the most fundamental method in preventing obesity and ill health. Even though the packaging of food comes with calorie and nutrition labels, it might not be very favorable for the reference of people. Thus, the scientists to help people started using AI-based techniques and methodologies to know the ways of determining their daily calorie intake of their food. This chapter proposes a review of various AI-based food calorie estimation methodologies in diet assessment which are suggested to help the normal people and patients so that normal people and doctors could succeed to fight against diet-based health conditions.*

DOI: 10.4018/978-1-6684-5141-0.ch015

## 1. INTRODUCTION

Increased percentage of fat in the human body leads to a severe condition termed as obesity. Now-a-day obesity is targeting not only adults but also children. There are many reasons why some people have difficulty losing weight. Generally, obesity can be due to inheritance, environment and physiological conditions, diet intake, physical lifestyle. Being Overweight and obese can lead a major factors of risk for certain chronic diseases, diabetes, cancer, musculoskeletal disorders, and cardiovascular diseases. Obese problems in childhood can also lead to adulthood disability or even premature death. Low and middle income countries are mostly deprived of nutrition and can lead to malnutrition. Being overweight or obese are non-communicable diseases, so can be preventable. Regular exercise and consumption of healthy food can help in preventing obesity and health-related issues which develop due to obesity. Through the express progress of social media, mobile networks, and the Internet of Things (IoT), individuals start commonly uploading, sharing, and recording food images 'recipes', 'cookery videos', and 'diet diaries', which is leading to the large-scale centered food dataset (Min et al., 2019). In prior years, the study had been performed on food from distinct aspects, like food choice (Nestle et al., 1998), perception of food (Sørensen et al., 2003) consumption of food (Pauly,1986), safety of food (Chen & Tao, 2001) and culture of food (Harper & Siller, 2015). Statistics obtained from WHO showed that the obesity problem has increased twice since 1975, and the problem of diabetes has almost increased four times since 1980 and blood pressure has increased twice since 1975 (Jung et al., 2021). The abbreviation of BMI is Body Mass Index which is used to find the body fat based on the individual's height and mass. There are many tools available online to compute the BMI. Table-1 shows the weight status based on BMI of a person according to WHO standard.

*Table 1. Weight-based status on BMI*

| BMI value | Categories |
|---|---|
| Less than 18.5 | Underweight |
| 18.5-24.9 | Healthy weight |
| 25-29.9 | Overweight |
| 30 and above | Obese |

The motive of this chapter is to know the different applications of nutrient science using Artificial Intelligence. Using accurate methods for the measurement of food and calorie intake are essential to fight against obesity. The users or patients can use these methods can utilize these intelligent methods to have knowledge about their calorie intake with nutritional information, hence preventing obesity-related health issues and malnutrition in children.

Now-a-day, smart devices play an extremely vital role in day-to-day life; hence it is nothing new to use smart phones for the treatment of diet-related health issues (Jia et al., 2009). In This chapter, we review different proposals on AI based calorie measurement systems which would help doctors and patients can become successful to fight against nutrition-based health conditions. The recent diet-based monitoring systems handle the challenges in automated assessment of diet consumption by using machine learning techniques. Confusion Matrix is an approach that is used in machine learning for the summarization of a classification model. The artificial neural networks, abbreviated as ANNs, are being used at present widely as the modeling method in the field of Artificial Intelligence, which are inspired by the structure of the natural neurons in the brain of human beings. These ANNs are mathematical based models designed for the process of calculating input signals through rows of processing elements, referred as artificial neurons, that are connected to each other by means of artificial synapses (Sak & Suchodolska, 2021)

Newly, the deep learning approaches have obtained extensive responsiveness because of excellence in performance aimed at image recognition & classification. This deep learning methodology is a section of the machine learning techniques and can train more no of productive neural networks. In contrast, compared to conventional approaches, deep learning methods demonstrate better performance when processing large scale datasets and have an excellence in classification potential (Goodfellow et al., 2014, Deng & Yu, 2014).

## 2. FOOD CATEGORIZATION CLASSIFIERS

### 2.1 The Traditional Methods in Machine Learning

Most of the classifiers used by various traditional techniques used for the domain of food image recognition includes Multiple Kernel Learning (MKL), Support Vector Machines (SVM) (Hoashi et al., 2010) and the K-Nearest Neighbour(KNN) (Chen & Ngo,2016). This is because of their excellence in performance in comparison to other classification methodologies.

### 2.2 Deep Learning Methods

Deep learning methods have obtained essential responsiveness in the area of food recognition. It is because of their outstanding classification performance in evaluation with traditional methodologies (Termritthikun et al., 2017, McAllister et al., 2018). CNN abbreviated as Convolutional Neural Network, Ensemble Net, DCNN abbreviated as Deep Convolutional Neural Network, and Inception-v3 are approximately the most leading techniques which are in use by existing methods for the recognition of food images.

## 3. FOOD CATEGORIZATION OF EVALUATION METRICS

The performance of automated food recognition systems depends highly on the correctness of mapping the food images into their appropriate categories. Hence, evaluating metrics play a vital role for the correctness of food recognition systems. It is observed that classifiers may accomplish well under one type of metric but can perform poorly below another type of metric. The general types of metrics which are used for better performance in comparisons are listed as Recall, precision, accuracy, and F1 score.

### 3.1 Recall

The Recall score leads to the identification of the food model's capability for correctly classification of food classes. It calculates out of the total positive classes of food what the percentage of predicted positives is. It gives better perception when the rate of false negatives goes high. Recall can be calculated as shown in Eq. (1).

$$Recall = \frac{TP}{(TP + TN)} \tag{1}$$

### 3.2 Precision

The precision score is well defined as how a model can be suitably able to predict the classified values that are positive. It means that, from all positive predicted food classes what is the percentage which is truly positive. Precision score is favourable only when the false positives cost is high. Precision can be computed as shown in Eq. (2).

$$Precision\ Score = \frac{TP}{(TP + FP)} \tag{2}$$

### 3.3 Accuracy

The Accuracy of any model would determine whether the model being used is capable to make the prediction of food classes correctly. Accuracy can be computed as shown in Eq. (2). However, the accuracy may not perform as expected for the imbalance dataset, hence it cannot be applied as major performance metric. Therefore, we can incorporate metrics like 'F1 score', 'Precision' and 'Recall' for providing better perception of results.

$$Accuracy = \frac{(TP + FN)}{(TP + FP + FN + TN)} \tag{3}$$

### 3.4 F1-Score

The F1-score is calculated as the harmonic mean of the recall score & precision- score. This considers false-positives, false-negatives both. Hence, it gives better performance on the imbalanced datasets. Accuracy can be computed is shown as Eq. (4).

$$F1\ Score = \frac{(2 * (Precision * Recall))}{(Precision + Recall)} \tag{4}$$

## 4. ANALYSIS OF DIFFERENT METHODS FOR FOOD NUTRITION AND CALORIES MEASUREMENT

### 4.1 Using image Recognition for Food Calories Analysis

This method creates feature-based vector which uses several features like texture, color, etc. and then classifies the food images applying SVM model. This study focused on the Thai food. For training the SVM, a set of example food images were grouped by using the type of food and the number of calories (Tammachat & Pantuwong, 2014).

### 4.2 Real-Time based Mobile Food-Recognition System

Since mostly the processes of image recognition methods are performed using a smartphone, this system does not send images to the server but runs in a real-time way on an ordinary smartphone. For recognizing the food items, an user draws a boundary boxes on the display first by touching the smartphone screen, and now the system will start the food item recognition using the boundary box for identifying. The recognition can be made more accurate by segmenting every food item region by 'GrubCut', and then extract a color histogram and 'SURF based' bag-of-features, then lastly will categorize the food items into one among the available 50 food classifications with the linear Support Vector Machine(SVM) and the fast too kernel (Oliveira et al., 2014).

### 4.3 Fruits and Vegetables Calorie Counter Using Convolutional Neural Networks

This technique uses the state-of-art deep learning methods for the extraction of features and classification. The Deep learning techniques, specifically CNN, has been mostly used model for variety of grouping related problems and have attained promising results. This trained model has attained 75% accuracy level for the classification of forty-three diverse varieties of fruits. The comparable type of methods could achieve only up to 70% using fewer classes. (Akbari Fard et al., 2016)

### 4.4 Foodcam: A Real-Time Food Recognition System On a Smartphone

This system is implemented as a standalone Android based mobile application for smartphones which uses multiple number of CPU cores efficiently for real-time recognition. The experiments resulted in achieving the 79.2% rate of classification for the top five categories for a 100 category food-based dataset with ground-truth bounding boxes were used with HOG and color patches with Fisher Vector coding as the image features (Kawano & Yanai, 2015)

### 4.5 Food Calorie Measurement using Deep Learning Neural Network

This method runs on smartphones, which allows the user click a photo of the food items and would measure automatically the amount of calorie intake. For identifying the food accuracy of system, the deep CNN is used for classifying the food images of 10000 high resolutions for the system training. The result shows that the accurateness of this method for the recognition of food of the range is 99% for a single portion (Pouladzadeh et al., 2016)

## 4.6 Precision Nutrition

The Precision nutrition is a new trend which has become famous now a day. There are no such universally agreed definitions of the terms named as personalized or precision nutrition (Bush et al., 2020). In some cases, these terminologies are used with close similarity (Zeisel, 2020), but they are used distinctly in many other cases (Chen & Ngo 2016). The research in Precision Nutrition aims using the personal information of people or groups of people to deliver personal nutrition advice. In recent research, the machine learning techniques, algorithms, and tools were applied for precision nutrition for different cases. The results of (Ordovas et al., 2018) have shown that there are 15 problems which are spread across 7 domains of health & nutrition. There are 4 machine learning methods that are available, termed as classification, regression, recommendation, and clustering, which mostly implement the supervised approach.

The Machine Learning methods can be used in the research of Precision Nutrition with may lead to high performance. Personal Nutrition can be implemented to facilitate precise nutritional advices to an individual, to facilitate the prediction of postprandial glycaemia (Mendes-Soares et al., 2019), triglycerides (Berry et al., 2020) and cancer (Shiao et al., 2018). The usage of sophisticated techniques like machine learning (ML) and deep learning (DL) can help in interpreting various factors which are of great use. Eye and hair color which are part of, them can also be extended as a response to nutrition (Bashiardes et al., 2018). There can also be some known kind of relationships between weight management and genetics (Goni et al, 2015) lactose (as in the instance of lactose intolerance) (Mattar et al.,2012), metabolic syndrome (Perez-Martinez et al., 2012) and more Zeisel, 2020). Mezgec et al. (2017) had developed a food & drink recognition system named as NutriNet, which used CNN architecture and made comparison with other CNNs and many solver types. Personalized nutritional approaches would help in the general health maintenance of athletes to maximize their sports performance (Guest et al.,2019). The DL (Deep learning) was mostly used for food recognition in the articles using imaging technology for food logging, although it also appeared in Metabolic Health (Sowah et al.,2020), Bodyweight (Curbelo Montañez et al., 2018), and the Nutritional Management of Chronic Diseases (Kim & Chung, 2020, Baek, 2019). This is mostly because of the fact that DL results in good performance in image recognition.

## 4.7 Food Recognition and Nutrition Estimation using Machine Learning

This is a machine learning based method which automatically performs the classification of food based images and can estimate the attributes of food accurately. Deep learning pattern comprising of a CNN abbreviated as convolutional neural network organizes the food into certain categories as a part of the prototype system training set. The prototype system is designed based on the client and server model. The server receives a request for detection of image from the client. The processing of the image is done at the server side. The design of the prototype is developed with three different software components, the first one includes pretrained convolutional neural network model used as a training module for classification, the second component is a training module of text data for attribute estimation, and the third component is a server side module. The experiment was done using machine learning training with a wide range of food categories, with each category including a variety of food images, to achieve high classification accuracy (Shen et al., 2020).

### 4.7.1 Pre-Processing Phase

To retrieve the essential physical characteristics from the images of distinct food items, each image is categorized into its respective class. Depending on different attributes, the food item category is decided. Machine-learning tool known as Word2Vec helps for the estimation of different word vector representations. Hence, as mentioned, food attributes and ingredients are extracted using the pretrained word2vec model.

### 4.7.2 System Architecture

The architecture consists of 3 modules, the first module is retrieval of text data, the second one is training text data, and the third module is convolutional neural network based classification model as shown in Figure-1. After the classification process completed by using the pretrained model, each food item name is passed to the first module for retrieval of text data. This required text information is pulled out from the Uniform Resource Locator extraction and Google search. And then HyperText Markup Language(HTML), is extracted from the webpage and by using HyperText Markup Language(HTML) parser, the stop words will be removed from the Python library. The text is now sent to the system for lemmatization process and stemming. Now, the text data forms a textual corpus which will be the input for the next module, i.e., training text data. The next step after extraction of the text corpus is to train the text data with word2vec. After the training and learning module of the system is successfully completed, the next module, i.e., CNN based classification module, starts by accepting input from user.

*Figure 1. Food recognition and nutrition estimation using machine learning model[39]*

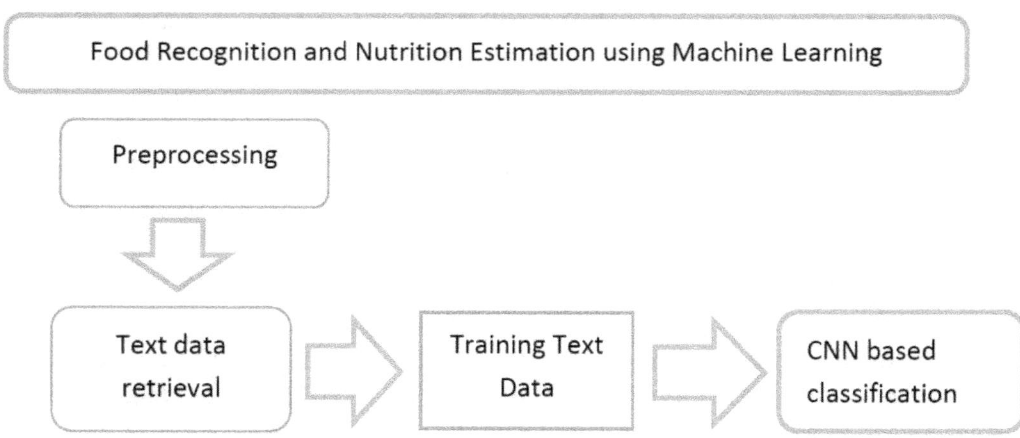


The above system will take the image of food from the consumer and includes appropriate classification, then the attributes are identified by the system. The proposed system which does classification along with extraction of attributes has achieved 85% accuracy.

### 4.7.3 Drawbacks and future enhancements

The Current system could not exactly identify and evaluate the physical images with different combinations. Hence, future work can be done to process a mixed food physical image by implementing image segmentation. The features and data sets have a greater impact on the revealing of results. The prevailing data sets are insufficient which comprise partial parameters like distinct light conditions, angle of camera and distinct backgrounds, etc. Best techniques should be implemented for the review different types of data sets (Pouladzadeh et al., 2015). It is essential to have a better understanding of the importance of calorie calculations. The literature (Farinella et al., 2014) used game technique to obtain essential information on foods and its calorie value. In future enhancement of calorie estimation of food items along with nutrition measurements deep learning techniques can be used.

As being obese is a major issue of concern now a day, people are more curious to know their food calorie count. The novel model discussed above gives information about the food calories. The system takes input as food images from the user and does the classification and then discloses the characteristics of food. The dataset consisting of common meal and sub-continental food is used. The Inception V-3 and V-4 models were used to identify the food items and using attribute estimation model the attributes are measured. The results of the model were enhanced through multicrop, data augmentation and some similar techniques. The proposed system could give an accuracy of 85%.

## 4.8 Design and Implementation of Food Nutrition Information System Using SURF and Fatsecret API

The Applications like Calorie Counter, Coach, Noom and Lose were been developed for and monitoring the needs of calorie and nutritional needs (Bayu et al., 2013, Sheikh, 2013). Still, these applications could not be used easily because the customers should know the food name before to use these applications. In this design, a faster and smarter process is developed as an Android based application that can display the nutritional information by taking the food photograph as input. The application is Android based augmented reality which can help the user in retrieving the diet information in an easier way. The nutritional information will be displayed in terms of number of calories, fat, carbohydrates, and proteins per serving of food. And this is all possible by just on the click of the food picture. The experiment done on this application for food recognition accuracy level are noted as 92%, and the average time that is required for the identification of the food name and its nutrients is noted as 9.295 seconds (Hariadi et al., 2015).

### 4.8.1 Systems Architecture

In the application, the users need to click the photo of the food item by using the phone's camera. After this, the results are displayed on the screen. The process would start once the user clicks the photograph of the food items from an Android based application. In the next step, Android based application will send the image directly to the server which recognizes the image and identifies the food's name. SURF abbreviated as Speeded-Up Robust Feature is used for the image recognition process. The SURF algorithm works fast and is accurate in detecting the similarity of the images (Hariadi et al., 2015), The application will connect to the FatSecret server thru Fat Server API and displays the nutritional information after identifying the food name. The food's name is passed as an input to the Fatsecret and then the Fatsecret will give the output as nutritional information. The database of FatSecret stores the nutritional information of the food items in the form of fat, calories, carbs, and proteins. The Experiments were led with two hundred images which consisted of 10 items with 20 dissimilar images per item (Bay et al.,2008). Figure 2 shows the design of the application by using SURF and Fatsecret API.

*Figure 2. Design of the application by using SURF and FatSecret API [45]*

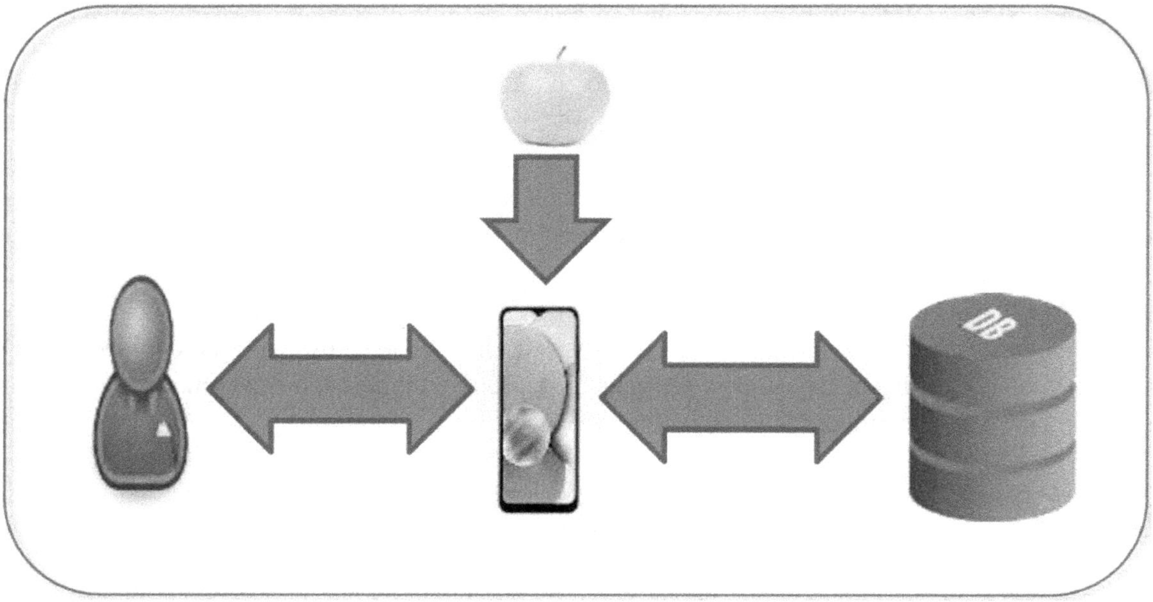


## 4.9 Artificial Intelligence Based Calorie Estimator

There is a higher risk of obesity and related diseases with increased levels of BMI. As maintaining healthy weight and lifestyle is a tedious task as we need to have proper awareness of calorie intake. People are advised by doctors to keep track of their diet by personal nutritionists. Therefore, based on this requirement in place of a personal nutritionist which everyone may not be afford, an AI based model can help in calorie estimation. Calorie estimators based on computer vision were introduced to estimate the calories from food images as shown in Figure 3 (Ganesh & Hemavathi, 2017).

*Figure 3. Tensor flow-based food recognition flow diagram [46]*

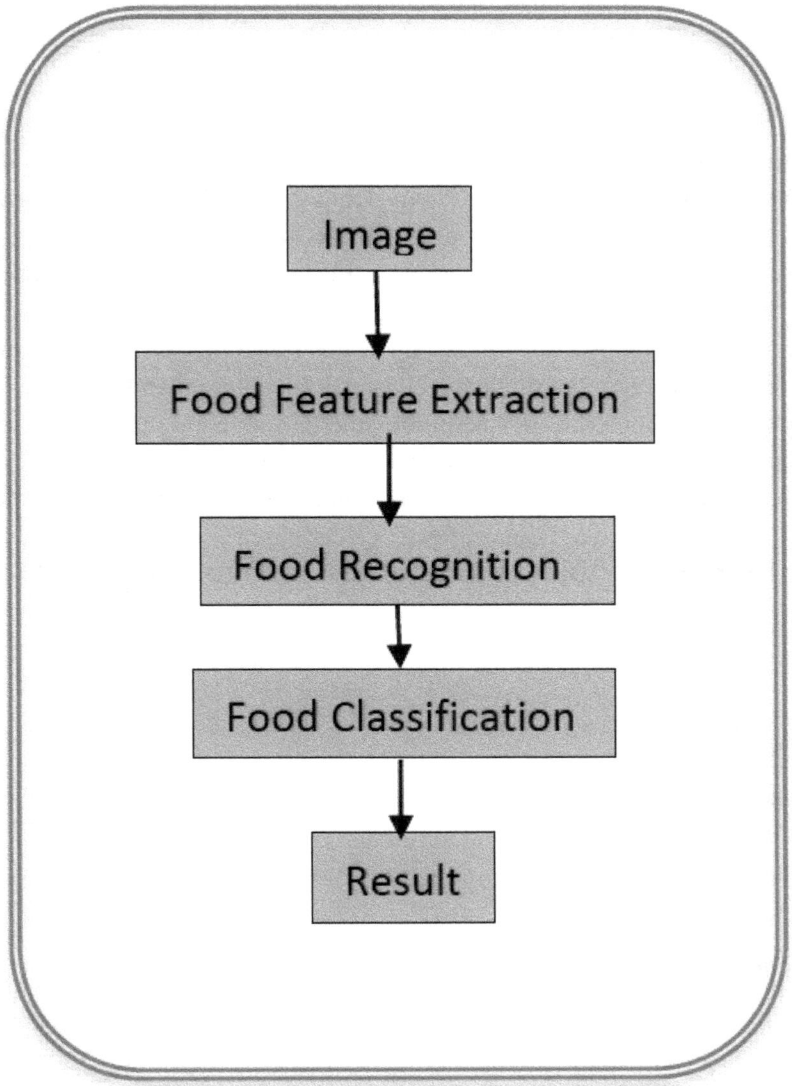


This is a food image calorie estimator technique using Artificial Intelligence based on deep learning methodology (Ganesh & Hemavathi, 2017). This AI based technique uses Indian cuisine food image as the dataset which contains the calorie information in detail. Object detection technique is one of the most important techniques used for the full estimation of calories. This system is an initially released AI based calorie estimator whose dataset is Indian Food cuisine. The system can detect the Indian cuisine food items; it finds the calories count and can track the person's diet by giving an estimate of approximate calories by considering their BMI values.

### 4.9.1 System Architecture

An object detection Application interface (API) based on called Opencv is used in the model to detect the food objects. The output from Opencv is transferred to Tensorflow for detection of contents of the food image. The Tensorflow is used in this model for simply identifying the food images of the model. However, Tensorflow is a deep learning library which is an open source designed for complex computation and building machine learning based systems from scratch. It has a pretrained model termed as inception, which has been classified with almost 1000 class images by the Google's imagenet.

For the proposed model, the inception model's final layer is taken on to retrain the system for identifying the Indian origin dishes and giving the results which are accurate and precise. When an image is fed as input to the directory of tensorflow, it calculates the probability value for the image. The features are extracted and the food image type is recognized using the feature map. After recognition of the food image, the classification is done. The final result obtained is passed to the calories estimation table for calculating the respective calories, now the amount of consumed calories can be analysed in an efficient manner so that the obesity and overweight problems related problems can be controlled in future.

## 4.10 AI Based Proposed Project Modules for Calorie Estimation

The AI based project modules for calorie estimation are:

1. Collection of DataSet
2. Model Training
3. Model testing
4. Calorie details Tracking
5. Estimation of calories.

*Table 2. Shows the food recognition percentage of single food items*

| SNO | Food item | % of recognition |
|---|---|---|
| I | Dosa | 100% |
| II | Idly | 99.5% |
| III | Cutlet | 87% |
| IV | Bonda | 93% |
| V | Poori | 100% |
| VI | Pongal | 88% |

## 4.11 A Comprehensive Survey of Image-Based Food Recognition and Volume Estimation Methods for Dietary Assessment

This survey explores a wide spectrum of vision-oriented methods which are specially designed for food item volume estimation and image recognition (Tahir & Loo, 2021). Practically, the food item identification process consists of these 4 tasks: retrieving food item images from the related food-based datasets, extracting of features using deep visual / handcrafted, selecting the related extracted features, and at the last stage, selection of the classification techniques by using either the traditional approach of machine learning or the deep learning model. This procedure is followed by the food ingredients classification which can provide in-depth insight of nutritional information.

## 5. CONCLUSION

In spite of flawless performance demonstrated by the 'state-of-the-art' approach, there still exists various challenges and limitations. The standard personalized nutrition-based framework methodology with the protection of patient's confidentiality can help in establishing preventive and predictive strategies for the advancement of health care and improved disease controlling management. There is a definite necessity for full-scale datasets for standard performance evaluation, such as using large datasets of food images improves the overall performance of the models. During the dealing of an open-ended, dynamic food dataset, the classifiers should be able to perform an open-ended continuous learning.

Yet, the prevailing methods include quite a lot of bottlenecks, which reduce the ability of food recognition when it deals with open-ended learning by means of methods for proposal of usage but can be prone to catastrophic disremembering. They usually tend to fail to recall the former knowledge gained from the images while they start learning new data. Such type of methods can only work greatly for fixed food image datasets. Moreover, the suggested techniques of food ingredients classification are still in the stage of stressing with performance concerns when they are applied to prepare and mixture food items.

Similarly, the automated food portion estimate techniques are classified into two kinds labelled as 'single-view' image approach, 'multi-view' image approach. Mostly the 'multi-view' image approach performs more accurately than the 'single view' approach but 'multi-view' image technique require more complex type of processing and images considered from altered angles, which results in the reduction of the rate of user retention. Most of the 'single-view' and 'multi-view' approaches need calibrated objects always, which make the use of these results difficult for the elderly patients. Hence, there is scope for inventive health care and dietetic based assessment application which would incorporate wearable devices with smartphones to improvise the research areas. Furthermore, the dietetic based assessment methods must be able to address these types of challenges to improve the ability of providing better insights for effective ways of health care and chronic illness avoidance.

## REFERENCES

Akbari Fard, M., Hadadi, H., & Tavakoli Targhi, A. (2016, April). Fruits and vegetables calorie counter using convolutional neural networks. In *Proceedings of the 6th international conference on digital health conference* (pp. 121-122). Academic Press.

Baek, J. W., Kim, J. C., Chun, J., & Chung, K. (2019). Hybrid clustering based health decision-making for improving dietary habits. *Technology and Health Care*, *27*(5), 459–472.

Bashiardes, S., Godneva, A., Elinav, E., & Segal, E. (2018). Towards utilization of the human genome and microbiome for personalized nutrition. *Current Opinion in Biotechnology*, *51*, 57–63.

Bay, H., Ess, A., Tuytelaars, T., & Van Gool, L. (2008). Speeded-up robust features (SURF). *Computer Vision and Image Understanding*, *110*(3), 346–359.

Bayu, M. Z., Arshad, H., & Ali, N. M. (2013). Nutritional information visualization using mobile augmented reality technology. *Procedia Technology*, *11*, 396–402.

Berry, S. E., Valdes, A. M., Drew, D. A., Asnicar, F., Mazidi, M., Wolf, J., ... Spector, T. D. (2020). Human postprandial responses to food and potential for precision nutrition. *Nature Medicine*, *26*(6), 964–973.

Bush, C. L., Blumberg, J. B., El-Sohemy, A., Minich, D. M., Ordovás, J. M., Reed, D. G., & Behm, V. A. Y. (2020). Toward the definition of personalized nutrition: A proposal by the American Nutrition Association. *Journal of the American College of Nutrition*, *39*(1), 5–15.

Chen, J., & Ngo, C. W. (2016, October). Deep-based ingredient recognition for cooking recipe retrieval. In *Proceedings of the 24th ACM international conference on Multimedia* (pp. 32-41). ACM.

Chen, J., & Ngo, C. W. (2016, October). Deep-based ingredient recognition for cooking recipe retrieval. In *Proceedings of the 24th ACM international conference on Multimedia* (pp. 32-41). Academic Press.

Chen, Z., & Tao, Y. (2001). Food safety inspection using "from presence to classification" object-detection model. *Pattern Recognition*, *34*(12), 2331–2338. doi:10.1016/S0031-3203(00)00169-2

Curbelo Montañez, C. A., Fergus, P., Montañez, A. C., Hussain, A., Al-Jumeily, D., & Chalmers, C. (2018, October). Deep Learning Classification of Polygenic Obesity using Genome Wide Association Study SNPs. In *2018 International Joint Conference on Neural Networks (IJCNN)*. IEEE.

Deng, L., & Yu, D. (2014). Deep learning: methods and applications. *Foundations and Trends in Signal Processing, 7*(3–4), 197-387.

Farinella, G. M., Moltisanti, M., & Battiato, S. (2014, October). Classifying food images represented as bag of textons. In *2014 IEEE International Conference on Image Processing (ICIP)* (pp. 5212-5216). IEEE.

Ganesh, V. D., & Hemavathi, A. D. (2017). Artificial Intelligence Based Calorie Estimator. *International Journal of Pure and Applied Mathematics*, *117*(15), 1201–1205.

Goni, L., Cuervo, M., Milagro, F. I., & Martínez, J. A. (2015). Future perspectives of personalized weight loss interventions based on nutrigenetic, epigenetic, and metagenomic data. *The Journal of Nutrition, 146*(4), 905S-912S.

Goodfellow, I., Bengio, Y., & Courville, A. (2016). *Deep learning*. MIT press.

Guest, N. S., Horne, J., Vanderhout, S. M., & El-Sohemy, A. (2019). Sport nutrigenomics: Personalized nutrition for athletic performance. *Frontiers in Nutrition*, *6*, 8.

Hariadi, R. R., Khotimah, W. N., & Wiyono, E. A. (2015, October). Design and implementation of food nutrition information system using SURF and FatSecret API. In *2015 International Conference on Advanced Mechatronics, Intelligent Manufacture, and Industrial Automation (ICAMIMIA)* (pp. 181-183). IEEE.

Harper, C., & Siller, M. (2015). OpenAG: A globally distributed network of food computing. *IEEE Pervasive Computing*, *14*(4), 24–27. doi:10.1109/MPRV.2015.72

Hoashi, H., Joutou, T., & Yanai, K. (2010, December). Image recognition of 85 food categories by feature fusion. In *2010 IEEE International Symposium on Multimedia* (pp. 296-301). IEEE.

Jia, W., Zhao, R., Yao, N., Fernstrom, J. D., Fernstrom, M. H., Sclabassi, R. J., & Sun, M. (2009, April). A food portion size measurement system for image-based dietary assessment. In *2009 IEEE 35th Annual Northeast Bioengineering Conference* (pp. 1-2). IEEE.

Jung, C. H., Son, J. W., Kang, S., Kim, W. J., Kim, H. S., Kim, H. S., Seo, M., Shin, H.-J., Lee, S.-S., Jeong, S. J., Cho, Y., Han, S. J., Jang, H. M., Rho, M., Lee, S., Koo, M., Yoo, B., Moon, J.-W., Lee, H. Y., ... Yoon, K. H. (2021). Diabetes fact sheets in Korea, 2020: An appraisal of current status. *Diabetes & Metabolism Journal*, *45*(1), 1–10. doi:10.4093/dmj.2020.0254 PMID:33434426

Kawano, Y., & Yanai, K. (2015). Foodcam: A real-time food recognition system on a smartphone. *Multimedia Tools and Applications*, *74*(14), 5263–5287.

Kim, J. C., & Chung, K. (2020). Knowledge-based hybrid decision model using neural network for nutrition management. *Information Technology Management*, *21*(1), 29–39.

McAllister, P., Zheng, H., Bond, R., & Moorhead, A. (2018). Combining deep residual neural network features with supervised machine learning algorithms to classify diverse food image datasets. *Computers in Biology and Medicine*, *95*, 217–233.

Mendes-Soares, H., Raveh-Sadka, T., Azulay, S., Ben-Shlomo, Y., Cohen, Y., Ofek, T., ... Nelson, H. (2019). Model of personalized postprandial glycemic response to food developed for an Israeli cohort predicts responses in Midwestern American individuals. *The American Journal of Clinical Nutrition*, *110*(1), 63–75.

Mezgec, S., & Koroušić Seljak, B. (2017). NutriNet: A deep learning food and drink image recognition system for dietary assessment. *Nutrients*, *9*(7), 657.

Min, W., Jiang, S., Liu, L., Rui, Y., & Jain, R. (2019). A survey on food computing. *ACM Computing Surveys*, *52*(5), 1–36. doi:10.1145/3329168

Nestle, M., Wing, R., Birch, L., DiSogra, L., Drewnowski, A., Middleton, S., & Economos, C. (1998). *Behavioral and social influences on food choice*. Academic Press.

Ordovas, J. M., Ferguson, L. R., Tai, E. S., & Mathers, J. C. (2018). Personalised nutrition and health. *BMJ (Clinical Research Ed.)*, 361.

Pauly, D. (1986). *A simple method for estimating the food consumption of fish populations from growth data and food conversion experiments*. Academic Press.

Perez-Martinez, P., Garcia-Rios, A., Delgado-Lista, J., Perez-Jimenez, F., & Lopez-Miranda, J. (2012). Metabolic syndrome: Evidences for a personalized nutrition. *Molecular Nutrition & Food Research*, *56*(1), 67–76.

Pouladzadeh, P., Kuhad, P., Peddi, S. V. B., Yassine, A., & Shirmohammadi, S. (2016, May). Food calorie measurement using deep learning neural network. In *2016 IEEE international instrumentation and measurement technology conference proceedings* (pp. 1-6). IEEE.

Pouladzadeh, P., Yassine, A., & Shirmohammadi, S. (2015, September). Foodd: food detection dataset for calorie measurement using food images. In *International Conference on Image Analysis and Processing* (pp. 441-448). Springer.

Sak, J., & Suchodolska, M. (2021). Artificial intelligence in nutrients science research: A review. *Nutrients*, *13*(2), 322. doi:10.3390/nu13020322 PMID:33499405

Sheikh, W. (2013). *Prototype of an image recognition based calorie recording application* [Ph.D. dissertation]. Heriot-Watt University.

Shen, Z., Shehzad, A., Chen, S., Sun, H., & Liu, J. (2020). Machine learning based approach on food recognition and nutrition estimation. *Procedia Computer Science*, *174*, 448–453.

Shiao, S. P. K., Grayson, J., Lie, A., & Yu, C. H. (2018). Personalized nutrition—Genes, diet, and related interactive parameters as predictors of cancer in multiethnic colorectal cancer families. *Nutrients*, *10*(6), 795.

Sørensen, L. B., Møller, P., Flint, A., Martens, M., & Raben, A. (2003). Effect of sensory perception of foods on appetite and food intake: A review of studies on humans. *International Journal of Obesity*, *27*(10), 1152–1166. doi:10.1038j.ijo.0802391 PMID:14513063

Sowah, R. A., Bampoe-Addo, A. A., Armoo, S. K., Saalia, F. K., Gatsi, F., & Sarkodie-Mensah, B. (2020). Design and development of diabetes management system using machine learning. *International Journal of Telemedicine and Applications*, 2020.

Tahir, G. A., & Loo, C. K. (2021, December). A Comprehensive Survey of Image-Based Food Recognition and Volume Estimation Methods for Dietary Assessment. *Health Care*, *9*(12), 1676.

Tammachat, N., & Pantuwong, N. (2014, October). Calories analysis of food intake using image recognition. In *2014 6th International Conference on Information Technology and Electrical Engineering (ICITEE)* (pp. 1-4). IEEE.

Termritthikun, C., Muneesawang, P., & Kanprachar, S. (2017). NU-InNet: Thai food image recognition using convolutional neural networks on smartphone. *Journal of Telecommunication Electronic and Computer Engineering*, *9*(2-6), 63–67.

Zeisel, S. H. (2020). Precision (personalized) nutrition: Understanding metabolic heterogeneity. *Annual Review of Food Science and Technology*, *11*, 71–92.


Zeisel, S. H. (2020). Precision (personalized) nutrition: Understanding metabolic heterogeneity. *Annual Review of Food Science and Technology*, *11*, 71–92.

# Compilation of References


Aaltonen, I., Salmi, T., & Marstio, I. (2018). Refining levels of collaboration to support the design and evaluation of human-robot interaction in the manufacturing industry. *Procedia CIRP*, *72*, 93–98. doi:10.1016/j.procir.2018.03.214

Abade, A., Ferreira, P., & Vidal, F. (2020). Plant Diseases recognition on images using Convolutional Neural Networks: A Systematic Review. *Computers and Electronics in Agriculture*, *185*, 106125. Advance online publication. doi:10.1016/j.compag.2021.106125

Abate, T., van Huis, A., & Ampofo, J. K. O. (2000). Pest Management Strategies in Traditional Agriculture: An African Perspective. *Annual Review of Entomology*, *45*(1), 631–659. doi:10.1146/annurev.ento.45.1.631 PMID:10761592

Abhilash, P. C., & Singh, N. (2009). Pesticide use and application: An Indian scenario. *Journal of Hazardous Materials*, *165*(1–3), 1–12. doi:10.1016/j.jhazmat.2008.10.061 PMID:19081675

Abiodun, O. I., Jantan, A., Omolara, A. E., Dada, K. V., Mohamed, N. A., & Arshad, H. (2018). State-of-the-art in Artificial Neural Network Applications: A Survey. *Heliyon*, *4*(11), e00938. doi:10.1016/j.heliyon.2018.e00938 PMID:30519653

Adek, R. T., & Ula, M. (2020). A Survey on The Accuracy of Machine Learning Techniques for Intrusion and Anomaly Detection on Public Data Sets. *2020 International Conference on Data Science, Artificial Intelligence, and Business Analytics (DATABIA)*. 10.1109/DATABIA50434.2020.9190436

Adeux, G., Vieren, E., Carlesi, S., Bàrberi, P., Munier-Jolain, N., & Cordeau, S. (2019). Mitigating crop yield losses through weed diversity. *Nature Sustainability*, *2*(11), 1018–1026.

Agarwal, M., Singh, A., Arjaria, S., Sinhad, A., & Gupta, S. (2019). ToLeD: Tomato Leaf Disease Detection using Convolution Neural Network. *Procedia Computer Science*, *167*, 293–301. doi:10.1016/j.procs.2020.03.225

Ageikina, I. I., Lazareva, E. G., Mikhailova, I. Y., & Semipyatny, V. K. (2020). Results designing and analysis when introducing new beverage identification criteria. *Food Systems*, *3*(3), 4–7.

Agnihotri, N. (2000). Pesticide Consumption in Agriculture in India - an Update. *Pesticide Research Journal*, *12*, 150–155.

Ahirwar, S., Swarnkar, R., Bhukya, S., & Namwade, G. (2019). Application of drone in agriculture. *International Journal of Current Microbiology and Applied Sciences*, *8*(1), 2500–2505. doi:10.20546/ijcmas.2019.801.264

Ahmad, A., & Dey, L. (2007). A k-mean clustering algorithm for mixed numeric and categorical data. *Data & Knowledge Engineering*, *63*(2), 503–527. doi:10.1016/j.datak.2007.03.016

Ahmad, F., Khaliq, A., Qiu, B., Sultan, M., & Ma, J. (2021). *Advancements of Spraying Technology in Agriculture*. Technology In Agriculture.

Ahmadi, P., Muharam, F. M., Ahmad, K., Mansor, S., & Seman, I. A. (2017). Early detection of Ganoderma basal stem rot of oil palm using artificial neural network spectral analysis. *Plant Disease*, *101*(6), 1009–1016. doi:10.1094/PDIS-12-16-1699-RE PMID:30682927

Ahmed, N., De, D., & Hussain, I. (2018). Internet of things (IoT) for smart precision agriculture and farming in rural areas. *IEEE Internet of Things Journal*, *5*(6), 4890–4899.

Ahumada, O., & Villalobos, J. R. (2009). Application of planning models in the agri-food supply chain: A review. *European Journal of Operational Research*, *196*(1), 1–20. doi:10.1016/j.ejor.2008.02.014

Aiello, L. M., Schifanella, R., Quercia, D., & Prete, L. D. (2019). Large-scale and high-resolution analysis of food purchases and health outcomes. *EPJ Data Science*, *8*(1), 14. doi:10.1140/epjds13688-019-0191-y

Akashe, M. M., Pawade, U. V., & Nikam, A. V. (2018). Classification of Pesticides: A Review. *International Journal of Research in Ayurveda and Pharmacy*, *9*, 144–150.

Akbari Fard, M., Hadadi, H., & Tavakoli Targhi, A. (2016, April). Fruits and vegetables calorie counter using convolutional neural networks. In *Proceedings of the 6th international conference on digital health conference* (pp. 121-122). Academic Press.

Akhtar, P., Tse, Y. K., Khan, Z., & Rao-Nicholson, R. (2016). Data-driven and adaptive leadership contributing to sustainability: Global agri-food supply chains connected with emerging markets. *International Journal of Production Economics*, *181*, 392–401. doi:10.1016/j.ijpe.2015.11.013

Al Ohali, Y. (2011). Computer vision based date fruit grading system: Design and implementation. *Journal of King Saud University - Computer and Information Sciences, 23*(1), 29–36. doi:10.1016/j.jksuci.2010.03.003

Alaimo, L. S., Fiore, M., & Galati, A. (2020). How the covid-19 pandemic is changing online food shopping human behaviour in Italy. *Sustainability*, *12*(22), 9594. doi:10.3390u12229594

Alam, M., Alam, M. S., Roman, M., Tufail, M., Khan, M. U., & Khan, M. T. (2020, April). Real-Time Machine-Learning Based Crop/Weed Detection and Classification for Variable-Rate Spraying in Precision Agriculture. In *2020 7th International Conference on Electrical and Electronics Engineering (ICEEE)* (pp. 273-280). IEEE.

Alamsyah, A., Purnama, I. K. E., Setijadi, E., & Purnomo, M. H. (2019). M.P.R. selection to the OLSR quality of service in MANET using minmax algorithm. *Iranian Journal of Electrical and Computer Engineering*, *9*(1), 417.

Albu-Salih, A.T., & Khudhair, H.A. (2021). ASR-FANET: An adaptive SDN-based routing framework for FANET. *International Journal of Electrical & Computer Engineering, 11*(5).

Alexandratos, N. (1995). *World Agriculture: towards 2010: An FAO Study*. Food & Agriculture Org.

Alexandratos, N. (Ed.). (1995). *World agriculture: towards 2010: an FAO study. FAO.*

Alexey. (n.d.). *darknet: YOLOv4 / Scaled-YOLOv4 / YOLO - Neural Networks for Object Detection (Windows and Linux version of Darknet).* Academic Press.

Ali, M., Deo, R. C., Downs, N. J., & Maraseni, T. (2018a). Multi-stage hybridized online sequential extreme learning machine integrated with Markov Chain Monte Carlo copula-Bat algorithm for rainfall forecasting. *Atmospheric Research*, *213*, 450–464. doi:10.1016/j.atmosres.2018.07.005

Al-Jarrah, O. Y., Yoo, P. D., Muhaidat, S., Karagiannidis, G. K., & Taha, K. (2015). Efficient machine learning for big data: A review. *Big Data Research*, *2*(3), 87–93. doi:10.1016/j.bdr.2015.04.001

Allard, M. W., Strain, E., Melka, D., Bunning, K., Musser, S. M., Brown, E. W., & Timme, R. (2016). Practical value of food pathogen traceability through building a whole-genome sequencing network and database. *Journal of Clinical Microbiology*, *54*(8), 1975–1983. doi:10.1128/JCM.00081-16 PMID:27008877

Allawi, M. F., Jaafar, O., Ehteram, M., Mohamad Hamzah, F., & El-Shafie, A. (2018). Synchronizing artificial intelligence models for operating the dam and reservoir system. *Water Resources Management*, *32*(10), 3373–3389. doi:10.100711269-018-1996-3

Almog, U. (2020). *YOLO V3 explained*. Towards Data Science. https://towardsdatascience.com/yolo-v3-explained-ff5b850390f?source=rss----7f60cf5620c9---4&ref=morioh.com&utm_source=morioh.com

Alom, M. Z., Taha, T. M., Yakopcic, C., Westberg, S., Sidike, P., Nasrin, M. S., Hasan, M., Van Essen, B. C., Awwal, A. A. S., & Asari, V. K. (2019). A State-of-the-Art Survey on Deep Learning Theory and Architectures. *Electronics (Basel)*, *8*(3), 292. doi:10.3390/electronics8030292

Alonso, J., Villa, A., & Bahamonde, A. (2015). Improved estimation of bovine weight trajectories using Support Vector Machine Classification. *Computers and Electronics in Agriculture*, *110*, 36–41. doi:10.1016/j.compag.2014.10.001

Altenburger, K. M., & Ho, D. E. (2019). When algorithms import private bias into public enforcement: The promise and limitations of statistical debiasing solutions. *Journal of Institutional and Theoretical Economics*, *175*(1), 98–122. doi:10.1628/jite-2019-0001

Althouse, B. M., Scarpino, S. V., Meyers, L. A., Ayers, J. W., Bargsten, M., Baumbach, J., Brownstein, J. S., Castro, L., Clapham, H., Cummings, D. A. T., Del Valle, S., Eubank, S., Fairchild, G., Finelli, L., Generous, N., George, D., Harper, D. R., Hébert-Dufresne, L., Johansson, M. A., ... Wesolowski, A. (2015). Enhancing disease surveillance with novel data streams: Challenges and opportunities. *EPJ Data Science*, *4*(1), 17. doi:10.1140/epjds13688-015-0054-0 PMID:27990325

Amara, J., Bouaziz, B., & Algergawy, A. (2017). A Deep Learning-based Approach for Banana Leaf Diseases Classification. In Datenbanksysteme für Business, Technologie und Web (BTW 2017) - Workshopband. Bonn: Gesellschaft für Informatik e.V.

Amemiya, T., Akiyama, K., Leow, C. S., Buayai, P., Makino, K., Mao, X., & Nishizaki, H. (2021). Development of a support system for judging the appropriate timing for grape harvesting. *2021 International Conference on Cyberworlds*. 10.1109/CW52790.2021.00040

Ananthanarayana, T., Ptucha, R., & Kelly, S. C. (2020). Deep Learning based Fruit Freshness Classification and Detection with CMOS Image sensors and Edge processors. *IS&T International Symposium on Electronic Imaging*, *2020*(12), 172–172. doi:10.2352/ISSN.2470-1173.2020.12.FAIS-172

Anas, D. F., Jaya, I., & Nurjanah. (2021). Design and implementation of fish freshness detection algorithm using deep learning. *IOP Conference Series. Earth and Environmental Science*, *944*(1), 012007. doi:10.1088/1755-1315/944/1/012007

Andrea, C. C., Daniel, B. B. M., & Misael, J. B. J. (2017, October). Precise weed and maize classification through convolutional neuronal networks. In *2017 IEEE Second Ecuador Technical Chapters Meeting (ETCM)* (pp. 1-6). IEEE.

Ankalaki, S., Chandra, N. & Majumdar, J. (2016). Applying Data Mining Approach and Regression Model to Forecast Annual Yield of Major Crops in Different District of Karnataka. *International Journal of Advanced Research in Computer and Communication Engineering, 5*(2), 25-29.

Arai, Kato, & Fujita. (2010). Assessment of operator stress induced by robot collaboration in assembly. *CIRP Annals – Manufacturing Technology, 59*, 5–8.

Arakeri, M. P., Kumar, B. V., Barsaiya, S., & Sairam, H. V. (2017). Computer vision based robotic weed control system for precision agriculture. *Proceedings of International Conference on Advances in Computing, Communications and Informatics*, 1201–1205. 10.1109/ICACCI.2017.8126005

Arif, S., Kumar, R., Abbasi, S., Mohammadani, K., & Dev, K. (2021). *Weeds Detection and Classification using Convolutional Long-Short-Term Memory*. Academic Press.

Arivazhagan, S. (2010). Fruit recognition using color and texture features. *Journal of Emerging Trends in Computing and Information Sciences*, *1*(2), 90–94.

Arivazhagan, S., & Ligi, S. V. (2018). Mango Leaf Diseases Identification Using Convolutional Neural Network. *International Journal of Pure and Applied Mathematics*, *120*, 11067–11079.

Arulbalachandran, D., Mullainathan, L., & Latha, S. (2017). Food Security and Sustainable Agriculture. *Sustainable Agriculture Towards Food Security*, 3–13. doi:10.1007/978-981-10-6647-4_1

Asad, M. H., & Bais, A. (2019). Weed detection in canola fields using maximum likelihood classification and deep convolutional neural network. *Information Processing in Agriculture*.

Ashok, V., & Vinod, D. S. (2014). Automatic quality evaluation of fruits using probabilistic neural network approach. In *2014 International Conference on Contemporary Computing and Informatics (IC3I)* (pp. 308-311). IEEE.

Aswini, N., & Uma, S. V. (2019). Obstacle avoidance and distance measurement for unmanned aerial vehicles using monocular vision. *Iranian Journal of Electrical and Computer Engineering*, *9*(5), 3504.

Athani, Tejeshwar, Patil, Patil, & Kulkarni. (n.d.). *Soil moisture monitoring using IoT enabled Arduino sensors with neural networks for improving soil management for farmers and predict seasonal rainfall for planning future harvest in North Karnataka*. Academic Press.

Aydemir, E., & Karagul, K. (2020). Solving a periodic capacitated vehicle routing problem using simulated annealing algorithm for a manufacturing company. *Brazilian Journal of Operations & Production Management*, *17*(1), 1–13. doi:10.14488/BJOPM.2020.011

Azarmdel, H., Jahanbakhshi, A., Mohtasebi, S. S., & Muñoz, A. R. (2020). Evaluation of image processing technique as an expert system in mulberry fruit grading based on ripeness level using artificial neural networks (ANNs) and support vector machine (SVM). In Postharvest Biology and Technology (Vol. 166, p. 111201). Elsevier BV. doi:10.1016/j.postharvbio.2020.111201

Badage, A. (2018). Crop Disease Detection using Machine Learning: Indian Agriculture. *International Research Journal of Engineering and Technology*, *5*(9).

Baek, J. W., Kim, J. C., Chun, J., & Chung, K. (2019). Hybrid clustering based health decision-making for improving dietary habits. *Technology and Health Care*, *27*(5), 459–472.

Bah, M. D., Hafiane, A., & Canals, R. (2017, November). Weeds detection in UAV imagery using SLIC and the hough transform. In *2017 Seventh International Conference on Image Processing Theory, Tools and Applications (IPTA)* (pp. 1-6). IEEE.

Bah, M. D., Hafiane, A., & Canals, R. (2019). CRowNet: Deep network for crop row detection in UAV images. *IEEE Access: Practical Innovations, Open Solutions*, *8*, 5189–5200.

Bah, M. D., Hafiane, A., Canals, R., & Emile, B. (2019, November). Deep features and One-class classification with unsupervised data for weed detection in UAV images. In *2019 Ninth International Conference on Image Processing Theory, Tools and Applications (IPTA)* (pp. 1-5). IEEE.

Baker, L., & Fan, H. (2017). Innovations of AlphaGo. *DeepMind Blog*. https://deepmind.com/blog/article/innovations-alphago

Balster, A., & Friedrich, H. (2019). Dynamic freight flow modelling for risk evaluation in food supply. *Transp. Res.E*, *121*, 4–22. doi:10.1016/j.tre.2018.03.002

Banko, M. (2001). Scaling to very very large corpora for natural language disambiguation. In *ACL '01: Proceedings of the 39th Annual Meeting on Association for Computational Linguistics* (pp. 26–33). Stroudsberg, PA: Assoc. Comput. Linguist. 10.3115/1073012.1073017

Bannerjee, G., Sarkar, U., Das, S., & Ghosh, L. (2018). Artificial intelligence in agriculture: A literature survey. *International Journal of Scientific Research in Computer Science Applications and Management Studies*, *7*(3), 1–6.

Bansal, S., Chowell, G., Simonsen, L., Vespignani, A., & Viboud, C. (2016). Big data for infectious disease surveillance and modeling. *The Journal of Infectious Diseases*, *214*(suppl 4), S375–S79. doi:10.1093/infdis/jiw400 PMID:28830113

Barrero, O., Rojas, D., Gonzalez, C., & Perdomo, S. (2016). Weed detection in rice fields using aerial images and neural networks. *Proceedings of the the 2016 XXI Symposium on Signal Processing, Images and Artificial Vision (STSIVA)*, 1–4. 10.1109/STSIVA.2016.7743317

Barrett, D. M., Beaulieu, J. C., & Shewfelt, R. (2010). Color, flavor, texture, and nutritional quality of fresh-cut fruits and vegetables: Desirable levels, instrumental and sensory measurement, and the effects of processing. *Critical Reviews in Food Science and Nutrition*, *50*(5), 369–389. doi:10.1080/10408391003626322 PMID:20373184

Barzegar, M., Zare, D., & Stroshine, R. L. (2015). An integrated energy and quality approach to optimization of green peas drying in a hot air infrared-assisted vibratory bed dryer. *Journal of Food Engineering*, *166*, 302–315. doi:10.1016/j.jfoodeng.2015.06.026

Bashiardes, S., Godneva, A., Elinav, E., & Segal, E. (2018). Towards utilization of the human genome and microbiome for personalized nutrition. *Current Opinion in Biotechnology*, *51*, 57–63.

Bay, H., Ess, A., Tuytelaars, T., & Van Gool, L. (2008). Speeded-up robust features (SURF). *Computer Vision and Image Understanding*, *110*(3), 346–359.

Bayu, M. Z., Arshad, H., & Ali, N. M. (2013). Nutritional information visualization using mobile augmented reality technology. *Procedia Technology*, *11*, 396–402.

Behmann, J., Mahlein, A. K., Rumpf, T., Römer, C., & Plümer, L. (2015). A review of advanced machine learning methods for the detection of biotic stress in precision crop protection. *Precision Agriculture*, *16*(3), 239–260. doi:10.100711119-014-9372-7

Behnck, L. P., Doering, D., Pereira, C. E., & Rettberg, A. (2015). A modified simulated annealing algorithm for SUAVs path planning. *IFAC-PapersOnLine*, *48*(10), 63–68. doi:10.1016/j.ifacol.2015.08.109

Ben Ayed, R., Ennouri, K., BenAmar, F., Moreau, F., Triki, M. A., & Rebai, A. (2017). Bayesian and phylogenic approaches for studying relationships among table olive cultivars. *Biochemical Genetics*, *55*(4), 300–313. doi:10.100710528-017-9802-0 PMID:28466379

Ben Ayed, R., & Hanana, M. (2021). Artificial Intelligence to Improve the Food and Agriculture Sector. *Journal of Food Quality*.

Ben-Ayed, R., Kamoun-Grati, N., & Rebai, A. (2013). An overview of the authentication of olive tree and oil. *Comprehensive Reviews in Food Science and Food Safety*, *12*(2), 218–227. doi:10.1111/1541-4337.12003

Bera. (2021). An application of operational analytics: for predicting sales revenue of restaurant. In *Machine Learning Algorithms for Industrial Applications* (pp. 209–235). Springer.

Berg, H., Manuweera, G., & Konradsen, F. (2017). *Global trends in the production and use of DDT for control of malaria and other vector-borne diseases* (Vol. 16). Malaria Journal.

Berry, S. E., Valdes, A. M., Drew, D. A., Asnicar, F., Mazidi, M., Wolf, J., ... Spector, T. D. (2020). Human postprandial responses to food and potential for precision nutrition. *Nature Medicine*, *26*(6), 964–973.

Bhardwaj, H., Tomar, P., Sakalle, A., & Sharma, U. (2021). *Artificial Intelligence and Its Applications in Agriculture With the Future of Smart Agriculture Techniques.* . doi:10.4018/978-1-7998-1722-2.ch002

Bhattacharyya, R., Ghosh, B. N., Mishra, P. K., Mandal, B., Rao, C. S., Sarkar, D., Das, K., Anil, K. S., Lalitha, M., Hati, K. M., & Franzluebbers, A. J. (2015). Soil degradation in India: Challenges and potential solutions. *Sustainability*, *7*(4), 3528–3570. doi:10.3390u7043528

Biffis, E., & Chavez, E. (2017). Satellite data and machine learning for weather risk management and food security. *Risk Analysis*, *37*(8), 1508–1521.

BK, C., Siddappa, M., & HS, J. (2020). Novel framework using dynamic passphrase towards secure and energy-efficient communication in Manet. *International Journal of Electrical & Computer Engineering, 10*(2).

Bochkovskiy, A., Wang, C.-Y., & Liao, H.-Y. M. (2020). *YOLOv4: Optimal speed and accuracy of object detection.* https://arxiv.org/abs/2004.10934

Bon, H. D., Huat, J., Parrot, L., & Sinzogan, A. (2014). Pesticide risks from fruit and vegetable pest management by small farmers in sub-Saharan Africa. A review. *Agronomy for Sustainable Development*, *34*(4), 723–736. doi:10.100713593-014-0216-7

Boris Boiarskii, H. H. (2019). Comparison of NDVI and NDRE indices to detect differences in vegetation and chlorophyll content. *J. Mech. Cont.& Math. Sci*, *1*(Special Issue), 20–29. doi:10.26782/jmcms.spl.4/2019.11.00003

Bosoon Park. (2015). *Hyperspectral Imaging Technology in Food and Agriculture*. Springer.

Bouazizi, M., & Ohtsuki, T. (2016). A pattern-based approach for sarcasm detection on Twitter. *Morbidity and Mortality Weekly Report*, *46*, 258–261.

Boudh, S., & Singh, J. S. (2018). Pesticide Contamination: Environmental Problems and Remediation Strategies. *Emerging and Eco-Friendly Approaches for Waste Management*, 245–269. doi:10.1007/978-981-10-8669-4_12

Brahimi, M., Arsenovic, M., Laraba, S., Sladojevic, S., Boukhalfa, K., & Moussaoui, A. (2018). Deep Learning for Plant Diseases: Detection and Saliency Map Visualisation. In J. Zhou & F. Chen (Eds.), *Human and Machine Learning. Human–Computer Interaction Series*. Springer. doi:10.1007/978-3-319-90403-0_6

Brea, F., Gimeneza, J. M., & Fachinottia, V. D. (2018). Prediction of wind pressure coefficients on building surfaces using Arti ðcial Neural Networks. *Energy and Building*, *158*, 1429–1441. doi:10.1016/j.enbuild.2017.11.045

Bucher, S., Ikeda, K., Broszus, B., Gutierrez, A., & Low, A. (2021). Adaptive Robotic Chassis (ARC): RoboCrop A Smart Agricultural Robot Toolset. *Interdisciplinary Design Senior Theses*. https://scholarcommons.scu.edu/idp_senior/69

Bu, F., & Wang, X. (2019). A smart agriculture IoT system based on deep reinforcement learning. *Future Generation Computer Systems*, *99*, 500–507. doi:10.1016/j.future.2019.04.041

Bujari, A., Calafate, C. T., Cano, J. C., Manzoni, P., Palazzi, C. E., & Ronzani, D. (2017). Flying ad-hoc network application scenarios and mobility models. *International Journal of Distributed Sensor Networks*, *13*(10). doi:10.1177/1550147717738192

Bünger, L. (2021). *Robotic waste sorting*. Worcester Polytechnic Institute. https://www.wpi.edu/Academics/Projects

Bush, C. L., Blumberg, J. B., El-Sohemy, A., Minich, D. M., Ordovás, J. M., Reed, D. G., & Behm, V. A. Y. (2020). Toward the definition of personalized nutrition: A proposal by the American Nutrition Association. *Journal of the American College of Nutrition*, *39*(1), 5–15.

Caferoglu, B. A., Elbir, Z., & Cihan, O. (2022). *GitHub repository*. https://github.com/BeratAsrin/Freshness_Detection_of_Agricultural_Products_Using_Artificial_Intelligence

Camuñas-Mesa, L. A., Barranco, B. L., & Gotarredona, T. S. (2019). Neuromorphic Spiking Neural Networks and Their Memristor-CMOS Hardware Implementations. *Materials (Basel)*, *12*(17), 2745. doi:10.3390/ma12172745 PMID:31461877

Carey, J. (2017). News Feature: On the brink of a recycling revolution? *Proceedings of the National Academy of Sciences of the United States of America*, *114*(4), 612–616. doi:10.1073/pnas.1620655114 PMID:28119494

Carvalho, F. P., Villeneuve, J. P., Cattini, C., Rendón, J., & Oliveira, J. M. (2009). Pesticide and PCB residues in the aquatic ecosystems of Laguna de Terminos, a protected area of the coast of Campeche, Mexico. *Chemosphere*, *74*(7), 988–995.

Castillo & Meliif. (1970). Automated quality control in the food industry combining artificial intelligence techniques with fractal theory. *WIT Transactions on Information and Communication Technologies, 10.*

Cauvery, N.K. (2020). Trust-based secure routing against lethal behavior of nodes in wireless adhoc network. *International Journal of Electrical & Computer Engineering, 10*(2).

Ceballos-Magaña, S. G., Jurado, J. M., Muñiz-Valencia, R., Alcázar, A., de Pablos, F., & Martín, M. J. (2012). Geographical authentication of tequila according to its mineral content by means of support vector machines. *Food Analytical Methods*, *5*(2), 260–265. doi:10.100712161-011-9233-1

Cerdeira-Pena, A., Carpente, L., & Amiama, C. (2017). Optimised forage harvester routes as solutions to a traveling salesman problem with clusters and time windows. *Biosystems Engineering*, *164*, 110–123. doi:10.1016/j.biosystemseng.2017.10.002

Champ, J., Mora-Fallas, A., Goëau, H., Mata-Montero, E., Bonnet, P., & Joly, A. (2020). Instance segmentation for the fine detection of crop and weed plants by precision agricultural robots. *Applications in Plant Sciences*, *8*(7), e11373.

Chandy, A. (2019). Pest Infestation Identification in Coconut Trees Using Deep Learning. Journal of Artificial Intelligence, *1*(01), 10–18. doi:10.36548/jaicn.2019.1.002

Chatterjee, S., Nandan, M., Ghosh, A., & Banik, S. (2022). DTNMA: Identifying Routing Attacks in Delay-Tolerant Network. In *Cyber Intelligence and Information Retrieval* (pp. 3–15). Springer. doi:10.1007/978-981-16-4284-5_1

Chen, J., & Ngo, C. W. (2016, October). Deep-based ingredient recognition for cooking recipe retrieval. In *Proceedings of the 24th ACM international conference on Multimedia* (pp. 32-41). Academic Press.

Chen, T., & Guestrin, C. (2016). XGBoost: a scalable tree boosting system. In *KDD '16: Proceedings of the 22nd ACMSIGKDD International Conference on Knowledge Discovery and Data Mining* (pp. 785–94). New York: Assoc. Comput. Mach. 10.1145/2939672.2939785

Cheng, B., & Matson, E. T. (2015, June). A feature-based machine learning agent for automatic rice and weed discrimination. In *International Conference on Artificial Intelligence and Soft Computing* (pp. 517-527). Springer. 10.1007/978-3-319-19324-3_46

Chen, J., & Ngo, C. W. (2016, October). Deep-based ingredient recognition for cooking recipe retrieval. In *Proceedings of the 24th ACM international conference on Multimedia* (pp. 32-41). ACM.

Chen, S., Huang, D., Nong, W., & Kwan, H. S. (2016). Development of a food safety information database for Greater China. *Food Control*, *65*, 54–62. doi:10.1016/j.foodcont.2016.01.002

Chen, W., & Zhao, D. (2013). *Path Planning for Spray Painting Robot of Workpiece Surfaces*. Robust Control with Engineering Applications.

Chen, Z., & Tao, Y. (2001). Food safety inspection using "from presence to classification" object-detection model. *Pattern Recognition*, *34*(12), 2331–2338. doi:10.1016/S0031-3203(00)00169-2

Chidinma-Mary-Agbai. (2020). Application of artificial intelligence (AI) in food industry. *GSC Biological and Pharmaceutical Sciences*, *13*(01), 171–178. doi:10.30574/gscbps.2020.13.1.0320

Chlingaryan, A., Sukkarieh, S., & Whelan, B. (2018). Machine learning approaches for crop yield prediction and nitrogen status estimation in precision agriculture: A review. *Computers and Electronics in Agriculture*, *151*, 61–69. doi:10.1016/j.compag.2018.05.012

Chowdary, U. V. & Venkataramana, K. (2018). Tomato Crop Yield Prediction using ID3. *International Journal of Innovative Research in Technology, 4*(10), 663-62.

Condé, B. C., Fuentes, S., Caron, M., Di Xiao, R. C., & Howell, K. S. (2017). Development of a robotic and computer vision method to assess foam quality in sparkling wines. *Food Control*, *71*, 383–392. doi:10.1016/j.foodcont.2016.07.020

Conesa-Muñoz, J., Pajares, G., & Ribeiro, A. (2016). Mix-opt: A new route operator for optimal coverage path planning for a fleet in an agricultural environment. *Expert Systems with Applications*, *54*, 364–378. doi:10.1016/j.eswa.2015.12.047

Connelly, R., Playford, C. J., Gayle, V., & Dibben, C. (2016). The role of administrative data in the big data revolution in social science research. *Social Science Research*, *59*, 1–12. doi:10.1016/j.ssresearch.2016.04.015 PMID:27480367

Craninx, M., Fievez, V., Vlaeminck, B., & De Baets, B. (2008). Artificial neural network models of the rumen fermentation pattern in dairy cattle. *Computers and Electronics in Agriculture*, *60*(2), 226–238. doi:10.1016/j.compag.2007.08.005

Curbelo Montañez, C. A., Fergus, P., Montañez, A. C., Hussain, A., Al-Jumeily, D., & Chalmers, C. (2018, October). Deep Learning Classification of Polygenic Obesity using Genome Wide Association Study SNPs. In *2018 International Joint Conference on Neural Networks (IJCNN)*. IEEE.

D N, V., & Choudhary, D. S. (2021). An AI solution for Soil Fertility and Crop Friendliness Detection and Monitoring. *International Journal of Engineering and Advanced Technology*, *10*(3), 172–175. doi:10.35940/ijeat.C2270.0210321

Dahikar, S. S., & Rode, S. V. (2014). Agricultural Crop Yield Prediction Using Artificial Neural Network Approach. International Journal of Innovative Research in Electrical, Electronics, Instrumentation and Control Engineering, 2(1), 683–686.

Dallman, T., Inns, T., Jombart, T., Ashton, P., Loman, N., Chatt, C., Messelhaeusser, U., Rabsch, W., Simon, S., Nikisins, S., Bernard, H., le Hello, S., Jourdan da-Silva, N., Kornschober, C., Mossong, J., Hawkey, P., de Pinna, E., Grant, K., & Cleary, P. (2016). Phylogenetic structure of European Salmonella Enteritidis outbreak correlates with national and international egg distribution network. *Microbial Genomics*, *2*(8), e000070. doi:10.1099/mgen.0.000070 PMID:28348865

Danti, A. (2012). Mean and range color features based identification of common Indian leafy vegetables. *International Journal of Signal Processing, Image Processing and Pattern Recognition*, *5*(3), 151–160.

Dasgupta, I., Saha, J., Venkatasubbu, P., & Ramasubramanian, P. (2020). AI Crop Predictor and Weed Detector Using Wireless Technologies: A Smart Application for Farmers. *Arabian Journal for Science and Engineering*, *45*(12), 11115–11127.

Davis, J. J., Boisvert, S., Brettin, T., Kenyon, R. W., Mao, C., Olson, R., Overbeek, R., Santerre, J., Shukla, M., Wattam, A. R., Will, R., Xia, F., & Stevens, R. (2016). Antimicrobial resistance prediction in PATRIC and RAST. *Scientific Reports*, *6*(1), 27930. doi:10.1038rep27930 PMID:27297683

Deepak Katkani, A. B. (2022). A Review on Applications and Utility of Remote Sensing and Geographic Information Systems in Agriculture and Natural Resource Management. *International Journal of Environment and Climate Change*.

Dehghan-Dehnavi, S., Fotuhi-Firuzabad, M., Moeini-Aghtaie, M., Dehghanian, P., & Wang, F. (2020). Estimating participation abilities of industrial customers in demand response programs: a two-level decision-making tree analysis. *2020 IEEE/IAS 56th Industrial and Commercial Power Systems Technical Conference (I&CPS)*, 1–8. 10.1109/ICPS48389.2020.9176817

Dela Cruz, J. R., Baldovino, R. G., Bandala, A. A., & Dadios, E. P. (2017). Water usage optimization of Smart Farm Automated Irrigation System using artificial neural network. *2017 5th International Conference on Information and Communication Technology (ICoIC7)*. 10.1109/ICoICT.2017.8074668

Deng, L., & Yu, D. (2014). Deep learning: methods and applications. *Foundations and Trends in Signal Processing*, *7*(3–4), 197-387.

Deng, J., Dong, W., Socher, R., Li, L.-J., Li, K., & Li, F.-F. (2009). ImageNet: A large-scale hierarchical image database. *2009 IEEE Conference on Computer Vision and Pattern Recognition*, 248–255. 10.1109/CVPR.2009.5206848

Deng, X., Cao, S., & Horn, A. L. (2021). Emerging applications of machine learning in food safety. *Annual Review of Food Science and Technology*, *12*(1), 513–538. doi:10.1146/annurev-food-071720-024112 PMID:33472015

Deng, X., den Bakker, H. C., & Hendriksen, R. S. (2016). Genomic epidemiology: Whole-genome-sequencing-powered surveillance and outbreak investigation of foodborne bacterial pathogens. *Annual Review of Food Science and Technology*, *7*(1), 353–374. doi:10.1146/annurev-food-041715-033259 PMID:26772415

Design of a Sensor to Detect Fruit Freshness. (2018). *International Journal of Scientific and Technological Research*, *4*(1), 1–6.

Devinney, K., Bekbay, A., Effland, T., Gravano, L., Howell, D., Hsu, D., O'Hallorhan, D., Reddy, V., Stavinsky, F., Waechter, H. N., & Gutelius, B. (2018). Evaluating Twitter for foodborne illness outbreak detection in New York City. *Online Journal of Public Health Informatics*, *10*(1), e120. doi:10.5210/ojphi.v10i1.8894

Dewi, T., Risma, P., & Oktarina, Y. (2020). Fruit sorting robot based on color and size for an agricultural product packaging system. *Bulletin of Electrical Engineering and Informatics*, *9*(4), 1438–1445. doi:10.11591/eei.v9i4.2353

Dharmaraj, V., & Vijayanand, C. (2018). Artificial Intelligence (AI) in Agriculture. *International Journal of Current Microbiology and Applied Sciences*, *7*(12), 2122–2128. doi:10.20546/ijcmas.2018.712.241

Dharminder, R. K. S., Kumar, V., Devedee, A. K., Mruthyunjaya, M., & Bhardwaj, R. (2019). The clean water: The basic need of human and agriculture. *International Journal of Chemical Studies*, *7*, 1994–1998.

Dharwal, R., & Kaur, L. (2016). Applications of Artificial Neural Networks: A Review. *Indian Journal of Science and Technology*, *9*(1), 1–8. doi:10.17485/ijst/2016/v9i47/106807

Di Vaio, A., Boccia, F., Landriani, L., & Palladino, R. (2020). Artificial intelligence in the agri-food system: Rethinking sustainable business models in the COVID-19 scenario. *Sustainability*, *12*(12), 4851.

Dogan, A., & Birant, D. (2021). Dogan, Alican, and DeryaBirant. "Machine learning and data mining in manufacturing. *Expert Systems with Applications*, *166*, 114060. doi:10.1016/j.eswa.2020.114060

Donepudi, P. K. (2014). Technology growth in shipping industry: An overview. *American Journal of Trade and Policy*, *1*(3), 137–142. doi:10.18034/ajtp.v1i3.503

Dos Santos Ferreira, A., Freitas, D. M., da Silva, G. G., Pistori, H., & Folhes, M. T. (2017). Weed detection in soybean crops using ConvNets. *Computers and Electronics in Agriculture*, *143*, 314–324. doi:10.1016/j.compag.2017.10.027

Doyle, M. P., Erickson, M. C., Alali, W., Cannon, J., Deng, X., Ortega, Y., Smith, M. A., & Zhao, T. (2015). The food industry's current and future role in preventing microbial foodborne illness within the United States. *Clinical Infectious Diseases*, *61*(2), 252–259. doi:10.1093/cid/civ253 PMID:25824814

Drouin, A., Giguère, S., Déraspe, M., Marchand, M., Tyers, M., Loo, V. G., Bourgault, A.-M., Laviolette, F., & Corbeil, J. (2016). Predictive computational phenotyping and biomarker discovery using reference-free genome comparisons. *BMC Genomics*, *17*(1), 754. doi:10.118612864-016-2889-6 PMID:27671088

Drury, B., & Roche, M. (2019). A survey of the applications of text mining for agriculture. *Computers and Electronics in Agriculture*, *163*, 104864. doi:10.1016/j.compag.2019.104864

Dubey, S. R., & Jalal, A. S. (2016). Apple disease classification using color, texture and shape features from images. *Signal, Image and Video Processing*, *10*(5), 819–826.

Du, C., & Sun, D. (2006). Learning techniques used in computer vision for food quality evaluation: A review. *Journal of Food Engineering*, *72*(1), 39–55. doi:10.1016/j.jfoodeng.2004.11.017

Dutta, S., Rakshit, S., & Chatterjee, D. (2020). Use of Artificial Intelligence in Indian Agriculture. *Food and Sci. Rep.*, *1*, 65–72. doi:10.13140/RG.2.2.24973.97762

Dyrmann, M., Karstoft, H., & Midtiby, H. S. (2016). Plant species classification using deep convolutional neural network. *Biosystems Engineering*, *151*, 72–80. doi:10.1016/j.biosystemseng.2016.08.024

Ebrahimi, M. A., Khoshtaghaza, M. H., Minaei, S., & Jamshidi, B. (2017). Vision-based pest detection based on SVM classification method. *Computers and Electronics in Agriculture*, *137*, 52–58. doi:10.1016/j.compag.2017.03.016

Effland, T., Lawson, A., Balter, S., Devinney, K., Reddy, V., Waechter, H. N., Gravano, L., & Hsu, D. (2018). Discovering foodborne illness in online restaurant reviews. *Journal of the American Medical Informatics Association: JAMIA*, *25*(12), 1586–1592. doi:10.1093/jamia/ocx093 PMID:29329402

Ehler, L. E. (2006). Integrated pest management (IPM): Definition, historical development and implementation, and the other IPM. *Pest Management Science*, *62*(9), 787–789. doi:10.1002/ps.1247 PMID:16786545

Ekwealor, K. U., Echereme, C. B., Ofobeze, T. N., & Okereke, C. N. (2019). Economic Importance of Weeds: A Review. *Asian Plant Research Journal*, *3*, 1–11. doi:10.9734/aprj/2019/v3i230063

Elferink, M., & Schierhorn, F. (2016). Global demand for food is rising. Can we meet it? *Harvard Business Review*, *7*(4), 1–7. https://www.researchgate.net/publication/302466629%0A

Eliasy, A., & Przychodzen, J. (2020). The role of AI in capital structure to enhance corporate funding strategies. *Array.*, *6*, 100017. Advance online publication. doi:10.1016/j.array.2020.100017

Eliazàt, A. (2020). *4 Ways AI is revolutionizing the food industry*. https://autome.me/4-ways-ai-is-revolutionizing-the-food-industry

Eli-Chukwu, N. C. (2019). Applications of Artificial Intelligence in Agriculture: A Review. *Eng. Technol. Appl. Sci. Res.*, *9*(4), 4377–4383. doi:10.48084/etasr.2756

Elijah, O., Rahman, T. A., Orikumhi, I., Leow, C. Y., & Hindia, M. N. (2018). An overview of Internet of Things (IoT) and data analytics in agriculture: Benefits and challenges. *IEEE Internet of Things Journal*, *5*(5), 3758–3773. doi:10.1109/JIOT.2018.2844296

Ellington, M. J., Ekelund, O., Aarestrup, F. M., Canton, R., Doumith, M., Giske, C., Grundman, H., Hasman, H., Holden, M. T. G., Hopkins, K. L., Iredell, J., Kahlmeter, G., Köser, C. U., MacGowan, A., Mevius, D., Mulvey, M., Naas, T., Peto, T., Rolain, J.-M., ... Woodford, N. (2017). The role of whole genome sequencing in antimicrobial susceptibility testing of bacteria: Report from the EUCAST Subcommittee. *Clinical Microbiology and Infection*, *23*(1), 2–22. doi:10.1016/j.cmi.2016.11.012 PMID:27890457

Esgario, J. G., de Castro, P. B., Tassis, L. M., & Krohling, R. A. (2021). *An app to assist farmers in the identification of diseases and pests of coffee leaves using deep learning. Information Processing in Agriculture*. doi:10.1016/j.inpa.2021.01.004

Espejo-Garcia, B., Mylonas, N., Athanasakos, L., Fountas, S., & Vasilakoglou, I. (2020). Towards weeds identification assistance through transfer learning. *Computers and Electronics in Agriculture*, *171*, 105306.

Esposito, M., Crimaldi, M., Cirillo, V., Sarghini, F., & Maggio, A. (2021). Drone and sensor technology for sustainable weed management: A review. *Chemical and Biological Technologies in Agriculture*, *8*(1), 1–11.

Estelles-Lopez. (2017). An automated ranking platform for machine learning regression models for meat spoilage prediction using multi-spectral imaging and metabolic profiling. *Food Research International*, *99*, 206–215.

Everingham, Y., Sexton, J., Skocaj, D. & Bamber, G. I. (2016). Accurate prediction of sugarcane yield using a random forest algorithm. *Agronomy for Sustainable Development, 36*(2). doi:10.1007/s13593-016-0364-z

Everingham, M., Van Gool, L., Williams, C. K. I., Winn, J., & Zisserman, A. (2009). The Pascal Visual Object Classes (VOC) Challenge. *International Journal of Computer Vision*, *88*(2), 303–338. doi:10.100711263-009-0275-4

Eyre, D. W., De Silva, D., Cole, K., Peters, J., Cole, M. J., Grad, Y. H., Demczuk, W., Martin, I., Mulvey, M. R., Crook, D. W., Walker, A. S., Peto, T. E. A., & Paul, J. (2017). WGS to predict antibiotic MICs for Neisseria gonorrhoeae. *The Journal of Antimicrobial Chemotherapy*, *72*(7), 1937–1947. doi:10.1093/jac/dkx067 PMID:28333355

F. A. O. Food. (2017). *The future of food and agriculture–Trends and challenges*. Annual Report, 296.

Faiçal, B. S., Costa, F. G., Pessin, G., Ueyama, J., Freitas, H., Colombo, A., Fini, P., Villas, H. L., Osório, F. S., Vargas, P. A., & Braun, T. (2014). The use of unmanned aerial vehicles and wireless sensor networks for spraying pesticides. *Journal of Systems Architecture*, *60*(4), 393–404.

Fainshtein, E., Serova, E., & Vorobyev, P. (2021). Contemporary research and analysis of food industry: case of Russian restaurant business network branch. In *Eurasian Business Perspectives: Proceedings of the 29th Eurasia Business and Economics Society Conference* (pp. 183–199). Springer International Publishing. 10.1007/978-3-030-65085-8_12

Fang, S., Ru, Y., Liu, Y., Hu, C., Chen, X., & Liu, B. (2021). *Route Planning of Helicopters Spraying Operations in Multiple Forest Areas* (Vol. 12). Forests.

Fan, Y., Qian, Y., Xie, F., & Soong, F. K. (2014). *TTS synthesis with bidirectional LSTM based recurrent neural networks. Proc. Interspeech.*

Faria, Alex dos Santos, Rocha, & Torres. (2012). Automatic classifier fusion for produce recognition. In *2012 25th SIBGRAPI Conference on Graphics, Patterns and Images* (pp. 252-259). IEEE.

Farinella, G. M., Moltisanti, M., & Battiato, S. (2014, October). Classifying food images represented as bag of textons. In *2014 IEEE International Conference on Image Processing (ICIP)* (pp. 5212-5216). IEEE.

Farooqui, N. A., & Ritika, R. (2019). An Identification and Detection Process for Leaves Disease of Wheat Using Advance Machine Learning Techniques. *Bioscience Biotechnology Research Communications*, *12*(4), 1081–1091. doi:10.21786/bbrc/12.4/31

Fathima, G.N. & Geetha, R. (2014). Agriculture Crop Pattern Using Data Mining Techniques. *International Journal of Advanced Research in Computer Science and Engineering, 4*(5), 781-786.

Feldgarden, M., Brover, V., Haft, D. H., Prasad, A. B., & Slotta, D. J. (2019). Validating the AMRFinder Tool and Resistance Gene Database by using antimicrobial resistance genotype-phenotype correlations in a collection of isolates. *Antimicrob. Agents Chemother., 63*(11).

Filip, M., Zoubek, T., Bumbalek, R., Cerny, P., Batista, C. E., Olsan, P., Bartos, P., Kriz, P., Xiao, M., Dolan, A., & Findura, P. (2020). Advanced computational methods for agriculture machinery movement optimization with applications in sugarcane production. *Agriculture*, *10*(10), 434. doi:10.3390/agriculture10100434

Filter, M., Appel, B., & Buschulte, A. (2015). Expert systems for food safety. *Current Opinion in Food Science*, *6*, 61–65. doi:10.1016/j.cofs.2016.01.004

Food and Agriculture Organization of the United Nations. (2019). *Agriculture and climate change Challenges and opportunities at the global and local level Collaboration on Climate-Smart Agriculture*. FAO. https://www.fao.org/3/CA3204EN/ca3204en.pdf

Food Waste. (n.d.). https://www.usda.gov/foodwaste/faqs/

Fraser, E. D. G., & Campbell, M. (2019). Agriculture 5.0: Reconciling Production with Planetary Health. *One Earth*, *1*(3), 278–280. doi:10.1016/j.oneear.2019.10.022

Friedlander, A., & Zoellner, C. (2020). Artificial Intelligence Opportunities to Improve Food Safety at Retail. *Food Protection Trends*, *40*(4), 272–278.

Friedman, J. H. (1997). On bias, variance, 0/1—Loss, and the curse-of-dimensionality. *Data Mining and Knowledge Discovery*, *1*(1), 55–77. doi:10.1023/A:1009778005914

Fu, Y. (2020). *Fruit Freshness Grading Using Deep Learning*. Auckland University of Technology.

Gamal ElMasry, N. M.-R. (2019). Recent Applications of Multispectral Imaging in Seed Phenotyping and Quality Monitoring—An Overview. *Sensors*.

Gandhi, N., Armstrong, L. J., Petkar, O., & Tripathy, A. K. (2016). Rice crop yield prediction in India using support vector machines. *13th International Joint Conference on Computer Science and Software Engineering (JCSSE)*, 1-5. 10.1109/JCSSE.2016.7748856

Ganesh, V. D., & Hemavathi, A. D. (2017). Artificial Intelligence Based Calorie Estimator. *International Journal of Pure and Applied Mathematics*, *117*(15), 1201–1205.

Ganganagowdar, N. V., & Siddaramappa, H. K. (2016). Recognition and classification of White Wholes (WW) grade cashew kernel using artificial neural networks. *Acta Scientiarum. Agronomy*, *38*(2), 145–155. doi:10.4025/actasciagron.v38i2.27861

Gao, J., French, A. P., Pound, M. P., He, Y., Pridmore, T. P., & Pieters, J. G. (2020). Deep convolutional neural networks for image-based Convolvulus sepium detection in sugar beet fields. *Plant Methods*, *16*(1), 1–12.

Garre, A., Fernandez, P. S., Brereton, P., Elliott, C., & Mojtahed, V. (2019). The use of trade data to predict the source and spread of food safety outbreaks: An innovative mathematical modelling approach. *Food Research International*, *123*, 712–721. doi:10.1016/j.foodres.2019.06.007 PMID:31285021

Garre, A., Ruiz, M. C., & Hontoria, E. (2020). Application of Machine Learning to support production planning of a food industry in the context of waste generation under uncertainty. *Operations Research Perspectives*, *7*, 100147. doi:10.1016/j.orp.2020.100147

Gates, M. C., Holmstrom, L. K., Biggers, K. E., & Beckham, T. R. (2015). Integrating novel data streams to support biosurveillance in commercial livestock production systems in developed countries: Challenges and opportunities. *Frontiers in Public Health*, *3*, 74. doi:10.3389/fpubh.2015.00074 PMID:25973416

Gaurav, V., Choudhary, S., Singh, A., & Agarwal, S. (2019). *Autonomous crop irrigation system using artificial intelligence. International Journal of Engineering and Advanced Technology* , 8).

Gebbers, A. V. I. (2010). Precision agriculture and food security. *Science, 327*(5967), 828–831.

Geetha, G., Samundeswari, S., Saranya, G., Meenakshi, K., & Nithya, M. (2020). Plant Leaf Disease Classification and Detection System Using Machine Learning. *Journal of Physics: Conference Series*, *1712*(1), 012012. doi:10.1088/1742-6596/1712/1/012012

Gertler, M., Czogiel, I., Stark, K., & Wilking, H. (2017). Assessment of recall error in self-reported food consumption histories among adults—particularly delay of interviews decrease completeness of food histories— Germany, 2013. *PLoS One*, *12*(6), e0179121. doi:10.1371/journal.pone.0179121 PMID:28640839

Ghosh, A., Ho, C. C., & Bestak, R. (2020). *Secured Energy-Efficient Routing in Wireless Sensor Networks Using Machine Learning Algorithm: Fundamentals and Applications*. Deep Learning Strategies for Security Enhancement in Wireless Sensor Networks. doi:10.4018/978-1-7998-5068-7.ch002

Gibson, D. J., Young, B. G., & Wood, A. J. (2017). Can weeds enhance profitability? Integrating ecological concepts to address crop-weed competition and yield quality. *Journal of Ecology*, *105*(4), 900–904.

Gibson, M. K., Forsberg, K. J., & Dantas, G. (2015). Improved annotation of antibiotic resistance determinants reveals microbial resistomes cluster by ecology. *The ISME Journal*, *9*(1), 207–216. doi:10.1038/ismej.2014.106 PMID:25003965

Global Digital Overview. (2020). *DataReportal – Global Digital Insights*. https://datareportal.com/global-digital-overview

Göbel, C., Langen, N., Blumenthal, A., Teitscheid, P., & Ritter, G. (2015). Cutting food waste through cooperation along the food supply chain. *Sustainability*, *7*(2), 1429–1445. doi:10.3390u7021429

Godfray, H. (2010). Food security: The challenge of feeding 9 billion people. *Science, 327*(5967), 812-818.

Godse, S.P., & Mahalle, P.N. (2018). A Computational Analysis of E.C.C. Based Novel Authentication Scheme in VANET. *International Journal of Electrical & Computer Engineering, 8*(6).

Golhani, K., Balasundram, S. K., Vadamalai, G., & Pradhan, B. (2018). A review of neural networks in plant disease detection using hyperspectral data. *Information Processing in Agriculture*, *5*(3), 354–371. doi:10.1016/j.inpa.2018.05.002

Gondchawar, N., & Kawitkar, R. S. (2016). IoT based smart agriculture. *International Journal of Advanced Research in Computer and Communication Engineering*, *5*(6), 838–842.

Goni, L., Cuervo, M., Milagro, F. I., & Martínez, J. A. (2015). Future perspectives of personalized weight loss interventions based on nutrigenetic, epigenetic, and metagenomic data. *The Journal of Nutrition, 146*(4), 905S-912S.

Goodfellow, I., Bengio, Y., & Courville, A. (2016). *Deep learning*. MIT press.

Gopi, R., Singh, S., & Raj, C. (2019). Status of Fusarium diseases of crop plants in North East India. *Indian Phytopathology, 72*(4), 637–646. doi:10.100742360-019-00148-3

Gothai, E., Natesan, P., Aishwariya, S., Aarthy, T. B., & Singh, G. B. (2020, March). Weed Identification using Convolutional Neural Network and Convolutional Neural Network Architectures. In *2020 Fourth International Conference on Computing Methodologies and Communication (ICCMC)* (pp. 958-965). IEEE.

Goyache, F., Bahamonde, A., Alonso, J., Lopez, S., del Coz, J. J., Quevedo, J. R., Ranilla, J., Luaces, O., Alvarez, I., Royo, L. J., & Diez, J. (2001). The usefulness of artificial intelligence techniques to assess subjective quality of products in the food industry. *Trends in Food Science & Technology, 12*(10), 370–381. doi:10.1016/S0924-2244(02)00010-9

Grabusts, P., Musatovs, J., & Golenkov, V. (2019). The application of simulated annealing method for optimal route detection between objects. *Procedia Computer Science, 149*, 95–101. doi:10.1016/j.procs.2019.01.112

Greis, N. P., & Nogueira, M. L. (2017). A data-driven approach to food safety surveillance and response. In Food Protection and Security: Preventing and Mitigating Contamination during Food Processing and Production. doi:10.1016/B978-1-78242-251-8.00005-9

Grewal, A. S., Singla, A., Kamboj, P., & Dua, J. S. (2017). Pesticide Residues in Food Grains, Vegetables and Fruits: A Hazard to Human Health. *Journal of Medicinal Chemistry and Toxicology, 2*(1), 40–46. doi:10.15436/2575-808X.17.1355

Guest, N. S., Horne, J., Vanderhout, S. M., & El-Sohemy, A. (2019). Sport nutrigenomics: Personalized nutrition for athletic performance. *Frontiers in Nutrition, 6*, 8.

Guh, R.-S. (2003). Integrating artificial intelligence into on-line statistical process control. *Quality and Reliability Engineering International, 19*(1), 1–20. doi:10.1002/qre.510

Guillier, L., Gourmelon, M., Lozach, S., Cadel-Six, S., Vignaud, M. L., Munck, N., Hald, T., & Palma, F. (2020). AB_SA: Accessory genes-based source attribution - tracing the source of Salmonella enterica Typhimurium environmental strains. *Microbial Genomics, 6*(7), mgen000366. doi:10.1099/mgen.0.000366 PMID:32320376

Gupta, A., Nelson, J. M., Barrett, T. J., Tauxe, R. V., Rossiter, S. P., Friedman, C. R., Joyce, K. W., Smith, K. E., Jones, T. F., Hawkins, M. A., Shiferaw, B., Beebe, J. L., Vugia, D. J., Rabatsky-Ehr, T., Benson, J. A., Root, T. P., & Angulo, F. J. (2004). Antimicrobial resistance among Campylobacter strains, United States, 1997–2001. *Emerging Infectious Diseases, 10*(6), 1102–1109. doi:10.3201/eid1006.030635 PMID:15207064

Haarnoja, T., Ha, S., Zhou, A., & Tan, J. (2018). *Learning to walk via deep reinforcement learning*. arXiv:1812.11103 [cs.LG]

Haglin, J. M., Jimenez, G., & Eltorai, A. E. M. (2019). Artificial neural networks in medicine. *Health Technology (Hong Kong), 9*(1), 1–6. doi:10.100712553-018-0244-4

Halevy, A., Norvig, P., & Pereira, F. (2009). The unreasonable effectiveness of data. *IEEE Intelligent Systems, 24*(2), 8–12. doi:10.1109/MIS.2009.36

Hall, A. J., Wikswo, M. E., Manikonda, K., Roberts, V. A., Yoder, J. S., & Gould, L. H. (2013). Acute gastroenteritis surveillance through the National Outbreak Reporting System, United States. *Emerging Infectious Diseases, 19*(8), 1305–1309. doi:10.3201/eid1908.130482 PMID:23876187

Hansen, M. F., Smith, M. L., Smith, L. N., Salter, M. G., Baxter, E. M., Farish, M., & Grieve, B. (2018). Towards on-farm pig face recognition using convolutional neural networks. *Computers in Industry*, *98*, 145–152. doi:10.1016/j.compind.2018.02.016

Hanson, A., Joy, A., & Francis, J. (2017). *Plant Leaf Disease Detection using Deep Learning and Convolutional Neural Network*. Academic Press.

Han, W., Jiang, F., & Zhu, Z. (2022). Detection of Cherry Quality Using YOLOV5 Model Based on Flood Filling Algorithm. *Foods*, *11*(8), 1127. doi:10.3390/foods11081127 PMID:35454714

Hariadi, R. R., Khotimah, W. N., & Wiyono, E. A. (2015, October). Design and implementation of food nutrition information system using SURF and FatSecret API. In *2015 International Conference on Advanced Mechatronics, Intelligent Manufacture, and Industrial Automation (ICAMIMIA)* (pp. 181-183). IEEE.

Hari, S. S., Sivakumar, M., Renuga, P., & Suriya, S. (2019). Detection of plant disease by leaf image using convolutional neural network. *2019 International Conference on Vision Towards Emerging Trends in Communication and Networking (ViTECoN)*, 1-5. 10.1109/ViTECoN.2019.8899748

Harper, C., & Siller, M. (2015). OpenAG: A globally distributed network of food computing. *IEEE Pervasive Computing*, *14*(4), 24–27. doi:10.1109/MPRV.2015.72

Harris, J. K., Hawkins, J. B., Nguyen, L., Nsoesie, E. O., Tuli, G., Mansour, R., & Brownstein, J. S. (2017). Using Twitter to identify and respond to food poisoning: The Food Safety STL Project. *Journal of Public Health Management and Practice*, *23*(6), 577–580. doi:10.1097/PHH.0000000000000516 PMID:28166175

Harris, J. K., Mansour, R., Choucair, B., Olson, J., & Nissen, C. (2014). Health department use of social media to identify foodborne illness—Chicago, Illinois, 2013–2014. *MMWR. Morbidity and Mortality Weekly Report*, *63*, 681–685. PMID:25121710

Harrison, C., Jorder, M., Stern, H., Stavinsky, F., & Reddy, V. (2014). Using online reviews by restaurant patrons to identify unreported cases of foodborne illness—New York City, 2012–2013. *MMWR. Morbidity and Mortality Weekly Report*, *63*, 441–445. PMID:24848215

Harvey, C. A., Rakotobe, Z. L., Rao, N. S., Dave, R., Razafimahatratra, H., Rabarijohn, R. H., Rajaofara, H., & MacKinnon, J. L. (2014). Extreme vulnerability of smallholder farmers to agricultural risks and climate change in Madagascar. *Philosophical Transactions of the Royal Society of London. Series B, Biological Sciences*, *369*(1639), 20130089. doi:10.1098/rstb.2013.0089 PMID:24535397

Hassaan, M. A., & Nemr, A. E. (2020). Pesticide's pollution: Classifications, human health impact, extraction and treatment techniques. *The Egyptian Journal of Aquatic Research*, *46*, 207–220.

Hastie, T., Tibshirani, R., & Friedman, J. (2009). *The Elements of Statistical Learning: Data Mining, Inference, and Prediction*. Springer. doi:10.1007/978-0-387-84858-7

He, G. (2002). *Destination-sequenced distance vector (DSDV) protocol*. Networking Laboratory, Helsinki University of Technology.

He, J., Baxter, S. L., Xu, J., Xu, J., Zhou, X., & Zhang, K. (2019). The practical implementation of artificial intelligence technologies in medicine. *Nature Medicine*, *25*(1), 30–36. doi:10.103841591-018-0307-0 PMID:30617336

Hendriksen, R. S., Bortolaia, V., Tate, H., Tyson, G. H., Aarestrup, F. M., & McDermott, P. F. (2019). Using genomics to track global antimicrobial resistance. *Frontiers in Public Health*, *7*, 242. doi:10.3389/fpubh.2019.00242 PMID:31552211

Hirsch, C., Bartocci, E., & Grosu, R. (2019). Capacitive soil moisture sensor node for IoT in agriculture and home. *Proc. IEEE 23rd International Symposium on Consumer Technologies*, 97–102.

Hoashi, H., Joutou, T., & Yanai, K. (2010, December). Image recognition of 85 food categories by feature fusion. In *2010 IEEE International Symposium on Multimedia* (pp. 296-301). IEEE.

Hochreiter, S., & Schmidhuber, J. (1997). Long short-term memory. *Neural Computation*, *9*(8), 1735–1780. doi:10.1162/neco.1997.9.8.1735 PMID:9377276

Hollemans, M. (2018, April 22). *MobileNet version 2.* Blog of Matthijs Hollemans about Machine Learning and AI. https://machinethink.net/blog/mobilenet-v2/

Hopfield, J. J. (1982). Neural networks and physical systems with emergent collective computational abilities. *Proceedings of the National Academy of Sciences of the United States of America*, *79*(8), 2554–2558. doi:10.1073/pnas.79.8.2554 PMID:6953413

Horn, A., & Friedrich, H. (2019). Locating the source of large-scale outbreaks of foodborne disease. *Journal of the Royal Society, Interface*, *16*(151), 20180624. doi:10.1098/rsif.2018.0624 PMID:30958197

Howard, A. G., Zhu, M., Chen, B., Kalenichenko, D., Wang, W., Weyand, T., Andreetto, M., & Adam, H. (2017). *Mobilenets: Efficient convolutional neural networks for mobile vision applications.* arXiv preprint arXiv:1704.04861.

How, M.-L., Chan, Y. J., & Cheah, S.-M. (2020). Predictive insights for improving the resilience of global food security using artificial intelligence. *Sustainability*, *12*(15), 6272.

Hughes, D., & Salathé, M. (2015). *An open access repository of images on plant health to enable the development of mobile disease diagnostics through machine learning and crowdsourcing.* ArXiv, abs/1511.08060.

Hu, K., Renly, S., Edlund, S., Davis, M., & Kaufman, J. (2016). Amodeling framework to accelerate food-borne outbreak investigations. *Food Control*, *59*, 53–58. doi:10.1016/j.foodcont.2015.05.017

Humphries, R. M., Abbott, A. N., & Hindler, J. A. (2019). Understanding and addressing CLSI breakpoint Revisions: A primer for clinical laboratories. *Journal of Clinical Microbiology*, *57*(6), e00203–e00219. doi:10.1128/JCM.00203-19 PMID:30971460

Huyghe, E., Verstraeten, J., & Van Kerckhove, A. (2017). Clicks as a healthy alternative to bricks: How online grocery shopping reduces vice purchases. *JMR, Journal of Marketing Research*, *54*(1), 61–74. doi:10.1509/jmr.14.0490

IBM Cloud Education. (n.d.). *What is overfitting?* Retrieved February 20, 2022, from https://www.ibm.com/cloud/learn/overfitting

Imran, A.Z.M., Hakim, M.L., Ahmed, M., Islam, M.T., & Hossain, E. (2021). Design of microstrip patch antenna to deploy unmanned aerial vehicle as U.E. in 5G wireless network. *International Journal of Electrical & Computer Engineering, 11*(5).

International Telecommunication Union (ITU). (2015). *ICT Facts and Figures – the World in 2015.* https://www.itu.int/en/ITU-D/Statistics/Documents/facts/ICTFactsFigures2015.pdf

Iqbal, J., Khan, Z. H., & Khalid, A. (2017). Prospects of robotics in food industry. *Food Science and Technology (Campinas)*, *37*(2), 159–165. doi:10.1590/1678-457x.14616

Jain. (2020). *Artificial Intelligence in Agriculture: Using Modern Day AI to Solve Traditional Farming Problems*. Data Science Bloagthon.

Jain, S. K. (2019). Water resources management in India-challenges and the way forward. *Current Science*, *117*(4), 569–576. doi:10.18520/cs/v117/i4/569-576

Jha, K., Doshi, A., Patel, P., & Shahd, M. (2019). A comprehensive review on automation in agriculture using artificial intelligence. *Artif. Intell. Agric.*, *2*, 1–12. doi:10.1016/j.aiia.2019.05.004

Jhawar, J. (2016). Orange sorting by applying pattern recognition on colour image. *Procedia Computer Science*, *78*, 691–697.

Jia, W., Zhao, R., Yao, N., Fernstrom, J. D., Fernstrom, M. H., Sclabassi, R. J., & Sun, M. (2009, April). A food portion size measurement system for image-based dietary assessment. In *2009 IEEE 35th Annual Northeast Bioengineering Conference* (pp. 1-2). IEEE.

Jiang, H., Zhang, C., Qiao, Y., Zhang, Z., Zhang, W., & Song, C. (2020). CNN feature based graph convolutional network for weed and crop recognition in smart farming. *Computers and Electronics in Agriculture*, *174*, 105450.

Jiang, P., Chen, Y., Liu, B., He, D., & Liang, C. (2019). Real-Time Detection of Apple Leaf Diseases Using Deep Learning Approach Based on Improved Convolutional Neural Networks. *IEEE Access: Practical Innovations, Open Solutions*, *7*, 59069–59080. doi:10.1109/ACCESS.2019.2914929

Jin, X., Che, J., & Chen, Y. (2021). Weed Identification Using Deep Learning and Image Processing in Vegetable Plantation. *IEEE Access: Practical Innovations, Open Solutions*, *9*, 10940–10950.

Jonathan, R. B., & Fisher, E. A.-F. (2017). Impact of satellite imagery spatial resolution on land use classification accuracy and modeled water quality. *Remote Sensing in Ecology and Conservation.*

José Paulo Molin, T. R. (2019). Sensor Systems For Mapping Soil Fertility Attributes: Challenges. *Engenharia Agrícola.*

Jung, C. H., Son, J. W., Kang, S., Kim, W. J., Kim, H. S., Kim, H. S., Seo, M., Shin, H.-J., Lee, S.-S., Jeong, S. J., Cho, Y., Han, S. J., Jang, H. M., Rho, M., Lee, S., Koo, M., Yoo, B., Moon, J.-W., Lee, H. Y., ... Yoon, K. H. (2021). Diabetes fact sheets in Korea, 2020: An appraisal of current status. *Diabetes & Metabolism Journal*, *45*(1), 1–10. doi:10.4093/dmj.2020.0254 PMID:33434426

Kader, A. A. (2008). Flavor quality of fruits and vegetables. *Journal of the Science of Food and Agriculture*, *88*(11), 1863–1868. doi:10.1002/jsfa.3293

Kakamoukas, G. A., Sarigiannidis, P. G., & Economides, A. A. (2020). FANETs in Agriculture-A routing protocol survey. Internet of Things, 100183.

Kakani, V., Nguyen, V. H., Kumar, B. P., Kim, H., & Pasupuleti, V. R. (2020). A critical review on computer vision and artificial intelligence in food industry. *Journal of Agriculture and Food Research*, *2*, 100033.

Kaloxylos, Eigenmann, & Teye, Politopoulou, Wolfert, Shrank, & Dillinger. (2012). Farm management systems and the Future Internet era. *Computers and Electronics in Agriculture*, *89*, 130–144.

Kamilaris, A., & Prenafeta-Boldú, F. X. (2018). Deep learning in agriculture: A survey. *Computers and Electronics in Agriculture*, *147*, 70–90. doi:10.1016/j.compag.2018.02.016

Kamyshova, G., Solovyev, D., Zatinatsky, S., Kolganov, D., & Terekhova, N. (2020). Improving the efficiency of circular irrigation machines based on models of neural network irrigation control. *E3S Web of Conferences*, 175. doi:10.1051/e3sconf/202017505007

Kang, Z. (2020). CagatayCatal, and BedirTekinerdogan. "Machine learning applications in production lines: A systematic literature review. *Computers & Industrial Engineering*, *149*, 106773.

Karlekar, A., & Seal, A. (2020). SoyNet: Soybean leaf diseases classification. *Computers and Electronics in Agriculture*, *172*, 105342. Advance online publication. doi:10.1016/j.compag.2020.105342

Kashyap, K. K., & Agrawal, A. (2018). Fanet: Survey on Design Challenges, Application Scenario and Communication Protocols. *IJRAR-International Journal of Research and Analytical Reviews*, *5*(4), 1024–1032.

Kashyap, P. K., Kumar, S., Jaiswal, A., Prasad, M., & Gandomi, A. H. (2021). Towards Precision Agriculture: IoT-Enabled Intelligent Irrigation Systems Using Deep Learning Neural Network. *IEEE Sensors Journal*, *21*(16), 17479–17491. doi:10.1109/JSEN.2021.3069266

Kate, K., Chaudhari, S., Prapanca, A., & Kalaganam, J. (2014). FoodSIS: a text mining system to improve the state of food safety in singapore. In *KDD '14: Proceedings of the 20th ACM SIGKDD International Conference on Knowledge Discovery and Data Mining*, (pp. 1709–18). New York: Assoc. Comput. Mach. 10.1145/2623330.2623369

Katiyar, S., & Farhana, A. (2021). Smart Agriculture: The Future of Agriculture using AI and IoT. *Journal of Computational Science*, *17*(10), 984–999. doi:10.3844/jcssp.2021.984.999

Katiyar, S., Khan, R., & Kumar, S. (2021). Artificial Bee Colony Algorithm for Fresh Food Distribution without Quality Loss by Delivery Route Optimization. *Journal of Food Quality*, *2021*, e4881289. doi:10.1155/2021/4881289

Katiyar, S., Khan, R., & Kumar, S. (2021). Artificial Bee Colony Algorithm for Fresh Food Distribution without Quality Loss by Delivery Route Optimization. *Journal of Food Quality*.

Kaufman, J., Lessler, J., Harry, A., Edlund, S., Hu, K., Douglas, J., Thoens, C., Appel, B., Käsbohrer, A., & Filter, M. (2014). A likelihood-based approach to identifying contaminated food products using sales data: Performance and challenges. *PLoS Computational Biology*, *10*(7), e1003692. doi:10.1371/journal.pcbi.1003692 PMID:24992565

Kaur, M., Gulati, H. & Kundra, H. (2014). Data Mining in Agriculture on Crop Price Prediction: Techniques and Applications. *International Journal of Computer Applications, 99*(12), 1-3.

Kawano, Y., & Yanai, K. (2015). Foodcam: A real-time food recognition system on a smartphone. *Multimedia Tools and Applications*, *74*(14), 5263–5287.

Kemiklioğlu, E., & Özen, Ö. (2018). Design of a Sensor to Detect Fruit Freshness. *International Journal of Scientific and Technological Research*, *4*(1), 1–6.

Khamayseh, Y., Yassein, M. B., & Abu-Jazoh, M. (2019). Intelligent black hole detection in mobile AdHoc networks. *Iranian Journal of Electrical and Computer Engineering*, *9*(3), 1968. doi:10.11591/ijece.v9i3.pp1968-1977

Khan, A., Ilyas, T., Umraiz, M., Mannan, Z. I., & Kim, H. (2020). Ced-net: Crops and weeds segmentation for smart farming using a small cascaded encoder-decoder architecture. *Electronics (Basel)*, *9*(10), 1602.

Khan, M. A., Khan, I. U., Safi, A., & Quershi, I. M. (2018). Dynamic routing in flying ad-hoc networks using topology-based routing protocols. *Drones (Basel)*, *2*(3), 27. doi:10.3390/drones2030027

Khan, M. A., Safi, A., Qureshi, I. M., & Khan, I. U. (2017). Flying ad-hoc networks (FANETs): A review of communication architectures, and routing protocols. *2017 First International Conference on Latest trends in Electrical Engineering and Computing Technologies (INTELLECT)*, 1-9. 10.1109/INTELLECT.2017.8277614

Khan, M. J., Zia, M. S., & Qasim, M. (2010). Use of pesticides and their role in environmental pollution. *World Academy of Science, Engineering and Technology*, *72*, 122–128.

Khan, R. U., Khan, K., Albattah, W., & Qamar, A. M. (2021). Image-Based Detection of Plant Diseases: From Classical Machine Learning to Deep Learning Journey. *Wireless Communications and Mobile Computing*, *2021*, 1–13. doi:10.1155/2021/5541859

Khan, R., Kumar, S., Dhingra, N., & Bhati, N. (2021). The Use of Different Image Recognition Techniques in Food Safety: A Study. *Journal of Food Quality*, *2021*, 1–10. Advance online publication. doi:10.1155/2021/7223164

Khan, R., Tyagi, N., & Chauhan, N. (2021). Safety of food and food warehouse using VIBHISHAN. *Journal of Food Quality*, *2021*, 1–12. Advance online publication. doi:10.1155/2021/1328332

Khare, V. R., Wang, F. Z., Wu, S., Deng, Y., & Thompson, C. (2008). Ad-hoc network of unmanned aerial vehicle swarms for search & destroy tasks. *Proceedings of the 4th International IEEE Conference Intelligent Systems*. 10.1109/IS.2008.4670440

Khoshnam, F., Namjoo, M., Golbakshi, H., & Dowlati, M. (2015). Physical and Mechanical Changes in Ripening Melon Fruits. *Yuzuncu Yil University Journal of Agricultural Sciences*, *26*(2), 135–144.

Kim, Burks, Qin, & Bulanon. (2009). Classification of grapefruit peel diseases using color texture feature analysis. *International Journal of Agricultural and Biological Engineering*, *2*(3), 41–50.

Kim, J. C., & Chung, K. (2020). Knowledge-based hybrid decision model using neural network for nutrition management. *Information Technology Management*, *21*(1), 29–39.

Kim, J. J., Kim, I., & Hwang, J. (2021). A change of perceived innovativeness for contactless food delivery services using drones after the outbreak of COVID-19. *International Journal of Hospitality Management*, *93*, 102758.

Kim, S., Lee, M., & Shin, C. (2018). IoT-based strawberry disease prediction system for smart farming. *Sensors (Basel)*, *18*(11), 4051. doi:10.339018114051 PMID:30463363

KiranMai, C., Krishna, M., & Reddy, A. V. (2006). Data Mining o f Geospatial Database for Agriculture Related Application. *Proceedings of Map India*, 83-96.

Kohavi, R., & John, G. H. (1997). Wrappers for feature subset selection. *Artificial Intelligence*, *97*(1-2), 273–324. doi:10.1016/S0004-3702(97)00043-X

Koh, H. K. (2010). A 2020 vision for healthy people. *The New England Journal of Medicine*, *362*(18), 1653–1656. doi:10.1056/NEJMp1001601 PMID:20445177

Kohli, P., Kumar, I., & Vimal, V. (2022). Plant Leaf Disease Identification Using Unsupervised Fuzzy C-Means Clustering and Supervised Classifiers. In R. Agrawal, J. He, E. Shubhakar Pilli, & S. Kumar (Eds.), *Cyber Security in Intelligent Computing and Communications* (pp. 281–293). Springer. doi:10.1007/978-981-16-8012-0_22

Kostik, V., Angelovska, B., Petreska, E. K., & Bauer, B. (2014). Determination of pesticide residues in plant-based foods from the Republic of Macedonia. *Journal of Food Nutrition Science*, *2*, 124–129.

Krishna, K. R. (2016). *Push Button Agriculture: Robotics, Drones, Satellite-Guided Soil and Crop Management*. Apple Academic Press.

Krizhevsky, A., Sutskever, I., & Hinton, G. E. (2012). ImageNet classification with deep convolutional neural networks. In Advances in Neural Information Processing Systems 25 (NIPS 2012). Neural Inf. Proc. Syst. Found.

Krumme, C., Llorente, A., Cebrian, M., Pentland, A. S., & Moro, E. (2013). The predictability of consumer visitation patterns. *Journal of the American Medical Association*, *312*, 117–118.

Kukreja, H., Bharath, N., Siddesh, C.S., & Kuldeep, S. (2016). An Introduction to Artificial Neural Network. International journal of advance research and innovative ideas in education, *1*(5), 27-30.

Kulkarni, S., Mandal, S. N., Sharma, G. S., & Mundada, M. R. (2018). Predictive analysis to improve crop yield using a neural network model. In 2018 International Conference on Advances in Computing, Communications and Informatics (ICACCI) (pp. 74-79). IEEE. 10.1109/ICACCI.2018.8554851

Kumar, A. & Kumar, A. (2015). Prediction of production of crops using K-Means and Fuzzy Logic. *International Journal of Computer Science and Mobile Computing, 4*(8), 44-56.

Kumar, A., Kumar, N. & Vats, V. (2018). Efficient crop yield prediction using machine learning algorithms. *International Research Journal of Engineering and Technology, 5*(6), 3151-3159.

Kumar, N. (2018). *Regularized Active Set Least Squares Algorithm for Nonnegative Matrix Factorization in Near Infrared Spectral Data* [PhD thesis]. http://hdl.handle.net/10603/233051

Kumar, T., & Prakash, N. (2020). Adoption of AI in agriculture: The game changer for Indian farmers. *International Conferences ICT, Society, and Human Beings*, 204-208.

Kumar, A. S., Suresh, G., Lekashri, S., Babu, L. G., & Manikandan, R. (2021). Smart Agriculture System with E–Carbage Using Iot. *International Journal of Modern Agriculture*, *10*(1), 928–931.

Kumar, I. (2020). Opportunities of Artificial Intelligence and Machine Learning in the Food Industry. *Journal of Cleaner Production*, 122574.

Kumar, V., Khan, M. R., & Walia, R. K. (2020). Crop Loss Estimations due to Plant-Parasitic Nematodes in Major Crops in India. *National Academy Science Letters*, *43*(5), 409–412. doi:10.100740009-020-00895-2

Kumar, Y., Kaur, K., & Singh, G. (2020). Machine Learning Aspects and its Applications Towards Different Research Areas. *2020 International Conference on Computation, Automation and Knowledge Management (ICCAKM)*, 150-156. 10.1109/ICCAKM46823.2020.9051502

Kurihara, J., Ishida, T., & Takahashi, Y. (2020). *Unmanned aerial vehicle (UAV)-based hyperspectral imaging system for precision agriculture and forest management. In Unmanned Aerial Vehicle: Applications in Agriculture and Environment*. Springer.

Kushwaha, A. K. & Bhattacharya, S. (2015). Crop yield prediction using Agro Algorithm in Hadoop. *International Journal of Computer Science and Information Technology & Security, 5*(2), 271-274.

Kuznetsova, A., Maleva, T., & Soloviev, V. (2020). Using YOLOv3 algorithm with pre- and post-processing for apple detection in fruit-harvesting robot. *Agronomy (Basel)*, *10*(7), 1016. doi:10.3390/agronomy10071016

Kuznetsova, A., Maleva, T., & Soloviev, V. (2021). YOLOv5 versus YOLOv3 for Apple Detection. In A. G. Kravets, A. A. Bolshakov, & M. Shcherbakov (Eds.), *Cyber-Physical Systems: Modelling and Intelligent Control. Studies in Systems, Decision and Control* (Vol. 338). Springer. doi:10.1007/978-3-030-66077-2_28

Lakew, D. S., Sa'ad, U., Dao, N. N., Na, W., & Cho, S. (2020). Routing in Flying Ad Hoc Networks: A Comprehensive Survey. *IEEE Communications Surveys and Tutorials*, *22*(2), 1071–1120. doi:10.1109/COMST.2020.2982452

Lazer, D., Pentland, A., Adamic, L., Aral, S., Barabási, A.-L., Brewer, D., Christakis, N., Contractor, N., Fowler, J., Gutmann, M., Jebara, T., King, G., Macy, M., Roy, D., & Van Alstyne, M. (2009). Computational social science. *Science*, *323*(5915), 721–723. doi:10.1126cience.1167742 PMID:19197046

Le Cun, Y., Bengio, Y., & Hinton, G. (2015). Deep learning. *Nature*, *521*(7553), 436–444. doi:10.1038/nature14539 PMID:26017442

LeCun, Y., Boser, B., Denker, J. S., Henderson, D., Howard, R. E., Hubbard, W., & Jackel, L. D. (1989). Backpropagation applied to handwritten zip code recognition. *Neural Computation*, *1*(4), 541–551. doi:10.1162/neco.1989.1.4.541

Lee, W., Lee, J. Y., Lee, J., Kim, K., Yoo, S., Park, S., & Kim, H. (2018). Ground control system based routing for reliable and efficient multi-drone control system. *Applied Sciences (Basel, Switzerland)*, *8*(11), 2027. doi:10.3390/app8112027

Leitold, D., Vathy-Fogarassy, A., & Abonyi, J. (2018). Network distance-based simulated annealing and fuzzy clustering for sensor placement ensuring observability and minimal relative degree. *Sensors (Basel)*, *18*(9), 3096. doi:10.339018093096 PMID:30223464

LeviR.RenegarN.SpringsS.ZamanT. (2019). Supply chain network analytics guiding food regulatory operational policy. SSRN. doi:10.2139/ssrn.3374620

Levow, G. (2006). Unsupervised and semi-supervised learning of tone and pitch accent. In *Proceedings of the Human Language Technology Conference of the North American Chapter of the Association for Computational Linguistics*, (pp. 224–31). New York: Assoc. Comput. Linguist. 10.3115/1220835.1220864

Li, Y., & Wang, Y. (2018, June). A multi-label image classification algorithm based on attention model. In *2018 IEEE/ACIS 17th International Conference on Computer and Information Science (ICIS)* (pp. 728-731). IEEE.

Liang, W. C., Yang, Y. J., & Chao, C. M. (2019, November). Low-Cost Weed Identification System Using Drones. In *2019 Seventh International Symposium on Computing and Networking Workshops (CANDARW)* (pp. 260-263). IEEE.

Libbrecht, M. W., & Noble, W. S. (2015). Machine learning applications in genetics and genomics. *Nature Reviews. Genetics*, *16*(6), 321–332. doi:10.1038/nrg3920 PMID:25948244

Li, J., Su, Z., Geng, J., & Yin, Y. (2018). Real-time detection of steel strip surface defects based on improved YOLO detection network. *IFAC-PapersOnLine*, *51*(21), 76–81. doi:10.1016/j.ifacol.2018.09.412

Lin, J., Shen, Z., Zhang, A., & Chai, Y. (2018). Blockchain and IoT based food traceability for smart agriculture. *Proceedings of the 3rd International Conference on Crowd Science and Engineering (ICCSE 2018)*, 3. 10.1145/3265689.3265692

Lin, X., Ruess, P., Marston, L., & Konar, M. (2019). Food flows between counties in the United States. *Environmental Research Letters*, *14*(8), 084011. doi:10.1088/1748-9326/ab29ae

Liu, S., Guo, L., Webb, H., Ya, X., & Chang, X. (2019). Internet of Things Monitoring System of Modern Eco-Agriculture Based on Cloud Computing. *IEEE Access: Practical Innovations, Open Solutions*, *7*, 37050–3705. doi:10.1109/ACCESS.2019.2903720

Liu, Z.-L., Nan, F., Zheng, X., Zielinska, M., Duan, X., Deng, L.-Z., Wang, J., Wu, W., Gao, Z.-J., & Xiao, H.-W. (2020). Color prediction of mushroom slices during drying using Bayesian extreme learning machine. *Drying Technology*, *38*(14), 1869–1881. doi:10.1080/07373937.2019.1675077

Li, Y., Ghosh, A., Zhang, Z., Yu, Z., & Wang, J. (2021). Sensors and Applications in Agricultural and Environmental Monitoring. *Journal of Sensors*, *2021*. doi:10.1155/2021/9841470

Li, Y., Metcalf, B. J., Chochua, S., Li, Z., Gertz, R. E. Jr, Walker, H., Hawkins, P. A., Tran, T., Whitney, C. G., McGee, L., & Beall, B. W. (2016). Penicillin-binding protein transpeptidase signatures for tracking and predicting β-lactam resistance levels in Streptococcus pneumoniae. *mBio*, *7*(3), e00756–e16. doi:10.1128/mBio.00756-16 PMID:27302760

López, E. M., García, M., Schuhmacher, M., & Domingo, J. L. (2008). A fuzzy expert system for soil characterization. *Environment International*, *34*(7), 950–958. doi:10.1016/j.envint.2008.02.005 PMID:18378310

Lottes, P., Khanna, R., Pfeifer, J., Siegwart, R., & Stachniss, C. (2017, May). UAV-based crop and weed classification for smart farming. In *2017 IEEE International Conference on Robotics and Automation (ICRA)* (pp. 3024-3031). IEEE.

Luchini, L. C., Peres, T. B., & deAndrea, M. M. (1999). Monitoring of Pesticides Residue in a cotton crop soil. *Journal of Environmental Science and Health*, *35*, 51–59.

Lupolova, N., Dallman, T. J., Holden, N. J., & Gally, D. L. (2017). Patchy promiscuity: Machine learning applied to predict the host specificity of Salmonella enterica and Escherichia coli. *Microbial Genomics*, *3*(10), e000135. doi:10.1099/mgen.0.000135 PMID:29177093

Lupolova, N., Dallman, T. J., Matthews, L., Bono, J. L., & Gally, D. L. (2016). Support vector machine applied to predict the zoonotic potential of E. coli O157 cattle isolates. *Proceedings of the National Academy of Sciences of the United States of America*, *113*(40), 11312–11317. doi:10.1073/pnas.1606567113 PMID:27647883

Lupolova, N., Lycett, S. J., & Gally, D. L. (2019). A guide to machine learning for bacterial host attribution using genome sequence data. *Microbial Genomics*, *5*(12), e000317. doi:10.1099/mgen.0.000317 PMID:31778355

Lu, X., Horn, A., Su, J., & Jiang, J. (2019). A universal measure for network traceability. *Omega*, *87*, 191–204. doi:10.1016/j.omega.2018.09.004

Macesic, N., Polubriaginof, F., & Tatonetti, N. P. (2017). Machine learning: Novel bioinformatics approaches for combating antimicrobial resistance. *Current Opinion in Infectious Diseases*, *30*(6), 511–517. doi:10.1097/QCO.0000000000000406 PMID:28914640

Machleb, J., Peteinatos, G. G., Kollenda, B. L., Andújar, D., & Gerhards, R. (2020). Sensor-based mechanical weed control: Present state and prospects. *Computers and Electronics in Agriculture*, *176*, 105638. doi:10.1016/j.compag.2020.105638

Maes, W. H., & Steppe, K. (2019). Perspectives for remote sensing with unmanned aerial vehicles in precision agriculture. *Trends in Plant Science*, *24*(2), 152–164. doi:10.1016/j.tplants.2018.11.007 PMID:30558964

Magomadov, V. S. (2019). Deep learning and its role in smart agriculture. *Journal of Physics: Conference Series*, *1399*(4), 044109. doi:10.1088/1742-6596/1399/4/044109

Maguire, F., Rehman, M. A., Carrillo, C., Diarra, M. S., & Beiko, R. G. (2019). Identification of primary antimicrobial resistance drivers in agricultural nontyphoidal Salmonella enterica serovars by using machine learning. *mSystems*, *4*(4), e00211–e00219. doi:10.1128/mSystems.00211-19 PMID:31387929

Maharana, A., Cai, K., Hellerstein, J., Hswen, Y., Munsell, M., Staneva, V., Verma, M., Vint, C., Wijaya, D., & Nsoesie, E. O. (2019). Detecting reports of unsafe foods in consumer product reviews. *JAMIA Open*, *2*(3), 330–338. doi:10.1093/jamiaopen/ooz030 PMID:31984365

Mahmud, M. S. A., Abidin, M. S. Z., & Mohamed, Z. (2018). Solving an agricultural robot routing problem with binary particle swarm optimization and a genetic algorithm. *International Journal of Mechanical Engineering and Robotics Research*, *7*(5), 521–527. doi:10.18178/ijmerr.7.5.521-527

Ma, K., Zhang, Y., & Trappe, W. (2005). Mobile network management and robust spatial retreats via network dynamics. *Proceedings of workshop on Resource Provisioning and Management in Sensor Networks (RPMSN05).*

Make sense. (n.d.). *Makesense.Ai*. Retrieved February 20, 2022, from https://www.makesense.ai

Manitz, J., Kneib, T., Schlather, M., Helbing, D., & Brockmann, D. (2014). Origin detection during food-borne disease outbreaks - a case study of the 2011 EHEC/HUS outbreak in Germany. *PLoS Currents*, *6*. Advance online publication. doi:10.1371/currents.outbreaks.f3fdeb08c5b9de7c09ed9cbcef5f01f2 PMID:24818065

Manjula, A., & Narsimha, G. (2016). Crop Yield Prediction with Aid of Optimal Neural Network in Spatial Data Mining: New Approaches. International Journal of Information and Computation Technology, 6(1), 25–33.

Mao, W., Hu, X., & Zhang, X. (2007, August). Weed detection based on the optimized segmentation line of crop and weed. In *International Conference on Computer and Computing Technologies in Agriculture* (pp. 959-967). Springer.

Marambe, B., & Silva, P. (2020). A sixty-day battle to tackle food security-response of the Sri Lankan government to the COVID-19 pandemic. *Sri Lanka Journal of Food and Agriculture*, *6*(1).

Marinello, F., Pezzuolo, A., Chiumenti, A., & Sartori, L. (2016). Technical analysis of unmanned aerial vehicles (drones) for agricultural applications. *Engineering for Rural Development, 15.*

Marvin, H. J., Janssen, E. M., Bouzembrak, Y., Hendriksen, P. J., & Staats, M. (2017). Big data in food safety: An overview. *Critical Reviews in Food Science and Nutrition*, *57*(11), 2286–2295. doi:10.1080/10408398.2016.1257481 PMID:27819478

Mathew, M. P., & Mahesh, T. Y. (2022). Leaf-based disease detection in bell pepper plant using YOLO v5. *SIViP*, *16*(3), 841–847. doi:10.100711760-021-02024-y

Mavani, N. R. (2021). Application of Artificial Intelligence in Food Industry—A Guideline. *Food Engineering Reviews*, 1–42.

McAllister, P., Zheng, H., Bond, R., & Moorhead, A. (2018). Combining deep residual neural network features with supervised machine learning algorithms to classify diverse food image datasets. *Computers in Biology and Medicine*, *95*, 217–233.

McDermott, P. F., Tyson, G. H., Kabera, C., Chen, Y., Li, C., Folster, J. P., Ayers, S. L., Lam, C., Tate, H. P., & Zhao, S. (2016). Whole-genome sequencing for detecting antimicrobial resistance in nontyphoidal Salmonella. *Antimicrobial Agents and Chemotherapy*, *60*(9), 5515–5520. doi:10.1128/AAC.01030-16 PMID:27381390

McNabb, D. E. (2019). Pathways to Sustainable Agriculture. Global Pathways to Water Sustainability, 185-199. doi:10.1007/978-3-030-04085-7_12

Mehdizadeh, S., Behmanesh, J., & Khalili, K. (2017). Using MARS, SVM, GEP and empirical equations for estimation of monthly mean reference evapotranspiration. *Computers and Electronics in Agriculture*, *139*, 103–114. doi:10.1016/j.compag.2017.05.002

Melander, B., Lattanzi, B., & Pannacci, E. (2015). Intelligent versus non-intelligent mechanical intra-row weed control in transplanted onion and cabbage. *Crop Protection (Guildford, Surrey)*, *72*, 1–8. doi:10.1016/j.cropro.2015.02.017

Mendes-Soares, H., Raveh-Sadka, T., Azulay, S., Ben-Shlomo, Y., Cohen, Y., Ofek, T., ... Nelson, H. (2019). Model of personalized postprandial glycemic response to food developed for an Israeli cohort predicts responses in Midwestern American individuals. *The American Journal of Clinical Nutrition*, *110*(1), 63–75.

Mendoza, F., Dejmek, P., & Aguilera, J. M. (2010). Gloss measurements of raw agricultural products using image analysis. *Food Research International (Ottawa, Ont.)*, *43*(1), 18–25. doi:10.1016/j.foodres.2009.08.003

Mezgec, S., & Koroušić Seljak, B. (2017). NutriNet: A deep learning food and drink image recognition system for dietary assessment. *Nutrients*, *9*(7), 657.

Miao, F., Zheng, S., & Tao, B. (2019, January). Crop Weed Identification System Based on Convolutional Neural Network. In *2019 IEEE 2nd International Conference on Electronic Information and Communication Technology (ICEICT)* (pp. 595-598). IEEE.

Milczarski, P. (2020). Machine Learning Application in Energy Consumption Calculation and Assessment in Food Processing Industry. In *International Conference on Artificial Intelligence and Soft Computing* (pp. 369-379). Springer.

Millan Escriva, D., & Laganiere, R. (2019). *OpenCV 4 Computer Vision Application Programming Cookbook: Build complex computer vision applications with OpenCV and C* (4th ed.). Packt Publishing.

Millar, K. D. L., Ebbs, S. D., Gibson, D. J., Wood, A. J., & Young, B. G. (2011). Evaluation of physiological parameters for the prediction of seed yield and quality for soybean (Glycine max) plants grown in the presence of weed competition. *Plant Biosystems*, *145*(1), 1–11. doi:10.1080/11263504.2010.514129

Millar, K., Gibson, D. J., Young, B. G., & Wood, A. J. (2007). Impact of interspecific competition on seed development and quality of five soybean cultivars. *Australian Journal of Experimental Agriculture*, *47*(12), 1455–1459. doi:10.1071/EA07034

Minaei-Bidgoli, B., Parvin, H., Alinejad-Rokny, H., Alizadeh, H., & Punch, W. F. (2014). Effects of resampling method and adaptation on clustering ensemble efficacy. *Artificial Intelligence Review*, *41*(1), 27–48. doi:10.100710462-011-9295-x

Min, W., Jiang, S., Liu, L., Rui, Y., & Jain, R. (2019). A survey on food computing. *ACM Computing Surveys*, *52*(5), 1–36. doi:10.1145/3329168

Mishra, S., Mishra, D., & Santra, G. H. (2016). Applications of Machine Learning Techniques in Agricultural Crop Production: A Review. *Indian Journal of Science and Technology*, *9*(38). Advance online publication. doi:10.17485/ijst/2016/v9i38/95032

Misra, N. N., Dixit, Y., Al-Mallahi, A., Bhullar, M. S., Upadhyay, R., & Martynenko, A. (2020). IoT, big data and artificial intelligence in agriculture and food industry. *IEEE Internet of Things Journal*, *1*, 1.

Misra, N. N., Dixit, Y., Al-Mallahi, A., Bhullar, M. S., Upadhyay, R., & Martynenko, A. (2020). IoT, big data and artificial intelligence in agriculture and food industry. *IEEE Internet of Things Journal*. Advance online publication. doi:10.1109/JIOT.2020.2998584

Mitcham, B., Cantwell, M., & Kader, A. (1996). Methods for Determining Quality of Fresh Commodities. *Perishables Handling Newsletter*, 1–5.

Moazzam, S. I., Khan, U. S., Tiwana, M. I., Iqbal, J., Qureshi, W. S., & Shah, S. I. (2019, October). A Review of Application of Deep Learning for Weeds and Crops Classification in Agriculture. In *2019 International Conference on Robotics and Automation in Industry (ICRAI)* (pp. 1-6). IEEE.

Mohanty, S. P., Hughes, D. P., & Salathé, M. (2016). Using Deep Learning for Image-Based Plant Disease Detection. *Frontiers in Plant Science*, *7*, 1419. doi:10.3389/fpls.2016.01419 PMID:27713752

Mohiuddin, M. A., Khan, S. A., & Engelbrecht, A. P. (2014). Simulated evolution and simulated annealing algorithms for solving multi-objective open shortest path first weight setting problem. *Applied Intelligence*, *41*(2), 348–365. doi:10.100710489-014-0523-3

Møller, F. T., Mølbak, K., & Ethelberg, S. (2018). Analysis of consumer food purchase data used for outbreak investigations, a review. *Eurosurveillance*, *23*(24), 1700503. doi:10.2807/1560-7917.ES.2018.23.24.1700503 PMID:29921346

Morales, I. R., Cebrián, D. R., Blanco, E. F., & Sierra, A. P. (2016). Early warning in egg production curves from commercial hens: A SVM approach. *Computers and Electronics in Agriculture*, *121*, 169–179. doi:10.1016/j.compag.2015.12.009

Moriguchi, K. (2020). Acceleration and enhancement of reliability of simulated annealing for optimizing thinning schedule of a forest stand. *Computers and Electronics in Agriculture*, *177*, 105691. doi:10.1016/j.compag.2020.105691

Morvan, X., Saby, N. P. A., Arrouays, D., Le Bas, C., Jones, R. J. A., Verheijen, F. G. A., Bellamy, P. H., Stephens, M., & Kibblewhite, M. G. (2008). Soil monitoring in Europe: A review of existing systems and requirements for harmonisation. *The Science of the Total Environment*, *391*(1), 1–12. doi:10.1016/j.scitotenv.2007.10.046 PMID:18063012

Mounashree, J. P., Sanjay, N., Sushmitha, B. S., Usha, B. G., & Shivamurthy, M. A. (2021). Automated weed removal system using convolutional neural network. *International Journal of Engineering Applied Sciences and Technology.*, *6*(2), 231–235.

Mukherjee, D., Chakraborty, S., Sarkar, I., Ghosh, A., & Roy, S. (n.d.). A Detailed Study on Data Centre Energy Efficiency and Efficient Cooling Techniques. *International Journal of Advanced Trends in Computer Science and Engineering, 9*(5), 9221-9242.

Munck, N., Leekitcharoenphon, P., Litrup, E., Kaas, R., Meinen, A., Guillier, L., Tang, Y., Malorny, B., Palma, F., Borowiak, M., Gourmelon, M., Simon, S., Banerji, S., Petrovska, L., Dallman, T. J., & Hald, T. (2020a). Four European Salmonella Typhimurium datasets collected to develop WGS-based source attribution methods. *Scientific Data*, *7*(1), 75. doi:10.103841597-020-0417-7 PMID:32127544

Munck, N., Njage, P. M. K., Leekitcharoenphon, P., Litrup, E., & Hald, T. (2020b). Application of whole-genome sequences and machine learning in source attribution of Salmonella Typhimurium. *Risk Analysis*, *40*(9), 1693–1705. doi:10.1111/risa.13510 PMID:32515055

Muruganandam, K., & Chauhan, U. (2021). Soil Moisture Sensor Nodes in IoT-Based Drip Irrigation System for Water Conservation. In Smart Sensors for Industrial Internet of Things. Internet of Things. Springer. doi:10.1007/978-3-030-52624-5_13

Muthukumaran, S., Ganesan, M., Dhanasekar, J., & Loganathan, G. B. (2021). Path Planning Optimization for Agricultural Spraying Robots Using Hybrid Dragonfly – Cuckoo Search Algorithm. *Journal of Agriculture Science*, *36*(1).

N, T. R., & Gupta, R. (2020). A Survey on Machine Learning Approaches and Its Techniques. *2020 IEEE International Students' Conference on Electrical, Electronics and Computer Science (SCEECS)*, 1–6. doi:10.1109/SCEECS48394.2020.190

Naik, S., & Patel, B. (2017). Thermal imaging with fuzzy classifier for maturity and size based non-destructive mango (MangiferaIndica L.) grading. In *2017 International Conference on Emerging Trends & Innovation in ICT (ICEI)* (pp. 15-20). IEEE.

Nair, D., Parte, A., Pokharkar, Y., & Pande, N. (2017). Recognition of Silverleaf Whitefly and Western Flower Thrips Via Image Processing and Artificial Neural Network. International Research Journal of Engineering and Technology, *4*(3), 1773–1776.

Nandi, C. S. (2016). A machine vision technique for grading of harvested mangoes based on maturity and quality. *IEEE Sensors Journal*, *16*(16), 6387–6396.

Naresh, M., & Munaswamy, P. (2019). Smart Agriculture System using IoT Technology. *International Journal of Recent Technology and Engineering, 7*(5).

Narvekar, M., & Fargose, P. (2015). Daily weather forecasting using artificial neural network. *International Journal of Computers and Applications*, *121*(22), 9–13. doi:10.5120/21830-5088

Nasierding, G., & Sajjanhar, A. (2013, December). Multi-label classification with clustering for image and text categorization. In *2013 6th International Congress on Image and Signal Processing (CISP)* (Vol. 2, pp. 869-874). IEEE. 10.1109/CISP.2013.6745287

Nepal, U., & Eslamiat, H. (2022). Comparing YOLOv3, YOLOv4 and YOLOv5 for Autonomous Landing Spot Detection in Faulty UAVs. *Sensors (Basel)*, *22*(2), 464. doi:10.339022020464 PMID:35062425

Nestle, M., Wing, R., Birch, L., DiSogra, L., Drewnowski, A., Middleton, S., & Economos, C. (1998). *Behavioral and social influences on food choice*. Academic Press.

Neto, O. C., Teixeira, A. S., Braga, A. P., Santos, C. C., & Leao, R. A. (2015). Application of artificial neural networks as an alternative to volumetric water balance in drip irrigation management in watermelon crop. Journal of the Brazilian Association of Agricultural Engineering, *35*(2), 266–279. doi:10.1590/1809-4430-Eng.Agric.v35n2p266-279/2015

Ngo, H.C., Hashim, U.R., Sek, Y.W., Kumar, Y.J., & Ke, W.S. (2019). Weeds Detection in Agricultural Fields using Convolutional Neural Network. *International Journal of Innovative Technology and Exploring Engineering*.

Nguyen, M., Brettin, T., Long, S. W., Musser, J. M., Olsen, R. J., Olson, R., Shukla, M., Stevens, R. L., Xia, F., Yoo, H., & Davis, J. J. (2018). Developing an in silico minimum inhibitory concentration panel test for Klebsiella pneumoniae. *Scientific Reports*, *8*(1), 421. doi:10.103841598-017-18972-w PMID:29323230

Nguyen, M., Long, S. W., McDermott, P. F., Olsen, R. J., Olson, R., Stevens, R. L., Tyson, G. H., Zhao, S., & Davis, J. J. (2019). Using machine learning to predict antimicrobial MICs and associated genomic features for nontyphoidal Salmonella. *Journal of Clinical Microbiology*, *57*(2), e01260–e18. doi:10.1128/JCM.01260-18 PMID:30333126

Niedbała, G. (2019). Application of Artificial Neural Networks for Multi-Criteria Yield Prediction of Winter Rapeseed. *Sustainability*, *11*(2), 533. doi:10.3390u11020533

Niehaus, D. (2016). *Tree Boosting with XGBoost—Why Does XGBoost Win "Every" Machine Learning Competition?* Nor. Univ. Sci. Technol.

Niehaus, K. E., Walker, T. M., Crook, D. W., Peto, T. E. A., & Clifton, D. A. (2014). Machine learning for the prediction of antibacterial susceptibility in Mycobacterium tuberculosis. In *IEEE-EMBS International Conference on Biomedical Health and Informatics* (pp. 618–21). Piscataway, NJ: IEEE. 10.1109/BHI.2014.6864440

Nisbet, R., Miner, G., & Elder, J. (2009). *Handbook of Statistical Analysis and Data Mining Applications*. Academic.

Nooraiyeen, A. (2020, July). Robotic Vehicle for Automated Detection of Leaf Diseases. *2020 IEEE International Conference on Electronics, Computing and Communication Technologies (CONECCT)*. 10.1109/CONECCT50063.2020.9198326

Norström, M., Kristoffersen, A. B., Görlach, F. S., Nygård, K., & Hopp, P. (2015). An adjusted likelihood ratio approach analysing distribution of food products to assist the investigation of foodborne outbreaks. *PLoS One*, *10*(8), e0134344. doi:10.1371/journal.pone.0134344 PMID:26237468

Noubir & Lin. (n.d.). Low-power DoS attacks in data wireless lans and countermeasures. *SIGMOBILE Mob. Comput.*

Nsoesie, E. O., Kluberg, S. A., & Brownstein, J. S. (2014). Online reports of foodborne illness capture foods implicated in official foodborne outbreak reports. *Preventive Medicine*, *67*, 264–269. doi:10.1016/j.ypmed.2014.08.003 PMID:25124281

Nunes, M. C. N., Emond, J. P., Rauth, M., Dea, S., & Chau, K. V. (2009). Environmental conditions encountered during typical consumer retail display affect fruit and vegetable quality and waste. *Postharvest Biology and Technology*, *51*(2), 232–241. doi:10.1016/j.postharvbio.2008.07.016

Obopile, M., Munthali, D., & Matilo, B. (2008). Farmers' knowledge, perceptions and management of vegetable pests and diseases in Botswana. *Crop Protection (Guildford, Surrey)*, *27*(8), 1220–1224. doi:10.1016/j.cropro.2008.03.003

Oldroyd, R. A., Morris, M. A., & Birkin, M. (2018). Identifying methods for monitoring foodborne illness: Review of existing public health surveillance techniques. *JMIR Public Health and Surveillance*, *4*(2), e57. doi:10.2196/publichealth.8218 PMID:29875090

Oliver, N., Lepri, B., Sterly, H., Lambiotte, R., Deletaille, S., De Nadai, M., Letouzé, E., Salah, A. A., Benjamins, R., Cattuto, C., Colizza, V., de Cordes, N., Fraiberger, S. P., Koebe, T., Lehmann, S., Murillo, J., Pentland, A., Pham, P. N., Pivetta, F., ... Vinck, P. (2020). Mobile phone data for informing public health actions across the COVID-19 pandemic life cycle. *Science Advances*, *6*(23), eabc0764. doi:10.1126ciadv.abc0764 PMID:32548274

Oluwole, O., & Cheke, R. A. (2009). Health and environmental impacts of pesticide use practices: A case study of farmers in Ekiti State, Nigeria. *International Journal of Agricultural Sustainability*, *7*(3), 153–163. doi:10.3763/ijas.2009.0431

Onishchuk, M. O. (2020). *Opto-mechanical sorting of municipal solid waste* (PhD dissertation). BHTY.

Ordovas, J. M., Ferguson, L. R., Tai, E. S., & Mathers, J. C. (2018). Personalised nutrition and health. *BMJ (Clinical Research Ed.)*, 361.

Osorio, K., Puerto, A., Pedraza, C., Jamaica, D., & Rodríguez, L. (2020). A Deep Learning Approach for Weed Detection in Lettuce Crops Using Multispectral Images. *AgriEngineering*, *2*(3), 471–488.

Oubbati, O. S., Atiquzzaman, M., Lorenz, P., Tareque, M. H., & Hossain, M. S. (2019). Routing in flying ad hoc networks: Survey, constraints, and future challenge perspectives. *IEEE Access: Practical Innovations, Open Solutions*, *7*, 81057–81105. doi:10.1109/ACCESS.2019.2923840

Palanisami, K., Kakumanu, K.R., Nagothu, U.S., & Ranganathan, C.R. (2019). Climate Change and Agriculture in India. *Climate Change and Future Rice Production in India*, 1–6. doi:10.1007/978-981-13-8363-2_1

Palti, J., & Cohen, Y. (1980). Downy mildew of Cucurbits (Pseudoperonospora Cubensis): The Fungus and its hosts, distribution, epidemiology and control. *Phytoparasitica*, *8*(2), 109–147. doi:10.1007/BF02994506

Pang, H., Mokhtari, A., Chen, Y., Oryang, D., Ingram, D. T., Sharma, M., Millner, P. D., & Van Doren, J. M. (2020). A predictive model for survival of Escherichia coli O157:H7 and generic E. coli in soil amended with untreated animal manure. *Risk Analysis*, *40*(7), 1367–1382. doi:10.1111/risa.13491 PMID:32378782

Panpatte, D. G. (2018). *Artificial Intelligence in Agriculture: An Emerging Era of Research. Intuitional Science.*

Pan, S. J., & Yang, Q. (2010). A survey on transfer learning. *IEEE Transactions on Knowledge and Data Engineering*, *22*(10), 1345–1359. doi:10.1109/TKDE.2009.191

Pan, Y. (2016). Heading toward artificial intelligence 2.0. *Engineering*, *2*(4), 409–413. doi:10.1016/J.ENG.2016.04.018

Parvin, H., Alinejad-Rokny, H., Minaei-Bidgoli, B., & Parvin, S. (2013). A new classifier ensemble methodology based on subspace learning. *Journal of Experimental & Theoretical Artificial Intelligence*, *25*(2), 227–250. doi:10.1080/0952813X.2012.715683

Parvin, H., MirnabiBaboli, M., & Alinejad-Rokny, H. (2015). Proposing a classifier ensemble framework based on classifier selection and decision tree. *Engineering Applications of Artificial Intelligence*, *37*, 34–42. doi:10.1016/j.engappai.2014.08.005

Patel, G. S., Rai, A., Das, N. N., & Singh, R. P. (Eds.). (2021). *Smart Agriculture: Emerging Pedagogies of Deep Learning, Machine Learning and Internet of Things* (1st ed.). CRC Press. doi:10.1201/b22627

Pathan, M., Patel, N., Yagnik, H., & Shah, M. (2020). Artificial Cognition for Applications in Smart Agriculture: A Comprehensive Review. *Artif. Intell. Agric.*, *4*, 81–95. doi:10.1016/j.aiia.2020.06.001

Patidar, S., Singh, U., & Sharma, S. K. (2020). Weed Seedling Detection Using Mask Regional Convolutional Neural Network. *International Conference on Electronics and Sustainable Communication Systems (ICESC)*, 311-316. 10.1109/ICESC48915.2020.9155701

Patil, G. G., & Banyal, R. K. (2019). Techniques of Deep Learning for Image Recognition. *2019 IEEE 5th International Conference for Convergence in Technology (I2CT)*, 1–5. 10.1109/I2CT45611.2019.9033628

Paul, M., Vishwakarma, S. K., & Verma, A. (2015). Analysis of Soil Behaviour and Prediction of Crop Yield using Data Mining approach. *International Conference on Computational Intelligence and Communication Networks*. 10.1109/CICN.2015.156

Pauly, D. (1986). *A simple method for estimating the food consumption of fish populations from growth data and food conversion experiments*. Academic Press.

Pawar, P., Turkar, V., & Patil, P. (2016). Cucumber disease detection using artificial neural network. *International Conference on Inventive Computation Technologies (ICICT)*, 1-5. 10.1109/INVENTIVE.2016.7830151

Pegorini, V., Zen Karam, L., Pitta, C. S. R., Cardoso, R., Da Silva, J. C. C., Kalinowski, H. J., Ribeiro, R., Bertotti, F. L., & Assmann, T. S. (2015). In Vivo Pattern Classification of Ingestive Behavior in Ruminants Using FBG Sensors and Machine Learning. *Sensors (Basel)*, *15*(11), 28456–28471. doi:10.3390151128456 PMID:26569250

PehlivanlıR.KolukırıkS.GörgülüE. (2021). Global Transformation and Differentiation III. figshare.

Péneau, S. (2005). *Freshness of Fruits and Vegetables: Concept and Perception*. Swiss Federal Institute of Technology Zurich.

People, H. (2010). *Healthy People 2020 Food Safety Objectives*. HHS. https://www. healthypeople.gov/2020/topics-objectives/topic/food-safety/objectives

Pereira, L. F. S. (2018). Predicting the ripening of papaya fruit with digital imaging and random forests. *Computers and Electronics in Agriculture*, *145*, 76–82.

Perez-Martinez, P., Garcia-Rios, A., Delgado-Lista, J., Perez-Jimenez, F., & Lopez-Miranda, J. (2012). Metabolic syndrome: Evidences for a personalized nutrition. *Molecular Nutrition & Food Research*, *56*(1), 67–76.

Pérez-Ortiz, M., Gutiérrez, P. A., Peña, J. M., Torres-Sánchez, J., López-Granados, F., & Hervás-Martínez, C. (2016, December). Machine learning paradigms for weed mapping via unmanned aerial vehicles. In 2016 IEEE symposium series on computational intelligence (SSCI) (pp. 1-8). IEEE. doi:10.1109/SSCI.2016.7849987

Pesesky, M. W., Hussain, T., Wallace, M., Patel, S., Andleeb, S., Burnham, C.-A. D., & Dantas, G. (2016). Evaluation ofmachine learning and rulesbased approaches for predicting antimicrobial resistance profiles in gram-negative bacilli from whole genome sequence data. *Frontiers in Microbiology*, *7*, 1887. doi:10.3389/fmicb.2016.01887 PMID:27965630

Phillips, L. (2006). Food and globalization. *Annual Review of Anthropology*, *35*(1), 37–57. doi:10.1146/annurev.anthro.35.081705.123214

Pinheiro, V., Pontes, R., & Furtado, V. (2015). A #hashtagtokenizer for Social Media Messages. *Int. J. Comput. Linguistics Appl.*, *6*, 141–158.

Ponnusamy, P., Roshan Ghias, A., Guo, C., & Sarikaya, R. (2020). Feedback-based self-learning in large-scale conversational ai agents. *Proceedings of the AAAI Conference on Artificial Intelligence*, 34(8), 13180–13187.

Popp, J., Pető, K., & Nagy, J. (2013). Pesticide productivity and food security. A review. *Agronomy for Sustainable Development*, *33*(1), 243–255. doi:10.100713593-012-0105-x

Pouladzadeh, P., Kuhad, P., Peddi, S. V. B., Yassine, A., & Shirmohammadi, S. (2016, May). Food calorie measurement using deep learning neural network. In *2016 IEEE international instrumentation and measurement technology conference proceedings* (pp. 1-6). IEEE.

Pouladzadeh, P., Yassine, A., & Shirmohammadi, S. (2015, September). Foodd: food detection dataset for calorie measurement using food images. In *International Conference on Image Analysis and Processing* (pp. 441-448). Springer.

Prajwala, T. M., Pranathi, A., Ashritha, K. S., Nagaratna, B. C., & Shasidhar, G. K. (2018). Tomato leaf disease detection using convolutional neural networks. In 2018 eleventh international conference on contemporary computing (IC3), 1-5. 10.1109/IC3.2018.8530532

Primicerio, J., Di Gennaro, S. F., Fiorillo, E., Genesio, L., Lugato, E., Matese, A., & Vaccari, F. P. (2012). A flexible unmanned aerial vehicle for precision agriculture. *Precision Agriculture*, *13*(4), 517–523. doi:10.100711119-012-9257-6

Priya, P., Muthaiah, U. & Balamurugan, M. (2018). Predicting yield of the crop using machine learning Algorithm. *International Journal of Engineering Sciences and Research Technology, 7*(4), 2277-2284.

Proakis, J. (2000). *Digital Communications*. McGraw-Hill.

Psirofonia, P., Samaritakis, V., Eliopoulos, P., & Potamitis, I. (2017). Use of unmanned aerial vehicles for agricultural applications with emphasis on crop protection: Three novel case-studies. *International Journal of Agricultural Science and Technology*, *5*(1), 30–39. doi:10.12783/ijast.2017.0501.03

Pujari, Devdas, Yakkundimath, & Byadgi. (2013). Grading and classification of anthracnose fungal disease of fruits based on statistical texture features. *International Journal of Advanced Science and Technology*, *52*(1), 121–132.

Pydipati, R., Burks, T. F., & Lee, W. S. (2006). Identification of citrus disease using color texture features and discriminant analysis. *Computers and Electronics in Agriculture*, *52*(1-2), 49–59.

Qin, J., Burks, T. F., Zhao, X., Niphadkar, N., & Ritenour, M. A. (2011). Multispectral detection of citrus canker using hyperspectral band selection. *Transactions of the ASABE*, *54*(6), 2331–2341. doi:10.13031/2013.40643

Quade, P., & Nsoesie, E. O. (2017). A platform for crowdsourced foodborne illness surveillance: Description of users and reports. *JMIR Public Health and Surveillance*, *3*(3), e42. doi:10.2196/publichealth.7076 PMID:28679492

Qureshi, K. N., Bashir, M. U., Lloret, J., & Leon, A. (2020). Optimized cluster-based dynamic energy-aware routing protocol for wireless sensor networks in agriculture precision. *Journal of Sensors*, *2020*, 2020. doi:10.1155/2020/9040395

Radke, A. M. (2013). A risk management-based evaluation of inventory allocations for make-to-order production. *CIRP Annals*, *62*(1), 459–462.

Ramakrishnan, M., & Nisha, S. A. A. (2015). Groundnut leaf disease detection and classification by using back propagation algorithm. In 2015 international conference on communications and signal processing (ICCSP), 964-968. 10.1109/ICCSP.2015.7322641

Ramesh, R., & Vardhan, B. (2015). Analysis of crop yield prediction using data mining techniques. *International Journal of Research in Engineering and Technology, 4*(1), 47-473.

Ramesh, S., & Hebbar, R. (2018, April). Plant Disease Detection Using Machine Learning. *International Conference on Design Innovations for 3Cs Compute Communicate Control (ICDI3C)*. 10.1109/ICDI3C.2018.00017

Ramjan, M. D., & Ansari, M. T. (2018). Factors affecting of fruits, vegetables and its quality. *Journal of Medicinal Plants Studies*, *6*(6), 16–18.

Ramos, P. J., Prieto, F. A., Montoya, E. C., & Oliveros, C. E. (2017). Automatic fruit count on coffee branches using computer vision. *Computers and Electronics in Agriculture*, *137*, 9–22. doi:10.1016/j.compag.2017.03.010

Rary, E., Anderson, S. M., Philbrick, B. D., Suresh, T., & Burton, J. (2020). Smart sanitation-biosensors as a public health tool in sanitation infrastructure. *International Journal of Environmental Research and Public Health*, *17*(14), 5146. doi:10.3390/ijerph17145146 PMID:32708793

Raut, R., Varma, H., Mulla, C., & Pawar, V. (2018). Soil Monitoring, Fertigation, and Irrigation System Using IoT for Agricultural Application. In Lecture Notes in Networks and Systems (pp. 67–73). doi:10.1007/978-981-10-5523-2_7

RavesaAkhter, S. A. (2021). Precision agriculture using IoT data analytics and machine learning. *Journal of King Saud University - Computer and Information Sciences*.

Razavian, A. S., Azizpour, H., Sullivan, J., & Carlsson, S. (2014). *CNN features off-the-shelf: an astounding baseline for recognition. In IEEE Conference on Computer Vision and Pattern Recognition Workshops*. CVPRW.

Real, J. L., Toril, J. U., Arriaza, J. A., & Valenciano, J. P. (2020). A Look at the Past, Present and Future Research Trends of Artificial Intelligence in Agriculture. *Agronomy (Basel)*, *10*(11), 1839. doi:10.3390/agronomy10111839

Rebetez, J., Satizábal, H., Mota, M., Noll, D., Büchi, L., Wendling, M., Cannelle, B., Pérez-Uribe, A., & Burgos, S. (2016). *Augmenting a convolutional neural network with local histograms - A case study in crop classification from high-resolution UAV imagery*. ESANN.

Redmon, J. (2015). *YOLO: Real-time object detection*. Retrieved February 20, 2022, from https://pjreddie.com/darknet/yolo/

Redmon, J. (2016). *Darknet: Convolutional neural networks*. Academic Press.

Redmon, J., & Farhadi, A. (2018). *YOLOv3: An Incremental Improvement*. https://arxiv.org/abs/1804.02767

Ren, S., He, K., Girshick, R., & Sun, J. (2015). *Faster r-cnn: Towards real-time object detection with region proposal networks*. arXiv preprint arXiv:1506.01497.

Rettore de Araujo Zanella, A., da Silva, E., & Carlos Pessoa Albini, L. (2020). Security challenges to smart agriculture: Current state, key issues, and future directions. *Array*, *8*, 100048. doi:10.1016/j.array.2020.100048

Riely, F., Mock, N., Cogill, B., Bailey, L., & Kenefick, E. (1999). *Food Security Indicators and Framework for Use in the Monitoring and Evaluation of Food Aid Programs, Nutrition Technical Assistance Project*. FANTA.

Rocha, Hauagge, & Wainer, & Goldenstein. (2010). Automatic fruit and vegetable classification from images. *Computers and Electronics in Agriculture*, *70* (1), 96-104.

Rockstrom, Williams, & Daily. (2017). Sustainable intensification of agriculture for human prosperity and global sustainability. *Ambio*, *46*(1), 4–17.

Rokach, L. (2010). Ensemble-based classifiers. *Artificial Intelligence Review*, *33*(1-2), 1–39. doi:10.100710462-009-9124-7

Rokunuzzaman, M., & Jayasuriya, H. P. W. (2013). Development of a low cost machine vision system for sorting of tomatoes. *Agricultural Engineering International: CIGR Journal*, *15*(1).

Rong, D., Rao, X., & Ying, Y. (2017). Computer vision detection of surface defect on oranges by means of a sliding comparison window local segmentation algorithm. *Computers and Electronics in Agriculture*, *137*, 59–68.

Rortais, A., Belyaeva, J., Gemo, M., van der Groot, E., & Linge, J. P. (2010). MedISys: An early-warning system for the detection of (re-)emerging food- and feed-borne hazards. *Food Research International*, *43*(5), 1553–1556. doi:10.1016/j.foodres.2010.04.009

Rowan, N. J., & Galanakis, C. M. (2020). Unlocking challenges and opportunities presented by COVID-19 pandemic for cross-cutting disruption in agri-food and green deal innovations: Quo Vadis? *The Science of the Total Environment*, 141362.

Rub, G., Kruse, R., Schneider, M., & Wagner, P. (2008). Estimation of neural network parameters for wheat yield prediction. In M. Bramer (Ed.), *Artificial Intelligence in Theory and Practice II, IFIP International Federation for Information Processing. Springer.*

Rudin, C. (2019). Stop explaining black box machine learning models for high stakes decisions and use interpretable models instead. *Nature Machine Intelligence*, *1*(5), 206–215. doi:10.103842256-019-0048-x PMID:35603010

Russakovsky, O., Deng, J., Su, H., Krause, J., Satheesh, S., Ma, S., Huang, Z., Karpathy, A., Khosla, A., Bernstein, M., Berg, A. C., & Fei-Fei, L. (2015). ImageNet Large Scale Visual Recognition Challenge. *International Journal of Computer Vision*, *115*(3), 211–252. doi:10.100711263-015-0816-y

Sabrol, H., & Kumar, S. (2016). Fuzzy and neural network based tomato plant disease classification using natural outdoor images. *Indian Journal of Science and Technology*, *9*(44), 1–8. doi:10.17485/ijst/2016/v9i44/92825

Sadiq, B. O., & Salawudeen, A. T. (2020). FANET optimization: A destination path flow model. *Iranian Journal of Electrical and Computer Engineering*, *10*(4), 4381–4389. doi:10.11591/ijece.v10i4.pp4381-4389

Sahingoz, O. (2014). Networking models in flying ad-hoc networks (FANETs): Concepts and challenges. *Journal of Intelligent & Robotic Systems*, *74*(1-2), 513–527. Advance online publication. doi:10.100710846-013-9959-7

Sahni, V., Srivastava, S., & Khan, R. (2021). Modelling techniques to improve the quality of food using artificial intelligence, Journal of Food Quality. *Journal of Food Quality*, *2021*, 1–10. Advance online publication. doi:10.1155/2021/2140010

Sahni, V., Srivastava, S., & Khan, R. (2021). Modelling Techniques to Improve the Quality of Food using Artificial Intelligence. *Journal of Food Quality*.

Saiz-Rubio, V., & Rovira-Más, F. (2020). From Smart Farming towards Agriculture 5.0: A Review on Crop Data Management. *Agronomy (Basel)*, *10*(2), 207. doi:10.3390/agronomy10020207

Sak, J., & Suchodolska, M. (2021). Artificial intelligence in nutrients science research: A review. *Nutrients*, *13*(2), 322. doi:10.3390/nu13020322 PMID:33499405

Salamí, E., Barrado, C., & Pastor, E. (2014). UAV flight experiments applied to the remote sensing of vegetated areas. *Remote Sensing*, *6*(11), 11051–11081. doi:10.3390/rs61111051

Saleem, S., & Rana, R. (2021). Ultrasonic Locust Repellent Circuit. *2021 International Conference on Computational Intelligence and Knowledge Economy (ICCIKE)*, 547-549. 10.1109/ICCIKE51210.2021.9410712

Salgotra, A. K., Manha, A. S., & Singh, P. D. (2018). Agricultural growth and productivity in India. *Internat. J. Appl. Soc. Sci.*, *5*, 192–199.

Samada, L. H., & Tambunan, U. S. F. (2020). Biopesticides as Promising Alternatives to Chemical Pesticides: A Review of Their Current and Future Status. *The Journal of Biological Sciences*, *20*(2), 66–76. doi:10.3844/ojbsci.2020.66.76

Sami Khanal, K. K. (2020). Remote Sensing in Agriculture—Accomplishments. *Remote Sensing*.

Sanchez, P. A., & Swaminathan, M. S. (2005). Public Health: Enhanced: Cutting World Hunger in Half. *Science*, *307*(5708), 357–359. doi:10.1126cience.1109057 PMID:15661994

Sandler, M., & Howard, A. (2018, April 3). *MobileNetV2: The Next Generation of On-Device Computer Vision Networks.* Google AI Blog. https://ai.googleblog.com/2018/04/mobilenetv2-next-generation-of-on.html

Sandler, M., Howard, A., Zhu, M., Zhmoginov, A., & Chen, L.-C. (2018). MobileNetV2: Inverted Residuals and Linear Bottlenecks. *2018 IEEE/CVF Conference on Computer Vision and Pattern Recognition.* 10.1109/CVPR.2018.00474

Sanjana Rao, G. P., Aditya Shastry, K., Sathyashree, S. R., & Sahu, S. (2021). Machine Learning based Restaurant Revenue Prediction. In V. Suma, N. Bouhmala, & H. Wang (Eds.), *Evolutionary Computing and Mobile Sustainable Networks* (pp. 363–371). Springer. doi:10.1007/978-981-15-5258-8_35

Sankararamakrishnan, N., Sharma, A. K., & Sanghi, R. (2005). Organochlorine and organophosphorous pesticide residues in ground water and surface waters of Kanpur. *Environment International, 31*(1), 113–120. doi:10.1016/j.envint.2004.08.001 PMID:15607785

Sarvini, T., Sneha, T., GS, S.G., Sushmitha, S., & Kumaraswamy, R. (2019, April). Performance Comparison of Weed Detection Algorithms. In *2019 International Conference on Communication and Signal Processing (ICCSP)* (pp. 843-847). IEEE.

Schmidt, R. H., & Piotter, H. M. (2020). The hygienic/sanitary design of food and beverage processing equipment. In *Food Safety Engineering*. Springer. doi:10.1007/978-3-030-42660-6_12

Schor, N., Bechar, A., Ignat, T., Dombrovsky, A., Elad, Y., & Berman, S. (2016). Robotic disease detection in greenhouses: Combined detection of powdery mildew and tomato spotted wilt virus. *IEEE Robotics and Automation Letters, 1*(1), 354–360. doi:10.1109/LRA.2016.2518214

Schwarz, Jürgen, & Sánchez. (2015). *29 Implementation of artificial intelligence into risk management decision-making processes in construction projects*. Universität der BundeswehrMünchen, InstitutfürBaubetrieb.

Shahriyar, S. A., Alam, K. M. R., Roy, S. S., & Morimoto, Y. (2018, December). An approach for multi label image classification using single label convolutional neural network. In *2018 21st international conference of computer and information technology (ICCIT)* (pp. 1-6). IEEE.

Shaikh, Memon, Mahoto, Zeadally, & Nebhen. (2022). Artificial Intelligence Best Practices in Smart Agriculture. *IEEE Micro, 42*(1), 17-24. . doi:10.1109/MM.2021.3121279

Shannon, S., Brennan, D., Hanna, D., Younger, Z., Hassan, J., & Breslin, G. (2018). The effect of a school-based intervention on physical activity and well-being: A non-randomised controlled trial with children of low socio-economic status. Sports medicine-open, 4(1), 1-12.

Sharma, A., Jain, A., Gupta, P., & Chowdary, V. (2021). *Machine Learning Applications for Precision Agriculture: A Comprehensive Review*. IEEE., doi:10.1002/9781119769323.ch8

Sharma, M., Carole, S., Paliyath, G., & Subramanian, J. (2008). Changes in Nutritional Quality of Fruits and Vegetables During Storage. In G. Paliyath, D. P. Murr, A. K. Handa, & S. Lurie (Eds.), *Postharvest Biology and Technology of Fruits Vegetables and Flowers* (pp. 443–466). Wiley-Blackwell.

Sharma, R. (2021). Artificial Intelligence in Agriculture: A Review. *International Conference on Intelligent Computing and Control Systems*.

Sharma, R., & Mishra, J. (2020). Review on Application of Drone Systems in Precision Agriculture. *Journal of Advanced Research in Electronics Engineering and Technology, 7*(2), 5–7.

Sharma, S., Gahlawat, V. K., Rahul, K., Mor, R. S., & Malik, M. (2021). Sustainable Innovations in the Food Industry through Artificial Intelligence and Big Data Analytics. *Logistics, 5*(4), 66.

Sharma, V., Kumar, R., & Kumar, N. (2018). DPTR: Distributed priority tree-based routing protocol for FANETs. *Computer Communications*, *122*, 129–151. doi:10.1016/j.comcom.2018.03.002

Shastry, A., Sanjay, H.A., & Bhanusree, E. (2016). Prediction of Crop Yield using Regression Analysis. *Indian Journal of Science and Technology, 9*(38), 1-5.

Sheikh, W. (2013). *Prototype of an image recognition based calorie recording application* [Ph.D. dissertation]. Heriot-Watt University.

Shen, Z., Shehzad, A., Chen, S., Sun, H., & Liu, J. (2020). Machine learning based approach on food recognition and nutrition estimation. *Procedia Computer Science*, *174*, 448–453.

Shiao, S. P. K., Grayson, J., Lie, A., & Yu, C. H. (2018). Personalized nutrition—Genes, diet, and related interactive parameters as predictors of cancer in multiethnic colorectal cancer families. *Nutrients*, *10*(6), 795.

Shobana, M., & Karthik, S. (2013). A Performance Analysis and Comparison of various Routing Protocols in MANET. *Proceedings of the 2013 International Conference on Pattern Recognition, Informatics and Mobile Engineering (PRIME)*. 10.1109/ICPRIME.2013.6496508

Shrivastava, V. K., & Pradhan, M. K. (2021). Rice plant disease classification using color features: A machine learning paradigm. *Journal of Plant Pathology*, *103*(1), 17–26. doi:10.100742161-020-00683-3

Shrivastava, V. K., Pradhan, M. K., Minz, S., & Thakur, M. P. (2019). Rice Plant Disease Classification Using Transfer Learning of Deep Convolution Neural Network. *International Archives of Photogrammetry and Remote Sensing*, 42(3) (W6), 631–635. doi:10.5194/isprs-archives-XLII-3-W6-631-2019

Simonyan, K., & Zisserman, A. (2014). Two-stream convolutional networks for action recognition in videos. *Advances in Neural Information Processing Systems*, *1-4*, 568–576.

Singh, R., Srivastava, S., & Mishra, R. (2020). AI and IoT Based Monitoring System for Increasing the Yield in Crop Production. *2020 International Conference on Electrical and Electronics Engineering (ICE3),* 301-305. 10.1109/ICE348803.2020.9122894

Singh, S. P., Shivam, & Kumar, I. (2022). Orange Fruit Recognition Using Neural Networks. In K. N. Das, D. Das, A. K. Ray, & P. N. Suganthan (Eds.), *Proceedings of the International Conference on Computational Intelligence and Sustainable Technologies* (pp. 417–427). Springer. 10.1007/978-981-16-6893-7_38

Singh, V., Sarwar, A. & Sharma, V. (2017). Analysis of soil and prediction of crop yield (Rice) using Machine Learning approach. *International Journal of Advanced Research in Computer Science, 8*(5), 1254-1259.

Singh, K., Kumar, S., & Kaur, P. (2016). Detection of Powdery Mildew Disease of Beans in India: A Review. Oriental Journal of Computer Science and Technology, *9*(3), 226–234. doi:10.13005/ojcst/09.03.08

Singh, K., & Verma, A. K. (2014). Applying OLSR routing in FANETs. *2014 IEEE International Conference on Advanced Communications, Control, and Computing Technologies*, 1212-1215.

Singh, K., & Verma, A. K. (2015, March). Experimental analysis of AODV, DSDV, and OLSR routing protocol for flying adhoc networks (FANETs). In *2015 IEEE International Conference on Electrical, Computer and Communication Technologies (ICECCT)* (pp. 1-4). IEEE. 10.1109/ICECCT.2015.7226085

Singh, K., & Verma, A. K. Flying adhoc networks concept and challenges. In M. D. B. A. Khosrow-Pour (Ed.), *Advanced Methodologies and Technologies in Network Architecture*. Mobile Computing, and Data Analytics.

Singh, S. P., Rawat, J., Mittal, M., Kumar, I., & Bhatt, C. (2022). Application of AI in SCM or Supply Chain 4.0. In S. L. Fernandes & T. K. Sharma (Eds.), *Artificial Intelligence in Industrial Applications: Approaches to Solve the Intrinsic Industrial Optimization Problems* (pp. 51–66). Springer International Publishing. doi:10.1007/978-3-030-85383-9_4

Sinha, J. P., Kushwaha, H. L., Kushwaha, D., Singh, N., & Purushottam, M. (2016). Prospect of Unmanned Aerial Vehicle (UAV) Technology for Agricultural Production Management. *International Journal of Emerging Trends in Engineering Research*, *8*(10), 6738–6741.

Si, Y. (2017). Potato tuber length-width ratio assessment using image analysis. *American Journal of Potato Research*, *94*(1), 88–93.

Sladojevic, S., Arsenovic, M., Anderla, A., Culibrk, D., & Stefanovic, D. (2016). Deep Neural Networks Based Recognition of Plant Diseases by Leaf Image Classification. *Computational Intelligence and Neuroscience*, *2016*, 1–11. doi:10.1155/2016/3289801 PMID:27418923

Smith, M. J. (2018). Getting value from artificial intelligence in agriculture. *Animal Production Science*, *60*(1), 46–54. doi:10.1071/AN18522

Sneha, A. J., & Chakravarthi, R. (2019). Neural network-based pest identification and control in cauliflower crop using sounds of pest. *International Journal of Innovative Technology and Exploring Engineering*, *8*(9), 3127–3134. doi:10.35940/ijitee.I8546.078919

Sobayo, R., Wu, H.-H., Ray, R., & Qian, L. 2018. Integration of Convolutional Neural Network and Thermal Images into Soil Moisture Estimation. *1st International Conference on Data Intelligence and Security (ICDIS)*, 207-210. 10.1109/ICDIS.2018.00041

Soltani-Fesaghandis, G., & Pooya, A. (2018). Design of an artificial intelligence system for predicting success of new product development and selecting proper market-product strategy in the food industry. *The International Food and Agribusiness Management Review*, *21*(7), 847–864. doi:10.22434/IFAMR2017.0033

Sørensen, L. B., Møller, P., Flint, A., Martens, M., & Raben, A. (2003). Effect of sensory perception of foods on appetite and food intake: A review of studies on humans. *International Journal of Obesity*, *27*(10), 1152–1166. doi:10.1038j.ijo.0802391 PMID:14513063

Sowah, R. A., Bampoe-Addo, A. A., Armoo, S. K., Saalia, F. K., Gatsi, F., & Sarkodie-Mensah, B. (2020). Design and development of diabetes management system using machine learning. *International Journal of Telemedicine and Applications*, 2020.

Spanaki, K. (2021). Disruptive technologies in agricultural operations: A systematic review of AI-driven AgriTech research. *Annals of Operations Research*, 1–34.

Srivastava, K., Pandey, P. C., & Sharma, J. K. (2020). An Approach for Route Optimization in Applications of Precision Agriculture Using UAVs (Vol. 4). Academic Press.

Srivastava, Shiney, & Shan. (2021). Plant Disease Prediction using Image Processing and Soft Computing Algorithms: A Review. *2021 International Conference on Computational Intelligence and Knowledge Economy (ICCIKE)*, 143-148. . doi:10.1109/ICCIKE51210.2021.9410728

Strategy, A. (2015). *SMARTer2030: ICT solutions for 21st century challenges. The Global eSustainability Initiative (GeSI)*. Tech. Rep. Brussels.

Suárez, P. L., Sappa, A. D., & Vintimilla, B. X. (2017). Infrared image colorization based on a tripletdcgan architecture. *IEEE Conference on Computer Vision and Pattern Recognition Workshops (CVPRW).*

Subeesh, A., & Mehta, C. R. (2021). Automation and digitization of agriculture using artificial intelligence and internet of things. *Artificial Intelligence in Agriculture*, *5*, 278–291. doi:10.1016/j.aiia.2021.11.004

Suchithra, M. S., & Pai, M. L. (2020). Improving the prediction accuracy of soil nutrient classification by optimizing extreme learning machine parameters. *Information Processing in Agriculture*, *7*(1), 72–82. doi:10.1016/j.inpa.2019.05.003

Su, J. X. (2017). Significant Remote Sensing Vegetation Indices: A Review of Developments and Applications. *Journal of Sensors.*

Sujatha, R., & Isakki, P. (2016). A study on crop yield forecasting using classification techniques. *International Conference on Computing Technologies and Intelligent Data Engineering (ICCTIDE'16),* 1-4.

Sun, Y., Liu, Y., Wang, G., & Zhang, H. (2017). Deep Learning for Plant Identification in Natural Environment. *Computational Intelligence and Neuroscience*, *2017*, 1–6. doi:10.1155/2017/7361042 PMID:28611840

Suresha, & Kumar, & Kumar. (2012). Texture features and decision trees based vegetables classification. *International Journal of Computer Applications*, *975*, 8878.

Surya Prabha, D., & Satheesh Kumar, J. (2015). Assessment of banana fruit maturity by image processing technique. *Journal of Food Science and Technology*, *52*(3), 1316–1327.

Su, Y., Xu, H., & Yan, L. (2017). Support vector machine-based open crop model (SBOCM): Case of rice production in China. *Saudi Journal of Biological Sciences*, *24*(3), 537–547. doi:10.1016/j.sjbs.2017.01.024 PMID:28386178

Szegedy, C., Vanhoucke, V., Ioffe, S., Shlens, J., & Wojna, Z. (2016, June). Rethinking the Inception Architecture for Computer Vision. *2016 IEEE Conference on Computer Vision and Pattern Recognition (CVPR)*, 2818–2826. 10.1109/CVPR.2016.308

Tahir, G. A., & Loo, C. K. (2021, December). A Comprehensive Survey of Image-Based Food Recognition and Volume Estimation Methods for Dietary Assessment. *Health Care*, *9*(12), 1676.

Tai, A. P. K., Martin, M. V., & Heald, C. L. (2014). Threat to future global food security from climate change and ozone air pollution. *Nature Climate Change*, *4*(9), 817–821. doi:10.1038/nclimate2317

Talaviya, T., Shah, D., Patel, N., Yagnik, H., & Shah, M. (2020). Implementation of artificial intelligence in agriculture for optimisation of irrigation and application of pesticides and herbicides. *Artificial Intelligence in Agriculture.*, *4*, 58–73. Advance online publication. doi:10.1016/j.aiia.2020.04.002

Tammachat, N., & Pantuwong, N. (2014, October). Calories analysis of food intake using image recognition. In *2014 6th International Conference on Information Technology and Electrical Engineering (ICITEE)* (pp. 1-4). IEEE.

Tang, J., Wang, D., Zhang, Z., He, L., Xin, J., & Xu, Y. (2017). Weed identification based on K-means feature learning combined with convolutional neural network. *Computers and Electronics in Agriculture*, *135*, 63–70. doi:10.1016/j.compag.2017.01.001

Tao, D., Yang, P., & Feng, H. (2020). Utilization of text mining as a big data analysis tool for food science and nutrition. *Comprehensive Reviews in Food Science and Food Safety*, *19*(2), 875–894. doi:10.1111/1541-4337.12540 PMID:33325182

Termritthikun, C., Muneesawang, P., & Kanprachar, S. (2017). NU-InNet: Thai food image recognition using convolutional neural networks on smartphone. *Journal of Telecommunication Electronic and Computer Engineering*, *9*(2-6), 63–67.

The future of Food and Agriculture: Trends and challenges. (2017). *FAO.* https://www.fao.org/3/i6583e/i6583e.pdf /

Tiwari, O., Goyal, V., Kumar, P., & Vij, S. (2019). An experimental set up for utilizing convolutional neural network in automated weed detection. *4th International Conference on Internet of Things: Smart Innovation and Usages (IoT-SIU)*, 1-6. 10.1109/IoT-SIU.2019.8777646

Toh, C. K. (1997). Associativity-based routing for ad hoc mobile networks. *Wireless Personal Communications*, *4*(2), 103–139. doi:10.1023/A:1008812928561

Topleva, S. A., & Prokopov, T. V. (2020). Integrated business model for sustainability of small and medium-sized enterprises in the food industry. *British Food Journal*, *122*(5), 1463–1483. doi:10.1108/BFJ-03-2019-0208

Tripathi, S., Shukla, S., Attrey, S., Agrawal, A., & Bhadoria, V. S. (2020). *Smart industrial packaging and sorting system. In Strategic System Assurance and Business Analytics*. Springer.

Tsang, S. (2019, March 23). *Review: Inception-v3 — 1st Runner Up (Image Classification) in ILSVRC 2015*. Medium. https://sh-tsang.medium.com/review-inception-v3-1st-runner-up-image-classification-in-ilsvrc-2015-17915421f77c

Tsoumakas, G. (2019). A survey of machine learning techniques for food sales prediction. *Artificial Intelligence Review*, *52*(1), 441–447.

Tudi, M., Ruan, H. D., Wang, L., Lyu, J., Sadler, R., Connell, D., Chu, C., & Phung, D. T. (2021). Agriculture Development, Pesticide Application and Its Impact on the Environment. *International Journal of Environmental Research and Public Health*, *18*(3), 1112–1112. doi:10.3390/ijerph18031112 PMID:33513796

Türkoğlu, M., & Hanbay, D. (2019). Plant disease and pest detection using deep learning-based features. *Turkish Journal of Electrical Engineering and Computer Sciences*, *27*(3), 1636–1651. doi:10.3906/elk-1809-181

Tyagi, A. C. (2016). Towards a second green revolution. *Irrigation and Drainage*, *65*(4), 388–389. doi:10.1002/ird.2076

Umamaheswari, S., Arjun, R., & Meganathan, D. (2018, October). Weed detection in farm crops using parallel image processing. In *2018 Conference on Information and Communication Technology (CICT)* (pp. 1-4). IEEE.

Upadhyay, M. (2020). Problems and Solutions to Low Agricultural Productivity in India. *Int. J. Adv. Multidiscip. Res.*, *3*, 40–49.

Vahdanjoo, M., Madsen, C. T., & Sørensen, C. G. (2020). Novel Route Planning System for Machinery Selection. Case: Slurry Application. *AgriEngineering*, *2*(3), 408–429. doi:10.3390/agriengineering2030028

Valdoria, J. C., Caballeo, A. R., Fernandez, B. I. D., & Condino, J. M. M. (2019, October). iDahon: An Android Based Terrestrial Plant Disease Detection Mobile Application Through Digital Image Processing Using Deep Learning Neural Network Algorithm. *2019 4th International Conference on Information Technology (InCIT)*. 10.1109/INCIT.2019.8912053

Valentino, F., Cenggoro, T. W., & Pardamean, B. (2021). A design of deep learning experimentation for fruit freshness detection. *IOP Conference Series. Earth and Environmental Science*, *794*(1), 012110. doi:10.1088/1755-1315/794/1/012110

Veenadhari, S., Misra, B., & Singh, S. (2014). Machine learning approach for forecasting crop yield based on climatic parameters. *International Conference on Computer Communication and Informatics*, 1-5. 10.1109/ICCCI.2014.6921718

Verheyen, K., & Adrianens, M. (2001). High resolution continuous soil classification using morphological soil profile descriptions. *Geoderma*, *101*(3-4), 31–48. doi:10.1016/S0016-7061(00)00088-4

Verma, M. (2015). *Energy Use in Global Food Production*. Springer International Publishing.

Waheed, A., Goyal, M., Gupta, D., Khanna, A., Hassanien, A. E., & Pandey, H. M. (2020). An Optimized Dense Convolutional Neural Network Model for Disease Recognition and Classification in Corn Leaf. *Computers and Electronics in Agriculture*, *175*, 105456. Advance online publication. doi:10.1016/j.compag.2020.105456

Walter, A., Finger, R., Huber, R., & Buchmann, N. (2017). Opinion: Smart farming is key to developing sustainable agriculture. *Proceedings of the National Academy of Sciences of the United States of America*, *114*(24), 6148–6150. doi:10.1073/pnas.1707462114 PMID:28611194

Wang, J., Yang, Y., Mao, J., Huang, Z., Huang, C., & Xu, W. (2016). Cnn-rnn: A unified framework for multi-label image classification. In *Proceedings of the IEEE conference on computer vision and pattern recognition* (pp. 2285-2294). 10.1109/CVPR.2016.251

Wang, S., & Zhang, Y. (2015). Fruit classification by wavelet-entropy and feedforward neural network trained by fitness-scaled chaotic ABC and biogeography-based optimization. *Entropy (Basel, Switzerland)*, *17*(8), 5711–5728.

Wang, X., Puri, V. M., & Demirci, A. (2020). Equipment Cleaning, Sanitation, and Maintenance. In *Food Engineering Series* (pp. 333–353). Springer. doi:10.1007/978-3-030-42660-6_13

Wang, X., Puri, V. M., & Demirci, A. (2020). *Equipment cleaning, sanitation, and maintenance. In Food Safety Engineering*. Springer.

Wardah, S., Djatna, T., Marimin, & Yani, M. (2020). New product development in coconut-based agro-industry: Current research progress and challenges. *IOP Conference Series. Earth and Environmental Science*, *472*(1), 012053. doi:10.1088/1755-1315/472/1/012053

Welcome to icrawler — icrawler 0.6.6 documentation. (n.d.). Retrieved February 20, 2022, from https://icrawler.readthedocs.io/en/latest/

Wen, S., & Huang, C. (2018). Delay-constrained routing based on stochastic model for flying ad hoc networks. *Mobile Information Systems*.

Withanage, D. P., & Damayanthi, B. W. R. (2019). Factors influencing the youths' interest in. *Agricultural Entrepreneurship in Sri Lanka*, *9*(12), 32–40.

Wu, J., Olesnikova, A., Song, C. H., & Lee, W. D. (2009). The Development and Application of Decision Tree for Agriculture Data. *IITSI*, 16-20. https://www.hilarispublisher.com/open-access/agriculture-role-on-indian-economy-2151-6219-1000176.pdf

Wu, Z., Chen, Y., Zhao, B., Kang, X., & Ding, Y. (2021). Review of Weed Detection Methods Based on Computer Vision. *Sensors (Basel)*, *21*(11), 3647.

Xiang, H., & Tian, L. (2006). Development of autonomous unmanned helicopter based agricultural remote sensing system. In *2006 ASAE Annual Meeting* (p. 1). American Society of Agricultural and Biological Engineers.

Xiang, H., & Tian, L. (2011). Development of a low-cost agricultural remote sensing system based on an autonomous unmanned aerial vehicle (UAV). *Biosystems Engineering*, *108*(2), 174–190. doi:10.1016/j.biosystemseng.2010.11.010

Xu, C., He, H., Hauser, S. C., & Gerling, G. J. (2019). Measurement of Touch Interaction Cues in Discriminating Soft Fruit. *Proceedings of the IEEE World Haptics Conference*.

Xu, J.-C., Zhang, M., Mujumdar, A. S., & Adhikari, B. (2017). Recent developments in smart freezing technology applied to fresh foods. *Critical Reviews in Food Science and Nutrition*, *57*(13), 2835–2843.

Xu, W., Trappe, W., & Zhang, Y. (2007). Channel surfing: defending wireless sensor networks from interference. *Proceedings of conference on Information Processing in Sensor Networks (IPSN)*, 499–508. 10.1109/IPSN.2007.4379710

Yao, J., Qi, J., Zhang, J., Shao, H., Yang, J., & Li, X. (2021). A Real-Time Detection Algorithm for Kiwifruit Defects Based on YOLOv5. *Electronics (Basel)*, *10*(14), 1711. doi:10.3390/electronics10141711

Yichun Xie, Z. S. (2008). Remote sensing imagery in vegetation mapping: A review. *Journal of Plant Ecology*, 9–23.

Yick, J., Mukherjee, B., & Ghosal, D. (2008). Wireless sensor network survey. *Computer Networks*, *52*(12), 2292–2330. doi:10.1016/j.comnet.2008.04.002

Yohanandan, S. (2020). *mAP (mean Average Precision) might confuse you!* Towards Data Science. https://towardsdatascience.com/map-mean-average-precision-might-confuse-you-5956f1bfa9e2

Yost, E., & Cheng, Y. (2021). Customers' risk perception and dine-out motivation during a pandemic: Insight for the restaurant industry. *International Journal of Hospitality Management*, *95*, 102889. doi:10.1016/j.ijhm.2021.102889

Yu, J., Schumann, A.W., Cao, Z., Sharpe, S.M., & Boyd, N.S. (2019). Weed detection in perennial ryegrass with deep learning convolutional neural network. *Frontiers in Plant Science, 10*, 1422.

Yuan, L. (2019). *Sustainability Aspects of ICT in Agriculture and Food Systems.* Academic Press.

Yu, J., Schumann, A. W., Cao, Z., Sharpe, S. M., & Boyd, N. S. (2019). Weed detection in perennial ryegrass with deep learning convolutional neural network. *Frontiers in Plant Science*, *10*, 1422.

Yu, Y., Xue, B., Chen, Z., & Qian, Z. (2019). Cluster tree topology construction method based on PSO algorithm to prolong the lifetime of ZigBee wireless sensor networks. *EURASIP Journal on Wireless Communications and Networking*, *2019*(1), 1–13. doi:10.118613638-019-1523-0

Zainal, N., Mohamood, N., Norman, M.F., & Sanmutham, D. (2019). Design and implementation of smart farming system for fig using connected-argonomics. *International Journal of Electrical & Computer Engineering, 9*(6).

Zakaria, M., Al-Shebany, M., & Sarhan, S. (2014). Artificial Neural Network: A Brief Overview. *International Journal of Engineering Research and Applications*, *4*, 7–12.

Zangina, U., Buyamin, S., Abidin, M. S. Z., & Mahmud, M. S. A. (2021). Agricultural rout planning with variable rate pesticide application in a greenhouse environment. *Alexandria Engineering Journal*, *60*, 3007–3020.

Zaniolo, L., & Marques, O. (2020). On the use of variable stride in convolutional neural networks. *Multimedia Tools and Applications*, *79*(19), 13581–13598. doi:10.100711042-019-08385-4

Zayan, S. A. (2019). *Impact of climate change on plant diseases and IPM strategies*. IntechOpen. doi:10.5772/intechopen.87055

Zeisel, S. H. (2020). Precision (personalized) nutrition: Understanding metabolic heterogeneity. *Annual Review of Food Science and Technology*, *11*, 71–92.

Zhang, A., Lipton, Z. C., Li, M., & Smola, A. J. (2021). *Dive into deep learning*. arXiv preprint arXiv:2106.11342.

Zhang, C., Li, M., & Guo, P. (2017). An interval multistage joint-probabilistic chance-constrained, programming model with left-hand-side randomness for crop area planning under uncertainty. *Journal of Cleaner Production*, *167*, 1276–1289. doi:10.1016/j.jclepro.2017.05.191

Zhang, C., Liu, J., Shang, J., & Cai, H. (2018). Capability of crop water content for revealing variability of winter wheat grain yield and soil moisture under limited irrigation. *The Science of the Total Environment*, *631*, 677–687. doi:10.1016/j.scitotenv.2018.03.004 PMID:29539596

Zhang, H., Wang, X., Memarmoshrefi, P., & Hogrefe, D. (2017). A survey of ant colony optimization based routing protocols for mobile ad hoc networks. *IEEE Access: Practical Innovations, Open Solutions*, *5*, 24139–24161. doi:10.1109/ACCESS.2017.2762472

Zhang, W. (2018). Global pesticide use: Profile, trend, cost/benefit and more. *Proc Int Acad Ecol Environ Scince*, *8*, 1–27.

Zhang, Y., Phillips, P., & Wang, S. (2016). Fruit classification by biogeography-based optimization and feedforward neural network. *Expert Systems: International Journal of Knowledge Engineering and Neural Networks*, *33*(3), 239–253.

Zhang, Y., & Wang, S. (2014). Fruit classification using computer vision and feedforward neural network. *Journal of Food Engineering*, *143*, 167–177.

Zhu, B., & Jiang, L. (2007). Gabor feature-based apple quality inspection using kernel principal component analysis. *Journal of Food Engineering*, *81*(4), 741–749.

# About the Contributors

**Rijwan Khan** has obtained Ph.D in Computer Engineering from Jamia Millia Islamia, New Delhi, India. He is presently working as Professor and Head Department of Computer Science and Engineering at ABES Institute of Technology (Affiliated to AKTU), Ghaziabad India. He has more than 50 publications in his area of research. He published 3 patents and 2 of his patents are granted. He is reviewer of different journal of Springer, Elsevier, ACM etc. He is author of 3 books. He is guest editor of SCI and Scopus indexed journals and editor of 3 Books.

* * *

**Firoz Ahmad** is working as a software engineer in Truminds Software Systems and he has completed his undergraduate degree majoring in Computer Engineering from Zakir Husain College of Engineering and Technology, Aligarh Muslim University.

**Berat Asrin Caferoglu** was born in November, 2000 in Turkey. Currently, he is a senior bachelor's student in Department of Electrical and Electronics Engineering in Marmara University.

**Usha Chauhan** is Professor in School of Electrical, Electronics and Communication Engineering, Galgotias University, NCR Delhi, India. She has received Ph. D degree in Image Processing. She has more than 22 years of experience working in teaching and research. Her current research areas are image processing, embedded system, artificial intelligence, internet of things, and machine learning. She is the author of over 10+ book chapters including some reputed publishers such as Springer, Elsevier, Wiley, CRC Pressl and 29+ international journals and conferences. She has published book chapters in Taylor & Francis Group (CRC Press), Weily, Springer, etc.

**Onur Cihan** was born in Ankara, Turkey. He received BS degrees in Control Engineering and Electrical Engineering from Istanbul Technical University in 2007. He got MS and PhD degrees in Electrical and Electronics Engineering from Bogazici University in 2009 and 2014, respectively. His PhD thesis was on convergence rate optimization of distributed consensus algorithms, which received the best PhD dissertation award from Bogazici University in 2014. From February 2014 until March 2015, he worked at Migros Research and Development Center, Turkey. In May 2015, he started working as a post doctoral researcher at Center for Systems Science, Yale University, New Haven, CT, before joining the Department of Electrical and Electronic Engineering at Marmara University in 2016. His research interests are in the areas of mathematical systems theory, multi-agent systems, distributed control and deep learning.

**Zeynep Elbir** was born in Amasya, Turkey. She is an undergraduate student in Bachelor's Degree of Electrical and Electronics Engineering in Marmara University.

**Ahona Ghosh** is a B.Tech., M.Tech. in Computer Science and Engineering and presently is an AICTE Doctoral Fellow in the Department of Computer Science and Engineering, Maulana Abul Kalam Azad University of Technology, West Bengal (In house). Before joining her Ph.D., she was an Assistant Professor in the Department of Computational Science at Brainware University, Barasat, Kolkata. She has published more than 30 papers in international conference proceedings, peer-reviewed international journals, and books with IEEE, Springer, Elsevier, IGI Global, CRC Press, etc. She is also a reviewer of different international journals. Her research interests include Machine Learning and the Internet of Things.

**Mahim Kumar** is currently pursuing B.Tech degree in electronics and communication engineering from Galgotias University, Greater Noida, India. His research interests include machine learning, image processing, internet of things and driver situation awareness in hazardous conditions.

**Nitendra Kumar** received a Ph. D (Mathematics) from Sharda University, Greater Noida, India and a Master of Science (Mathematics and Statistics) from Dr Ram Manohar Lohia Avadh University, Faizabad, India. Currently, he is working as an Assistant Professor at the IIMT College of Engineering, Greater Noida with interests in the Wavelets and its Variants, Data Mining, Inverse Problems, Epidemic Modeling, Fractional Derivatives and Statistical Methods. He has more than 10 years of experience in his research areas. Dr Kumar has published many research papers in reputed journals and also published 6 books on engineering mathematics. He contributes to the research community by undertaking various volunteer activities in the capacity of editor for two edited books, Guest Editor of reputed journals of Taylor and Francis.

**Caprio Mistry** is an undergraduate student of Brainware University who belongs to the Department of Electronics and communication engineering. Currently, he is pursuing his Bachelor of Technology in Electronics and Communication Engineering, 4 years degree (7semester Final Year). He had done a review project on the implementation of UAV networks for Agriculture and irrigation systems on another hand he had also done a review on IoT and modern agricultural aspects as his minor project. He has an interest in reversible computing implementation on Device-level designs and low-power VLSI. He had demonstrated his final year 1st stage project on how reversible logic works for heat dissipation and device speed minimization for low power VLSI circuits. He has an interest in FinFET and RibbonFET technology on the aspect of leakage current minimization on recent device modeling for low power VLSI.

**Kajol Mittal**, currently working as an Assistant Professor in department of computer science and engineering, ABES Institute of Technology and pursuing Ph.D. from JIIT, Noida. Her areas of interest are Machine Learning, Virtual Reality, IoT Security.

**Ayesha Naaz** is working as Professor in Department of Electronics and Communication Engineering Department of Muffakham Jah College of Engineering and Technology, Banjara Hills, Hyderabad and has 20+ years of teaching experience. She obtained her Ph. D degree from Osmania University, Hyderabad, Masters form JNTU, Hyderabad with Digital Systems and Computer Electronics as specialization.

She has guided 26 masters Projects and an equivalent number of undergraduate projects. She is also a recognized Ph. D supervisor of Osmania University. Her research interests include Signal Processing, Array signal processing, Image Processing, Audio signal Processing, Machine Learning, Microstrip Antennas, Digital Electronics.

**Sucharitha Nagulapally** is working as Assistant Professor of Electronics and Communication Engineering Department of Muffakham Jah College of Engineering and Technology, Banjara Hills, Hyderabad. She completed her B.Tech (ECE) from Vignana Jyothi Institute of Engineering and Technology in 2004 and M.Tech in Systems and Signal Processing (SSP) from JNTU Hyderabad in 2013 and Pursuing Ph.D from JNTU Hyderabad. She has 16 years of teaching experience and her research interests are in Signal Processing, Image Processing, Neural Networks, Electronics, IC technology, and Machine Learning.

**Rakesh Mohan Pujahari** is presently working as Assistant Professor ABES Institute of Technology of Ghaziabad, Uttar Pradesh, India. He is M. Sc., M. Phil. and Ph.D. degree to his credit. He has published many research papers both in National and International Journals of high repute like Elsevier, Springer etc. His research interests are Machine Learning, Artificial Intelligence, renewable energy and sustainable development . He is a life member of Solar energy society of India a unit of International solar energy society, Microscopy society of India and Odisha physical society. His recently Published chapters are: 1. Federated Learning for IoT Applications (EAI/Springer Innovations in Communication and Computing)( ISBN: 978-3-030-85559-8) Chapter 14: Quantum Federated learning for wireless communication https://link.springer.com/chapter/10.1007/978-3-030-85559-8_14 1. 2. Application of Machine Learning in Agriculture Intelligent Farming System through Weather Forecast Support.

**Mritunjay Rai** is working as an Assistant Professor in Department of Electronics and Communication at SRM University, Lucknow-Deva Road, Barabanki, UP. He is pursuingPh.D. from IIT-ISM Dhanbad. He has completed his B.Tech in ECE from Sri Ramswaroop Memorial College of Engineering and Management, Lucknow (SRMCEM) in 2009 and Master of Engineering (distinction) in Instrumentation and Control from Birla Institute of Technology-Mesra, Ranchi in 2013. His areas of interest lie in image processing, speech processing, Artificial Intelligence, machine learning, deep learning, Internet of Things (IoT)and robotics & automation. He has more than 9 years of working experience in research as well as academic. In addition, he has guided several UG and PG projects. He has published many research articles in reputed Journals published by IEEE, Inderscience, etc. He has reviewed many research papers of Journals and international/national conferences.

**Archana Sharma** is currently working in ABES Institute of Technology, Ghaziabad, as an assistant Professor, Department of Computer Science and Engineering. She has 17 Years of teaching experience. Her areas of Interest are Artificial Intelligence, Machine Learning, Optimization, etc.

**Shaurya Pratap Singh** is currently pursuing B.Tech degree in electronics and communication engineering from Galgotias University, Greater Noida, India. His research interests include machine learning, image processing, internet of things and driver situation awareness in hazardous conditions.

**Kavita Srivastava** is currently working as an Assistant Professor in the Department of Computer Science, Institute of Information Technology & Management affiliated with GGSIPU, New Delhi,

India. She obtained her M.Tech degree in IT in 2007 from USICT, GGSIPU, Delhi, India. She has an interest in Machine Learning, Deep Learning, Wireless Sensor Networks, the Internet of Things, and Image Processing.

**Padmesh Tripathi** is working as an Associate Professor of Mathematics at IIMT College of Engineering, Greater Noida, UP. He received his Ph.D. degree from Sharda University, Greater Noida (India). He completed his Master's and Bachelor's degrees from University of Allahabad, Prayagraj. He has been teaching since last 20 years at graduate and post-graduate levels. He has published many research papers in reputed journals and presented many papers in national/international conferences. He has been granted funds from prestigious organizations like: Cambridge University, UK (twice); University of California, Los Angeles, USA; University of Eastern Finland, Kuopio, Finland; INRIA, Sophia Antipolis, France, etc. and he visited these organizations. In addition, he has been granted funds from CIMPA (UNESCO) France to participate in a research school at Erbyl, Iraq. He has been member of Society of Industrial and Applied Mathematics (SIAM), Philadelphia, USA (2011-19). He is life member of Science and Engineering Institute (SCIEI), Los Angeles, USA; Indian Society of Industrial and Applied Mathematics (ISIAM), India; Ramanujan Mathematical Society, India; Purvanchal Academy of Sciences, India; International Association of Engineers, UK. He is member of EURO working group on continuous optimization (EUROPT), Italy; Society for Foundations of Computational Mathematics, USA. His area of research interests are inverse problems, signal processing, optimization, data science, AI, deep learning, etc.

# Index


## A

agriculture 1-6, 8, 10-12, 14-18, 24-29, 50, 52, 55-61, 63-64, 67, 69-78, 86-87, 89-95, 99, 101, 103, 109-113, 127-133, 136, 140-145, 148-150, 165-166, 169-173, 177, 180, 184-190, 202, 204-206, 208, 210-215, 217, 235, 250-251, 253, 255-258, 260-264, 270-274
Agronomic Engineering 190, 204
artificial intelligence 1-3, 8, 10-11, 14-16, 18-19, 24, 29, 33, 52, 57-58, 74-76, 78, 87, 89-92, 94, 109, 115, 127-128, 169, 171-172, 185-191, 194-202, 204, 210-214, 218, 236-237, 241-245, 247-248, 250, 252-258, 267-270, 272, 274, 277-278, 284-285, 288, 290
automation 10-11, 58, 74, 77, 87, 94-95, 110-111, 171, 184, 187, 190-192, 204, 206, 213, 241, 257, 263, 288

## C

Challenges in Food Industry 190
cobots 241, 243
computer vision 28-29, 35-36, 49, 51-52, 67, 74-76, 87, 91-92, 96, 98, 101, 103, 111, 149-150, 165-167, 200-201, 203, 211-212, 214-215, 235, 257, 262-263, 274, 284, 288
convolutional neural network 33, 109-111, 148, 150, 156, 165-167, 169, 175, 181, 184-189, 264, 278, 281-282
COVID-19 1, 34, 49, 202, 206, 211, 214, 240, 251, 256

## D

Decision Tree 112-114, 116, 120, 124, 128, 213, 220
Deep Learning 2, 6, 15, 17-20, 24, 27-30, 33-36, 49-50, 52, 55-56, 67, 75, 78, 91, 95-101, 104, 108, 110-111, 143, 148, 150-151, 154, 165-169, 172-173, 185, 187, 189, 191, 199, 203, 215, 221, 242, 245, 248, 257-259, 268, 273, 278, 280-281, 283, 285-289
Deep neural network 101, 148, 155, 165, 180

## F

farming 3, 5, 10-11, 15, 17-18, 55-58, 63, 67, 69, 74-77, 84, 87, 90, 92, 96, 108, 110, 112, 114, 130, 140-142, 144, 147, 149, 169, 171, 177, 184, 188-189, 204, 242, 250, 257-258, 261, 263
Flying AdHoc Network (FANET) 129
Food Distribution 93, 127, 206, 269, 273
food industry 2-4, 187, 190-191, 196-200, 203, 207-208, 211-214, 234, 241-243, 245-246, 249-258, 265-266, 272, 274
food processing 91, 194-195, 199-200, 206, 210, 213, 217, 236, 242-243, 245, 247-248, 257, 266-267, 269, 272
food production 1-2, 18, 52, 58, 78, 89, 190, 197, 200, 203-204, 206, 241-242, 250-251, 257
food quality 2, 78, 90, 93-94, 127-128, 191, 200, 208, 211, 235, 241-242, 251, 258, 269, 273-274
food safety 2-3, 14, 127, 191, 194-196, 199-201, 211-212, 216-217, 228-237, 239, 241-242, 247-248, 254-255, 257-258, 269-270, 273-274, 288
food sector 190-191, 196, 199-200, 202, 210, 241, 244, 250, 253, 257-258, 262-263, 268-269
food wastage 32-33
freshness 2, 29-34, 37-38, 44-54

## G

GAN 17, 20-21, 24
genomes 216, 222, 224-227, 229
Gradient Boosting Regression 113

## I

ICT-Enabled Food System 207

image processing 7, 16, 24, 35-36, 49-50, 55-58, 64-67, 70, 96, 98-101, 108-111, 150, 168, 177, 182-183, 188, 198, 209, 211, 214, 258, 264, 288
Inception V3 148, 151
Industrial Automation 190, 257, 288
Internet of Things 2, 11, 15, 18, 27, 57, 74-75, 90-94, 130, 138, 143, 187, 189, 194, 213, 253, 255, 263, 277
irrigation system 3, 15, 129, 133, 142, 179, 186, 263-264, 274

## L

Linear Regression 41, 70, 112-113, 117, 120, 124, 135, 172
local area index 17

## M

machine learning 1-2, 4, 14-15, 18-20, 25, 35, 52, 55, 57-58, 67-72, 75, 78, 91, 95-101, 103, 106, 109, 111-117, 126-128, 143, 150, 155, 165-167, 169, 171-172, 186-187, 189-193, 196-200, 202, 210-214, 216-218, 220-229, 231-232, 238-240, 242-245, 247, 252, 256-265, 267, 269-274, 277-278, 281-282, 286-287, 289-290
MobileNet V2 148, 165
multispectral image 55-56, 64, 67

## N

neural networks,CNN 17
novel data streams 216, 229, 233, 235

## O

object detection 29, 33, 35-37, 39-40, 43-53, 95-96, 98, 111, 285-286
online grocery shopping 29, 32, 237

## P

Pesticide Management 74-75, 169
pesticides 16, 18, 57, 60, 74-81, 83, 85-94, 96, 100, 129, 132, 149, 169-171, 264
Plant Disease Detection 24, 27, 148, 150-151, 164, 166-168, 182, 186
Polynomial Regression 112-113, 117, 121, 125
precision agriculture 10-11, 14, 26, 28, 55-61, 64, 69, 71-72, 92-94, 99, 109, 130, 140, 142, 144-145, 169, 171, 184, 186-187, 250, 274
Public Health 94, 167, 216-217, 222, 234-237, 240, 255, 274

## R

Random Forest 24, 57, 70-71, 98, 112-113, 116, 119, 123, 126, 150, 200, 208, 220, 227, 251
remote sensing 25, 55-61, 63, 65-66, 72-73, 133, 144-146, 150, 189
Ridge regression 113, 117, 119, 123
Ripeness 32, 46, 50, 52
RNN 17, 20-22, 24, 27, 101-102, 104, 179

## S

smart farming 11, 17-18, 110, 130, 142, 144, 147, 188-189, 257, 263
Smart pharming 129
sustainability 2, 14-15, 18, 49, 94, 108, 169, 171, 184-185, 187-188, 206-208, 211-212, 215, 242, 251, 256, 263, 274

## T

text data 216, 229-231, 233, 281-282
toxins 1
traceability 1-2, 15, 233, 238
trade data 216, 229, 232, 235
transactional data 216, 219, 231

## U

Unmanned Aerial Vehicle (UAV) 99, 129, 146

## V

Vegetation Index 55, 62-63

## W

weed detection 7, 77, 87, 95-103, 108-111, 184, 189, 254

## Y

YOLO 29, 33, 35-36, 44, 48-52, 95

www.ingramcontent.com/pod-product-compliance
Lightning Source LLC
LaVergne TN
LVHW060417300926
849162LV00006B/1454

*9781668451410*